AF893234

Green Energy and Technology

More information about this series at http://www.springer.com/series/8059

Aurora Monge-Barrio
Ana Sánchez-Ostiz Gutiérrez

Passive Energy Strategies for Mediterranean Residential Buildings

Facing the Challenges of Climate Change and Vulnerable Populations

Springer

Aurora Monge-Barrio
School of Architecture
University of Navarra
Pamplona
Spain

Ana Sánchez-Ostiz Gutiérrez
School of Architecture
University of Navarra
Pamplona
Spain

ISSN 1865-3529 ISSN 1865-3537 (electronic)
Green Energy and Technology
ISBN 978-3-319-69882-3 ISBN 978-3-319-69883-0 (eBook)
https://doi.org/10.1007/978-3-319-69883-0

Library of Congress Control Number: 2017960200

© Springer International Publishing AG 2018
This work is subject to copyright. All rights are reserved by the Publisher, whether the whole or part of the material is concerned, specifically the rights of translation, reprinting, reuse of illustrations, recitation, broadcasting, reproduction on microfilms or in any other physical way, and transmission or information storage and retrieval, electronic adaptation, computer software, or by similar or dissimilar methodology now known or hereafter developed.
The use of general descriptive names, registered names, trademarks, service marks, etc. in this publication does not imply, even in the absence of a specific statement, that such names are exempt from the relevant protective laws and regulations and therefore free for general use.
The publisher, the authors and the editors are safe to assume that the advice and information in this book are believed to be true and accurate at the date of publication. Neither the publisher nor the authors or the editors give a warranty, express or implied, with respect to the material contained herein or for any errors or omissions that may have been made. The publisher remains neutral with regard to jurisdictional claims in published maps and institutional affiliations.

Printed on acid-free paper

This Springer imprint is published by Springer Nature
The registered company is Springer International Publishing AG
The registered company address is: Gewerbestrasse 11, 6330 Cham, Switzerland

To Robert, Eduardo and Juan,

Aurora Monge-Barrio

To my family, for their unconditional support,

Ana Sánchez-Ostiz Gutiérrez

Preface

During these last two decades in Europe, efforts are being made for improving energy efficiency in buildings, aiming to reduce energy demand and consumption, which also bring a reduction in associated greenhouse gas emissions and a mitigation of global warming. Residential buildings' consumption in Southern Europe is mostly related to winter conditioning (heating); however, summer necessities (cooling) are on the rise. This is due to more severe temperature conditions, to users' increasing thermal comfort expectations and to the technical development of air conditioning systems. Bearing in mind the need of buildings to face warmer future conditions, we must not forget present cold winter climate conditions in many locations of Southern Europe. Hence, we should build and renovate for both present and future, being conscious of a building's whole life.

The south of Europe presents a great climatic variation and complexity. In relation to buildings, evermore warming conditions, heatwaves and hot spells, together with a mid-low adaptation capability, make this zone a hotspot for climate change impacts. At present, adaptation to climate change is a low priority in the political and administrative agenda, but is a need that cannot wait to be tackled. On the other hand, residential buildings, as opposed to tertiary buildings, face the challenge of securing comfortable conditions to a varied and changing population, of a marked intergenerational character, and whose vulnerability is a priori unknown. We understand for vulnerable a population with physical limitations (age, illness, etc.) and/or socio-economic problems (energy poverty), who can suffer unacceptable indoor environmental conditions, not only from a well-being but also from a health point of view.

Therefore, the construction and rehabilitation of residential buildings in the Mediterranean area face important challenges against very different and time changing climatic and occupational situations. There is a wide consensus in the need of empowering to maximum passive architectonic strategies, which can help buildings and their occupants adapt to exterior climatic conditions with a minimum energy consumption, with high-efficiency systems and the use of renewable energy, guaranteeing adequate comfort conditions.

These passive measures have to be very ambitious optimizing every resource in the building, both in summer and in winter conditions, being these strategies different and in many cases contradictory. They must be implemented from the beginning, whether it is a new project or a renovation, and must also be considered during execution and use. They must begin from the location on the site, form, volume and orientation of the building, and reach the thermal envelope's constructive details.

The building's envelope has a leading role in passive strategies. Its thermal resistance, airtightness and control of ventilation will be a priority in winter conditions, but also in summer, although with a different incidence in the effectiveness of the measures. The control of solar radiation throughout the year, mainly in the optimisation of orientations and shadings, is of a great relevance. Other measures like night ventilation and thermal mass control are very effective in summer, from a very simple to a very sophisticated technological level, both in the system and in its control and regulation. These strategies however, although well known and of a long tradition in the Mediterranean zone, are not sufficiently valued in many cases, either by the designer or by the user, mostly because of the ease of use and immediate response of active systems. Southern Europe presents an architecture more prepared, in general, for warmer conditions than other European areas, but at the same time, from the moment that technological systems are cheaper and easier to use, that popular knowledge that has buildings built and used without additional energy for comfortable interior conditions is being lost.

We can differentiate between those locations where the most severe climatic conditions make active systems essential, and those that do not normally count with them. In this way, we will have two approaches to tackle the effectiveness of passive strategies in buildings, that is, from the reduction of energy consumption of both heating and cooling, to lowering the risk of having inadequate interior conditions that may compromise users' health, especially in the summer, when not counting with active systems with which to restore thermal conditions.

Besides, the study of mitigation and adaptation to climate change must not be carried out only at building level, but must contemplate the neighbourhood and city scale in order to establish a diagnostic of the present situation of residential buildings and their contribution to energy consumption and CO_2 emissions. This will allow us to establish efficient renovation measures for the thermal envelope and evaluate the repercussion it can have in the group of residential buildings built without adequate thermal properties. At the same time, the existing knowledge of social and economic realities in neighbourhoods or city zones will permit the implementation and prioritizing of plans and actuation strategies that will make possible the detection and prevention of cases of energy poverty.

This book aims to tackle these important challenges in the residential architecture of the Southern European Mediterranean area. It is divided into two separate parts.

The first one introduces, puts into context and establishes the problematic and objectives related to typical climate and climate change, as well as the concepts of vulnerable population and the way of achieving acceptable conditions related to occupants' comfort and health. The optimized answer buildings must give should be mainly based in passive conditioning strategies. These strategies, together with high efficiency and renewable energy systems, are on the way to a zero energy consumption building.

The second part of the book will delve deeper into those strategies at an occupant, building and urban level, differentiating strategies for summer and winter conditions but looking for a balance optimized in a given building during a whole year. The study tackles from bibliographic reviews, results of monitorings and tests of case studies, to energy simulations of these case studies in the locations representative of the different climatic zones in the area considered. Approximations to these strategies in the residential buildings of Southern Europe aim to protect all the population from variable climate and vulnerability conditions, through measures that will allow buildings to be considered *Climate Ready*.

Pamplona, Spain

Aurora Monge-Barrio
Ana Sánchez-Ostiz Gutiérrez

Acknowledgements

First, we would like to thank our partners at SAVIArquietctura Research Group (University of Navarra, Spain) for their support during the writing of this book, and also for the time and efforts they have shared with us in the search for a more sustainable architecture. They are: Purificación González, Silvia Domingo, Germán Ramos, Carlos Fernández-Bandera, Joaquín Torres, Juan Echeverría, Miren Juaristi, Elena Aparicio, María Fernández-Vigil, Cristina Güell, Juan Carlos Gamero, Jorge San Miguel y Ana Castillejo. Special thanks to Jorge San Miguel and Ana Castillejo for their essential help during the monitorization process, to Cristina Güell for her review and assistance in the preparation of this book, to Miren Juaristi for her valuable appreciations and to Ainhoa Arriazu, María Dolores Juri and María Alecha for their appreciated collaboration. Second, to Robert Jones and Ann Hanningan for their help with the translation of the texts. Finally, to Springer for giving us the opportunity of publishing this book.

Thanks to all.

Contents

Chapter 1
Introduction. Resilience to Climate Change in the Built Environment in Southern Europe

1.1 Facing Climate Change

Climate change is a fact widely recognized by the international scientific community. The most important contributions are summarized in reports done periodically by the Intergovernmental Panel on Climate Change, IPCC (IPCC et al., 2013). According to this source '*Warming of the climate system is unequivocal, and since the 1950s, many of the observed changes are unprecedented over decades to millennia*' and '*It is extremely likely that human influence has been the dominant cause of the observed warming since the mid-20th century*'. First, the Kyoto Protocol (UNFCCC, 1998), and lately the agreement reached in the XXI 2015 United Nations Climate Change Conference COP21/CMP11 (UNFCCC, 2015), supposes an international compromise in the reduction of the greenhouse gas emissions, with the objective of limiting by 2100 global warming by 2 °C. The Scientific Community considers that this is the threshold beyond which there may be a risk of abrupt and irreversible changes, with dangerous and possibly catastrophic impacts. Actually, the average temperature is 0.85 °C higher, with an unprecedented increase in temperatures in the last three decades, with new records every year in global average temperatures (WMO, n.d.). So action or continuing action is imperative, forcing the pace of measures for mitigation and adaptation to climate change (IPCC, 2014). The magnitude of future climate change and its impact depend on the effectiveness of global climate mitigation efforts (EEA, 2016).

The European Union (EU) has established objectives to progressively reduce greenhouse gas emissions from now to 2050, aiming to put the EU in the path to a transformation into a low-carbon economy (European Commission, n.d.). Previous to this, a 2020 climate and energy package and a 2030 climate and energy framework have been established, with specific objectives for reductions in greenhouse gas emissions (from 1990 levels), share of renewables, and improvement in energy efficiency. The sectors implicated in the 2020 target are housing, agriculture, waste and transport (excluding aviation). However, all sectors

© Springer International Publishing AG 2018

A. Monge-Barrio and A. Sánchez-Ostiz Gutiérrez, *Passive Energy Strategies for Mediterranean Residential Buildings*, Green Energy and Technology,
https://doi.org/10.1007/978-3-319-69883-0_1

responsible for Europe's emissions are expected to contribute to a 2050 low-carbon economy including power generation and industry. Directive 2010/31/EU *on the energy performance of buildings* includes well-known key targets (EPDB, 2010), and all countries have to report their progress every year in order to successfully achieve the proposed European goals.

Fossil fuel combustion is a cause of both local air pollutants (especially particulates, ozone, methane, nitrogen oxides and sulphur dioxide) and greenhouse gases, both directly affecting people's health, but sources of renewable energy (photovoltaic, solar thermal, wave and wind power) do not appear to have adverse effects on health (Haines, Kovats, Campbell-Lendrum, & Corvalan, 2006).

In IPCC's Europe-based study, observed climate trends and future climate projections show regionally varying changes in temperature and precipitation in Europe, with projected rises in temperature throughout Europe, increasing precipitation in Northern Europe and decreasing in Southern Europe, with a marked increase in high-temperature extremes, meteorological droughts and heavy precipitation events variation across Europe (Kovats et al., 2014). In the report, the European region has been divided into five subregions derived from aggregating the Environmental Zones developed by Metzger (Metzger, Bunce, Jongman, Mücher, & Watkins, 2005): Alpine, Atlantic, Northern, Continental and Southern. A comprehensive assessment of the impacts in subregions by 2050 is shown in the report, focused on energy (including energy production and annual energy consumption), transport, settlements (focused on floods), tourism, human health (including mortality and morbidity related to heatwaves), social and cultural impacts, and environmental quality. All subregions are vulnerable to some impacts from climate change but these impacts differ significantly between the subregions. Vulnerability and adaptive capacity differ between them. Among them, Southern Europe is particularly vulnerable to climate change, as multiple sectors will be adversely affected, including energy and population health (Kovats et al., 2014).

The European Environment Agency (EEA) also studies the impacts and vulnerabilities of climate change by biogeographical regions: Arctic, Atlantic, Mountain, Coastal zones and regional seas, boreal, continental, and Mediterranean Regions (EEA, 2016), being the delimitation of the last one (Fig. 1.1) almost equal although with slight differences to the Southern Europe delimitated by IPCC.

According to the EEA, the Mediterranean Region is a hotspot of climate change impacts, having a very high number of economic sectors severely affected. The observed and projected climate change and impacts related to the built environment are the following (EEA, 2016):

- Large increase in heatwaves and increase in mortality from heatwaves.
- Decrease in potential for energy production and an increase in cooling demand.
- Decrease in precipitation and river flow, increasing risk of droughts, and increase in competition between different water users.

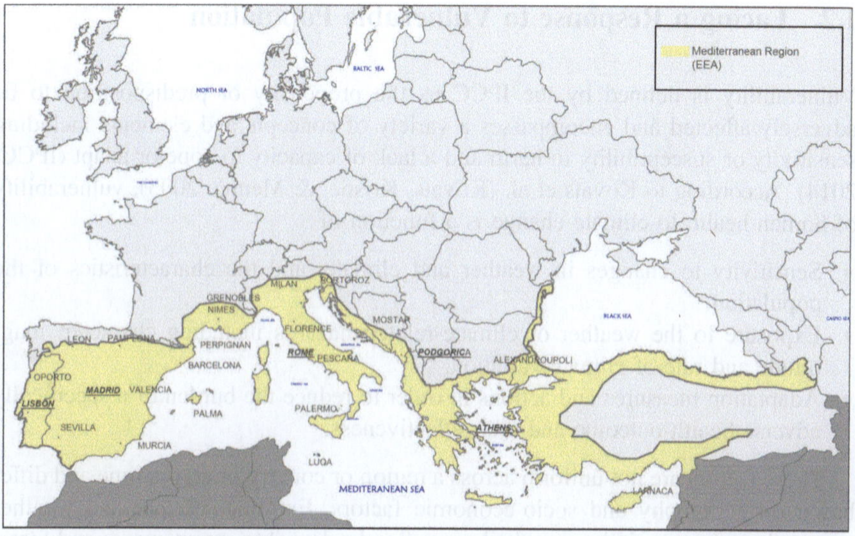

Fig. 1.1 European Mediterranean Region. Adapted from EEA Source (EEA, 2016)

Other changes and impacts summarized by the EEA for the Mediterranean Region are increasing risk of biodiversity loss, increasing risk of forest fires, increasing water demand for agriculture, decrease in crop yields, increasing risks for livestock production, expansion of habitats for southern disease vectors, decrease in summer tourism and potential increase in other seasons, increase in multiple climatic hazards, most economic sectors negatively affected, and high vulnerability to spillover effects of climate change from outside Europe.

Some impacts of climate change can actually be beneficial. In the Mediterranean zone, a reduction in heating energy demand in most locations is expected. However, there is a consensus that, on balance, benefits are not expected to outweigh the risk of negative effects (EEA, 2016).

On the other hand, adaptation is defined by the IPCC as the process of adjustment to actual or expected climate (and its effects), and in human systems, adaptation seeks to moderate or avoid harm, and even exploit beneficial opportunities. The capacity to adapt in Europe is high compared to other regions, but there are important differences in impacts and in capacity to respond between and within the European subregions, the countries around the Mediterranean having the lowest capacity (Kovats et al., 2014).

Adaptation appears, at present, to be a relatively low priority issue for city planners and governors in Europe (Carter, 2011), although climate change mitigation strategies, policies and actions are progressing at all governance levels (EEA, 2016). There is a need for better monitoring and evaluation of local and national adaptation and mitigation responses to climate change, and this includes evaluating the effectiveness of adaptation measures, over a range of timescales (Kovats et al., 2014; EEA, 2016).

1.2 Facing a Response to Vulnerable Population

Vulnerability is defined by the IPCC as the propensity or predisposition to be adversely affected and encompasses a variety of concepts and elements including sensitivity or susceptibility to harm and a lack of capacity to cope or adapt (IPCC, 2014). According to Kovats et al. (Kovats, Kristie, & Menne, 2003), vulnerability of human health to climate change is a function of

- Sensitivity to changes in weather and climate, and the characteristics of the population.
- Exposure to the weather or climate-related hazards including character, magnitude and rate of climate variation.
- Adaptation measures and actions in order to reduce the burden of a specifically adverse health outcome and their effectiveness.

These factors are not uniform across a region or country or across time and differ based on geography and socio-economic factors. Extreme weather and weather events (heatwaves, windstorms, hail, river floods, droughts, storm surges and forest fires) have adverse social and health effects (as well as significant impacts in multiple economic sectors), although this depends on multiple factors that determine human vulnerability (EEA, 2016). Heatwaves are the deadliest extreme weather event in Europe, and there are regional differences, being Southern Europe particularly affected as the EEA summarizes (Fig. 1.2).

Everyone is affected by climate change, but the effect of its impact on human health depends largely on people's vulnerability and also on their exposure to climatic hazards, and there are important variations within and across European regions (EEA, 2016). Identified population groups vulnerable to heatwaves and hot

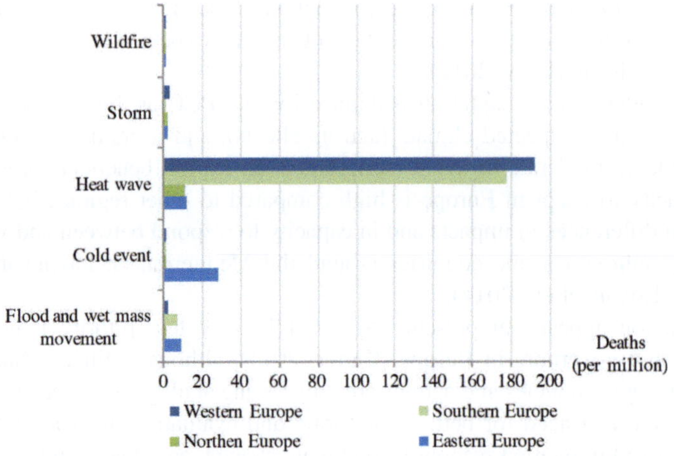

Fig. 1.2 Deaths (per million) due to extreme weather events, by European subregions from 1991 to 2015. Adapted from data summarized by the EEA according to the EM-DAT international disaster data base, Eurostat and WHO. Southern Europe includes Western Asia (EEA, 2016)

weather are the elderly, infants and children, people with chronic diseases, people taking certain medications, people whose socio-economic status may make them more vulnerable (ethnicity, occupation, education and social isolation), and people in certain occupations that require working in hot conditions, particularly under heavy physical activity (Matthies, Bickler, Marin, & Hales, 2008). These groups must therefore be specially identified and considered, due to their vulnerability and difficulties to adapt.

Actually, in the European Union, the group of population over 65 years supposes a 13.6% of the total, according to Eurostat. Although in projections forecasted for 2080, it is foreseen that the European population will diminish, the group of the elderly is expected to increase, especially relevant among them the group over 80 (EUROSTAT, n.d.).

Apart from this, energy poverty is understood as a temporary condition consequence of energy vulnerability, the home not receiving the adequate quantity of energy services (Hills, 2011). Although having mild climatic conditions, the Mediterranean countries in the south of Europe have a higher than average percentage of population in the EU, who cannot keep their homes warm enough in winter conditions (Bouzarovski, 2013). According to the same study, 30% of the total population of those countries is unable to keep their home cool enough in summer conditions.

Residential buildings have the peculiarity, in relation to office or educational buildings that the population inhabiting them is not homogeneous but intergenerational from young children to the elderly, and there may be ill or disabled people, precisely the ones who spend the most time in the home. There is also a common agreement in academic literature on the prevalence of high rates of energy poverty among the elderly, the handicapped and long-term ill households (Bouzarovski, 2013).

Finally, on a global scale, adverse impacts of future climate change are projected to outweigh beneficial ones in human health. Heatwaves and extreme cold spells are associated with increases in mortality and morbidity, especially in vulnerable populations. Although cold-related mortality is projected to go down due to better quality living conditions, there is inconclusive evidence if this will be also due to projected warming (EEA, 2016).

1.3 *Climate-Ready* Buildings

Resilience is defined by the IPCC as the capacity of social, economic and environmental systems to cope with a hazardous event or trend or disturbance, responding or reorganizing in ways that maintain their essential function, identity and structure, while also maintaining the capacity for adaptation, learning and transformation (IPCC, 2014). Resilience to very extreme events varies widely by sector and by country (Kovats et al., 2014), depending not only on the magnitude of the hazard but also in the capacity for adaptation.

In Southern Europe, buildings in most locations are designed attending to temperate climates, with energy needs for conditioning in winter and, in some locations, in summer. But buildings that were originally designed for certain thermal conditions will need to be used in warmer climates in the future (Matthies et al., 2008), so their ability as climate moderators could be compromised (Roaf, Crichton, & Nicol, 2009). The primary goal of buildings' adaptive capacity is to reduce future vulnerability to climate variability and change, so *coping capacity* describes what could be implemented now to minimize negative effects of climate variability and change (Kovats et al., 2003).

New buildings and rehabilitations projects should incorporate this determining factor in their design without forgetting the coldest conditions to deal with in the present moment, therefore looking for a design that includes both current and future needs along the whole year. Passive energy measures should be incorporated into the design from the beginning of the project.

Having in mind the residential stock already built, the focus on rehabilitation seems evident, because it increases residential buildings' lifetime. Rehabilitation nevertheless involves a great complexity from every point of view, both technical and socio-economic. Rehabilitation's starting point is a building, with some characteristics that will very much condition the design. Actuations may include treating the envelope to optimize good designs, or the reconfiguration of spaces or shapes of the building, with a higher implementation cost.

Passive measures are part of the buildings architecture and constructive design, and in the south of Europe, there is the complexity of having to respond to both winter (currently predominant in most locations) and summer conditions. Passive measures to respond to underheated and overheated seasons sometimes merge, but other times they can be quite different and even opposed, so the design of optimal solutions for the whole year and for future changing and warming conditions constitutes a real challenge (Fig. 1.3).

Fig. 1.3 Examples of new and traditional passive architecture in Southern Europe. Oropesa del Mar (Spain) and Santorini (Greece)

Passive design will be limited to point where, over certain thresholds, it will not be possible to guarantee adequate indoor conditions, neither in summer nor in winter. Energy consumption then will be necessary in order to ensure people's health and well-being. Passive cooling measures alone are unlikely to be sufficient to address adaptation in all regions and types of buildings, and retrofitting current housing stock can be expensive (Kovats et al., 2014).

Nonetheless, although in the future the help of energy consumption for cooling will be required, a good design leaning on passive measures will reduce energy requirements to the minimum, to ensure adequate conditions both during summer in general and under extreme heat-related events. Increased energy demand, especially in Southern Europe, requires additional power generation capacity, underused during the rest of the year and entailing higher supply costs (Kovats et al., 2014).

Lastly, it is important to ensure that responses to heatwaves do not contribute to exacerbate the problem of warming conditions. It would be easy to assume that the solution to protect population is the widespread use of air conditioning, although it supposes the increase of energy consumption and greenhouse gas emissions. In addition, air conditioning could not protect socio-economic vulnerable populations since they cannot afford energy costs (Matthies et al., 2008).

1.4 Synergies and Co-benefits of Mitigation Strategies

Efforts that contribute to the mitigation of emissions related to buildings that contribute to global warming have multiple benefits and added synergies. The European Union, promoting energy efficiency in buildings (among other sectors), also aims to increase its energy security, reducing dependence on imported energy and contributing to achieving a European Energy Union. Creation of jobs and increase in green growth, making Europe more competitive, are other expected benefits (Climate action 2020, n.d.).

Contributing to the eradication of energy poverty is one of the main synergies with mitigation efforts, through the improvements of the energy efficiency of residential buildings conducting to less energy consumption (Ürge-vorsatz & Herrero, 2012; Santamouris, 2016).

This measure, as opposed to subsidies and direct help to energy costs in households, which supposed a short-term measure (sometimes necessary, but ought to be considered temporary), contributes in the medium and long terms to improve people's life conditions by reducing the need for energy consumption and therefore related emissions.

These efforts in mitigation through improvements in energy efficiency in buildings also bring health benefits to all the population, through reduced air pollution and by reducing concentrations of ozone or NOx, particularly in summer and under heatwave conditions. Other issues related to energy consumption, energy poverty and local climate change are well described by Santamouris review (Santamouris, 2016).

1.5 Conclusions

The Mediterranean Area of Southern Europe is widely recognized as a hotspot of climate change impacts. One of the main impacts to people is on health, mainly caused by heatwaves. Everybody is affected by these new conditions, although vulnerable populations suffer the greatest impacts, and have less capacity to adapt.

Climate-ready residential buildings for future warming scenarios without forgetting current conditions constitute a real challenge in architectural design both for new and refurbished buildings. In temperate climates, such as the Mediterranean ones, buildings must attend to winter and summer conditions, and passive measures incorporated in the architecture will contribute to reduce or minimize overheating or energy needed for heating or cooling the spaces.

References

Bouzarovski, S. (2013). Energy poverty in the European Union : Landscapes of vulnerability, 276–289. http://doi.org/10.1002/wene.89.

Carter, J. G. (2011). Climate change adaptation in European cities. *Current Opinion in Environmental Sustainability, 3*(3), 193–198. https://doi.org/10.1016/j.cosust.2010.12.015.

Climate action 2020. (n.d.). 2020 climate & energy package | Climate action. Retrieved March 16, 2017 from https://ec.europa.eu/clima/policies/strategies/2020_en.

EEA. (2016). Climate change, impacts and vulnerability in Europe 2016: An indicator-based report.

EPDB. (2010). Directive 2010/31/EU of the European Parliament and of the Council of 19 May 2010 on the energy performance of buildings (recast). *Official Journal of the European Union*, 13–35. http://doi.org/10.3000/17252555.L_2010.153.eng.

European Commission. (n.d.). 2050 low-carbon economy | Climate action. Retrieved March 16, 2017 from https://ec.europa.eu/clima/policies/strategies/2050_en.

EUROSTAT. (n.d.). Eurostat. Retrieved April 7, 2017 from http://ec.europa.eu/eurostat.

Haines, A., Kovats, R. S., Campbell-Lendrum, D., & Corvalan, C. (2006). Climate change and human health: Impacts, vulnerability and public health. *Public Health, 120*(7), 585–596. https://doi.org/10.1016/j.puhe.2006.01.002.

Hills, J. (2011). Fuel poverty: The problem and its measurement. Retrieved from http://www.decc.gov.uk/en/content/cms/funding/fuel_poverty/hills_review/hills_review.aspx.

IPCC. (2014). *Climate change 2014: Impacts, adaptation, and vulnerability* (Climate Change 2014: Impacts, Adaptation and Vulnerability—Contributions of the Working Group II to the Fifth Assessment Report). 1–32. http://doi.org/10.1016/j.renene.2009.11.012.

IPCC, Stocker, T. F., Qin, D., Plattner, G.-K., Tignor, M. M. B., Allen, S. K., … Midgley, P. M. (2013). *Climate change 2013—The physical science basis. Intergovernmental panel on climate change.* http://doi.org/10.1038/446727a.

Kovats, S., Kristie, L. E., & Menne, B. (2003). Methods of assessing human health vulnerability and public health adaptation to climate change. *Health and Global Environmental Change. Series No.1.*

Kovats, R. S., Valentini, R., Bouwer, L. M., Georgopoulou, E., Jacob, D., Martin, E., … Soussana, J. (2014). *Climate change 2014: Impacts, adaptation, and vulnerability. Part B: Regional aspects. Contribution of working group II to the fifth assessment report of the intergovernmental panel on climate change. Europe*, 1267–1326.

Matthies, F., Bickler, G., Marin, N., & Hales, S. (2008). *Heat–health action plans*. WHO Europe, 45. Retrieved from http://www.euro.who.int/__data/assets/pdf_file/0006/95919/E91347.pdf.

Metzger, M. J., Bunce, R. G. H., Jongman, R. H. G., Mücher, C. A., & Watkins, J. W. (2005). A climatic stratification of the environment of Europe. *Global Ecology and Biogeography, 14* (6), 549–563. https://doi.org/10.1111/j.1466-822X.2005.00190.x.

Roaf, S., Crichton, D., & Nicol, F. (2009). *Adapting buildings and cities for climate change—A 21st century survival guide second edition*. http://dx.doi.org/10.1016/B978-1-85617-720-7.00001-2.

Santamouris, M. (2016). Innovating to zero the building sector in Europe: Minimising the energy consumption, eradication of the energy poverty and mitigating the local climate change. *Solar Energy, 128,* 61–94. https://doi.org/10.1016/j.solener.2016.01.021.

UNFCCC. (1998). Kyoto protocol to the United Nations framework Kyoto protocol to the United Nations framework. *Review of European Community and International Environmental Law, 7,* 214–217. https://doi.org/10.1111/1467-9388.00150.

UNFCCC. (2015). Report of the Conference of the Parties on its twenty-first session, held in Paris from 30 November to 13 December 2015. *Addendum-Part Two: Action Taken by the Conference of the Parties, 1194* (January), 1–36. Retrieved from http://unfccc.int/resource/docs/2015/cop21/eng/10a01.pdf.

Ürge-vorsatz, D., & Herrero, S. T. (2012). Building synergies between climate change mitigation and energy poverty alleviation, (October). http://doi.org/10.1016/j.enpol.2011.11.093.

WMO. (n.d.). Retrieved from https://public.wmo.int/en/our-mandate/climate.

Chapter 2
Climate Conditions and Future Scenarios in Southern Europe

2.1 Climate Conditions and Climatic Stratification

Climate classifications help understand the main characteristics of climate, which the built environment has to deal with, in order to be able to design the most adequate strategies for the minimum consumption of energy. Southern Europe presents different characteristics and associated conditioning needs to those of the rest of Europe. Climate is mild and wet during the winter and hot and dry during the summer, however the main characteristic is the high variability in seasonal mean temperature and total precipitation (Cartalis, 2016).

2.1.1 Southern or Mediterranean Europe

The Southern European subregion defined by the Intergovernmental Panel on Climate Change, IPCC (Kovats et al., 2014) or the Mediterranean Region defined by the European Environment Agency, EEA (2016) are geographical and ecological regions that explain, better than political boundaries, climate change and its impacts (Fig. 2.1).

Both regions have only slight differences in their delimitations. Subregions by the IPCC were accomplished following Environmental Stratification of Europe (EnS), developed by Metzger (Metzger, Bunce, Jongman, Mücher, & Watkins, 2005a).

EnS consists of 84 strata, aggregated into 13 Environmental Zones, based on the selection of 20 relevant environmental variables of climate, geomorphology, oceanity and northing, and geology and soil, with 1 km^2 resolution. These variables are altitude; slope and northing (latitude); oceanity; monthly minimum and maximum temperature; precipitation and percentage sunshine in the months of January,

© Springer International Publishing AG 2018

A. Monge-Barrio and A. Sánchez-Ostiz Gutiérrez, *Passive Energy Strategies for Mediterranean Residential Buildings*, Green Energy and Technology,
https://doi.org/10.1007/978-3-319-69883-0_2

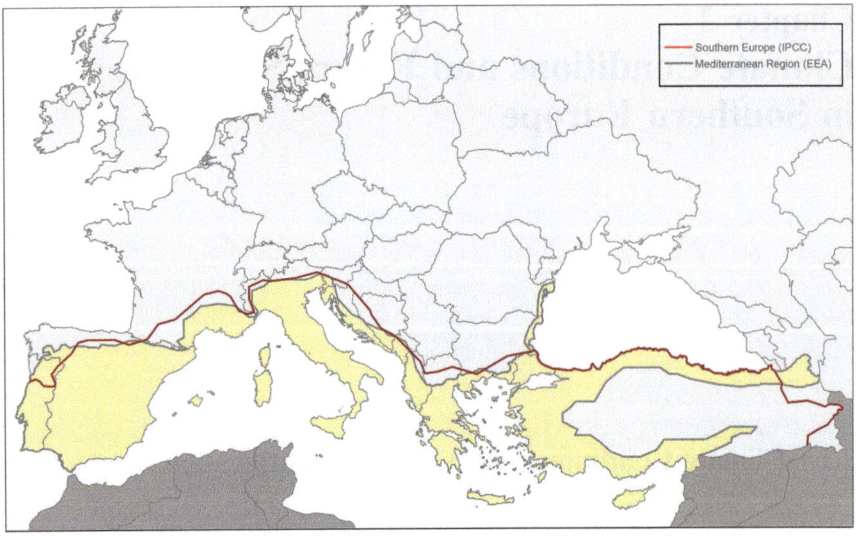

Fig. 2.1 Map including Southern European limits, defined by IPCC (Kovats et al., 2014) and the Mediterranean Region defined by EEA (2016). Adapted from both sources

April, July and October (and precipitation in November), in order to reflect overall seasonal climate variations.

The data used for climate purposes was 1971–2000 climate series, calculated as 30-year averages, from the CRU-TS1.2 dataset, developed by the Climatic Research Unit at the University of East Anglia. With the analysis of the three principal components, 88% of the variation could be explained and therefore clustered with ISODATA clustering routine into two main classes, Northern and Southern or Mediterranean Europe (Fig. 2.2). These three principal components are temperature gradient, oceanity gradient and precipitation pattern. While the first one explains 65% of the variation, the second and third ones explain 15 and 8% of the variation, respectively.

Northern Europe was clustered into 40 strata, covering 70% of Europe, and Southern Europe was clustered into 30 strata, that covering 30% of Europe, recognizes the greater variability of the latter. Detailed information of each stratum is available online (Metzger et al., 2005b).

The Environmental Zones that constitute the Southern or Mediterranean Region are mainly the following and are shown in Fig. 2.3 (Lusitain Zones, LUS strata, are also partially included depending on the delimitation):

- MDM. Mediterranean Mountains (11 strata);
- MDN. Mediterranean North (10 strata) and
- MDS. Mediterranean South (9 strata).

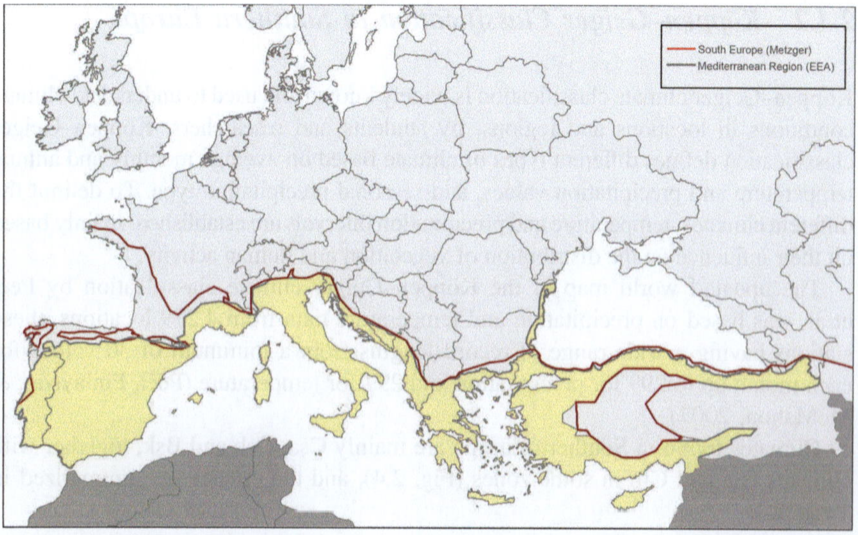

Fig. 2.2 Northern and Southern Europe delimitated only by temperature gradient, oceanity gradient, and precipitation pattern and the Mediterranean Region described by the EEA and based on Metzger (Metzger et al., 2005a). Adapted from both sources

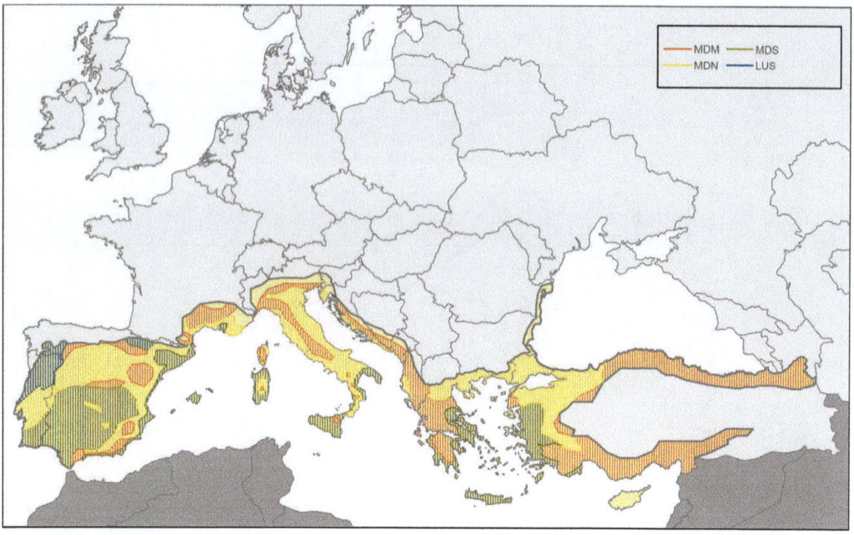

Fig. 2.3 Maps including Metzger climatic zones in Southern Europe, based on and adapted from Metzger (Metzger et al., 2005a)

2.1.2 *Köppen–Geiger Classification in Southern Europe*

Köppen–Geiger climate classification is widely known and used to understand climate conditions in locations and regions, by students and researchers. Köppen–Geiger classification defines different types of climate based on average monthly and annual temperature and precipitation values, and seasonal precipitation type. To delimit the different climates, temperature and precipitation intervals are established, mainly based on their influence on the distribution of vegetation and human activity.

The updated world map of the Köppen–Geiger climate classification by Peel et al. was based on precipitation and temperature data from 4,279 locations, these stations having a wide range of record lengths from a minimum of 30 values for each month up to 299 for precipitation and 297 for temperature (Peel, Finlayson, & McMahon, 2007).

Climates found in Southern Europe are mainly Csa, Csb and Bsk, together with climates Cfa and Cfb in some zones (Fig. 2.4), and the criteria are summarized in Table 2.1:

- Csa and Csb, recognized as Mediterranean climates, are dry and temperate climates. Csa is typical Mediterranean with hot summers, and Csb is oceanic Mediterranean with warm summers. Csa climates are found in coastal areas, like Palma (Balearic Islands, Spain), or continentalized, like Madrid (Spain). Csb climates, with oceanic influence, are a transition between Csa and Cfb, in places like Oporto (Portugal) or Leon (Spain).

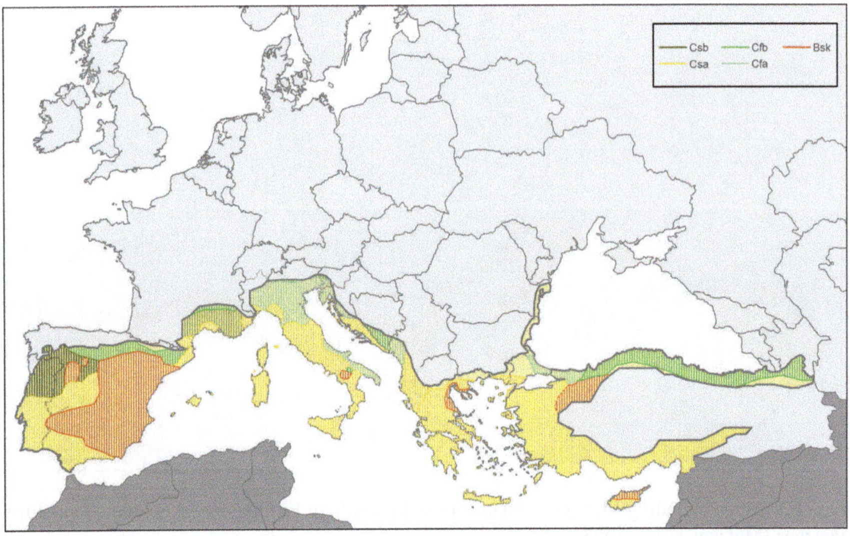

Fig. 2.4 Map of climatic zones according to Köppen–Geiger in Southern Europe, based on and adapted from Peel et al. (2007)

Table 2.1 Köppen-Geiger classification. Summary of symbols and defining criteria applied for the climates of the Mediterranean European Region. Extracted and adapted from Peel et al. (2007)

1st	2nd	3rd	Description	Criteria
B			Arid	MAP < 10 × $P_{threshold}$
	S		•Steppe	MAP • 5 × $P_{threshold}$
		k	–Cold	MAT < 18
C			Temperate	T_{hot} > 10 & 0 < T_{cold} < 18
	s		•Dry summer	P_{sdry} < 40 & P_{sdry} < $P_{wwet}/3$
	f		–Without dry season	Not (C_s) or C_w)
		a	–Hot Summer	$T_{hot} \geq 22$
		b	–Warm Summer	Not (a) & $T_{mon10} \geq 4$

Notes
MAP mean annual precipitation
MAT mean annual temperature
T_{hot} temperature of the hottest month
T_{cold} temperature of the coldest month
T_{mon10} number of months where the temperature is above 10
P_{sdry} precipitation of the driest month in summer
P_{wwet} precipitation of the wettest month in summer
$P_{thershold}$ varies according to the following rules (if 70% of MAP occurs in winter then)
$P_{threshold}$ 2 × MAT, if 70% of MAP occurs in summer then $P_{threshold}$=2 × MAT + 28, otherwise
$P_{threshold}$ 2 × MAT + 14
Summer is defined as the warmer 6-month period from April to September
Winter is defined as the cooler 6-month period from October to March

- Cfa (subtropical) and Cfb (oceanic) are temperate climates without a dry season, that differ in summer temperatures, Cfa is found in Milan (Italy) and Perpignan (France), and Cfb in Mostar (Bosnia and Herzegovina) and Portoroz (Slovenia).
- BSk (dry Mediterranean) is a semi-arid and steppe climate. It is principally found in the Iberian Peninsula, in Alicante, Murcia or Valencia.

Finally, Mediterranean climates Csa and Csb can be found in Africa (in Beirut and Tanger or in Cape Town), Australia (Perth) and America (Los Angeles and San Francisco, or Santiago de Chile).

The whole-of-record approach of Peel et al., assumes that data from one period is comparable with data from any other period, but as the authors recognize this assumption can be proved wrong by global or local trends, like the recent observed warming of global surface temperature, or, locally, cities affected with Urban Heat Island, UHI. However, at a broad level of climate typologies, this climate classification has been found to be relatively insensitive to temperature trends, although sensitivity is likely to be greater in the transition zones between climate types (Peel et al., 2007).

Location	Country	Koppen-Geiger	Altitude	Latitude	Mean Annual Temp	Mean Annual Temp 2050	Temp of the coldest month	Temp of the coldest month 2050	Temp of the hottest month	Temp of the hottest month 2050
Athens	GRC	Csa	15	37°54'	18,2	20,3	10,7	11	28	31,4
Sevilla	ESP	Csa	31	37°25'	18,6	21,1	10,2	11,9	27,8	31,5
Larnaca	CYP	Csa	2	34°52'	20	21,5	11,6	13,2	27,7	30,2
Murcia	ESP	BSk	62	38°0'	18,6	21,2	10,4	12,4	27,7	31,8
Luqa	MLT	Csa	91	35°50'	19	20,6	13	14	26,9	29,2
Palermo	ITA	Csa	117	38°5'	18,4	20	11,5	12,1	26,7	29,2
Podgorica	MNE	Cfa	33	42°22'	15,1	17,9	5,5	7,8	26,5	31,2
Florence	ITA	Cfb	38	43°47'	15,2	17,8	7	9,1	25,8	29,5
Valencia	ESP	BSk	62	39°30'	17,4	19,9	10,1	12	25,7	29,7
Mostar	BIH	Cfb	108	43°19'	14,6	17,4	3,8	6,2	25,6	30,2
Madrid	ESP	Csa	582	40°27'	14,4	17,2	5	6,9	25,5	30,4
Palma	ESP	Csa	7	39°32'	16,7	18,7	9,5	11,2	25,4	28,2
Lisbon	PRT	Csa	114	38°46'	17,4	19,4	11,2	12,7	25	27,4
Alexandroupoli	GRC	Csa	3	40°50'	14,5	17,3	5,2	7,4	25	29,7
Milan	ITA	Cfa	103	45°25'	13,6	16,3	2,2	4,1	24,3	28,6
Rome	ITA	Csa	3	41°47'	15,9	18	8,5	10,3	24,2	27,4
Pescara	ITA	Cfa	11	42°25'	14,9	17,4	6,6	8,6	24,1	28,1
Barcelona	ESP	Csa	6	41°16'	16,1	18,3	9,3	11,1	24	27,9
Perpignan	FRA	Cfa	47	42°43'	15,6	17,8	8,8	10,3	24	28,1
Nimes	FRA	Csa	62	43°52'	14,8	17	6	8	23,9	27,7
Portoroz	SVN	Cfb	95	45°31'	13,8	16,6	3,8	6,3	23,6	28,3
Pamplona	ESP	Cfb	453	42°46'	12,3	14,4	4,3	5,7	20,7	24,6
Grenobles	FRA	Cfb	386	45°20'	11,1	13,7	1,4	3,6	20,1	24,7
Leon	ESP	Csb	914	42°34'	10,5	12,1	2,1	3,6	19,3	21,7
Oporto	PRT	Csb	77	41°13'	14,4	16,3	9,4	11	19,2	22,2

Fig. 2.5 Selected locations in Southern Europe. Mean annual temperature, mean temperature of the coldest and hottest month for current and 2050 scenario

2.1.3 Representative Locations of Southern European Climates

Climate data used in the present book, with a regional approach and mainly focused on the efficiency of the built environment, are mainly *typical year weather files* from ASHRAE, IWEC2, with a general period of 1983–2008 (Huang, 2011). Therefore, some variations over updated Köppen–Geiger map developed by Peel et al. could be found in the assignation of the location classification for this study. Selected locations are presented in Fig. 2.5 in order to understand the differences within the locations of the Southern European Region.

In this figure, differences in mean annual temperature can be found first, from 10.5 °C in Leon (914 m altitude) and 11.1 °C in Grenobles (386 m altitude) to 19 °C in Luqa and 18.6 °C in Murcia and Seville, respectively. Taking into account the severity of the seasons, in winter, Leon and Grenobles continue having the coldest temperatures (1.4 and 2.1 °C) while Larnaca and Palermo have the hottest (11.6 and 11.5 °C). In summer conditions, the milder locations are Oporto and Leon (19.2 and 19.3 °C), while the warmer conditions are given in Athens (28 °C), Sevilla (27.8 °C), and Larnaca and Murcia (27.7 °C). It is also important to highlight that while there are locations with only one very remarkable season, there are other locations with great climatic severity in both seasons, for example, Mostar, Podgorica or Madrid.

Therefore, Southern Europe presents common and identifiable characteristics, principally based on gradient temperature criteria (Metzger et al., 2005a), but also important differences in the annual and monthly hottest and coldest temperatures, which translates into different needs and strategies for building conditioning.

2.2 Observed and Projected Climate Change

In this section, those observed and projected climate change impacts that have major incidence in the built environment in the Mediterranean Europe are exposed. For the aim of this book, these impacts are variability in average temperatures (warming) and an increase in extreme temperature events.

The Mediterranean Region has been identified as one of the most prominent climate response hotspots (Giorgi, 2006), and as such it is recognized and treated by the EEA and the IPCC. A hotspot was defined in the study as a *region for which potential climate change impacts on the environment or different activity sectors can be particularly pronounced, or whose climate is especially responsive to global change.*

2.2.1 Observed Climate Change

The average temperature in Europe has continued to increase with different regional and seasonal warming rates, being the decadal average temperature over land area for 2002–2011, 1.3° ± 0.11 °C above the 1850–1899 average, as is reviewed in the regionalized report for Europe by IPCC (Kovats et al., 2014).

Mediterranean climate variability both in the atmosphere and in the ocean was studied as part of the CIRCE project, and published in the first of two volumes of The Regional Assessment of Climate Change in the Mediterranean, edited by Navarra & Tubiana, RACCM (Ulbrich et al., 2013). With data from 1951 to 2005, the estimated changes indicate statistically significant Mediterranean summer temperature increase and a reduction in winter precipitation in some areas. Land surrounding the Mediterranean Sea has been warming during most of the twentieth century, experiencing during the period 1951–2000 a warming trend of about 0.1 °C per decade, the largest trend occurring over the Iberian Peninsula and the western part of North Africa (up to 0.2 °C/decade, considering mean summer temperatures) (Gualdi et al., 2013).

Since 1950, high-temperature extremes in Europe (hot days, tropical nights and heatwaves) have become more frequent, while low-temperature extremes (cold spells and frost days) have become less frequent on average (Kovats et al., 2014). *Mega-heatwaves* such as the 2003 and the 2010 events likely broke the 500-year-long seasonal temperature records over approximately 50% of Europe (Barriopedro, Fischer, Luterbacher, Trigo, & García-Herrera, 2011).

A Mediterranean-wide study of changes in temperature extremes over the last 50 years was also undertaken by CIRCE, encompassing intensity, frequency and duration, and a prominent increase was found in the whole Mediterranean Region (Ulbrich et al., 2013).

Furthermore, following the same source, a progressive and important drying of land surface since 1900, consistent with the increase in surface air temperature and the decrease in precipitation patterns was found, together with a rise of about 150 mm in the sea level, and considering the last two decades, also a positive salinity trend in the ocean layers. On the other hand, extreme winds related to cyclones and cut-off lows show largely negative trends.

2.2.2 Projected Climate Change

There is a range of possible emission scenarios to assess future impacts. Climate models show significant agreement for all emissions scenarios in warming all over Europe. The majority of scenarios are based on projections in the range of 1–4 °C in the global mean temperature per century, and Europe may experience higher rates of warming as is reviewed in the regionalized report for Europe by IPCC (Kovats et al., 2014). Following the report, the strongest warming projected is in Southern Europe in summer and Northern Europe in winter.

According to the CIRCE project, a substantial warming for 2021–2050 in comparison to the reference period of 1961–1990 might affect the Mediterranean Region under an A1B emission scenario. It might suppose an increase in temperatures of almost 1.5 °C in winter and almost 2 °C in summer, and might occur even in the first decades of the period (Gualdi et al., 2013). Locally, changes can be even warmer, being consistent with a current warming trend in the Iberian Peninsula and the western part of North Africa.

Climate projections in Europe show a marked increase in high-temperature extremes (mainly heatwaves and warm spells), droughts and heavy precipitation events, with variations across Europe, although there are small or no changes in general in wind speed extremes (Kovats et al., 2014). There is a general high confidence concerning changes in high-temperature extremes although there are large differences depending on the emissions scenario.

For the whole Mediterranean Region, an increase in very hot days and nights, longer warm spells and more intense and frequent heatwaves are projected (Gualdi et al., 2013). In the CIRCE project, changes in temperature extremes are presented for the period 2021–2050 as compared to 1961–1990, for summer and winter under the A1B emissions scenario. For land areas, the largest increases in the number of very hot days are projected in summer, and especially over the Iberian Peninsula (up to around 20–24 days, which is consistent with the mean maximum of about 3.5 °C). On the other hand, probability of a summer experiencing mega-heatwaves will increase by a factor of 5–10 until 2050 in Europe due to the 2003 and 2010 heatwaves (Barriopedro et al., 2011).

The number of very cold days and very cold nights beside cold spells duration is projected to decrease, and as an example, in the south-west of the Iberian Peninsula, the decrease in number of very cold nights will be 4 days (Gualdi et al., 2013).

2.2.3 *Predicted Climate Data: Scenarios Used in This Study*

There are a lot of improvements in research of models to explain and predict climate change, and the majority of the countries are developing their own projections and research. These models are translated to the typical meteorological weather files (epw or TMY files) to be used in building energy modelling software only in some countries, and to our knowledge, there are no updated scenarios in such format for the Mediterranean countries, with a common methodology or scenario.

For this reason, the Climate Change World Weather File Generator CCWorldWeatherGen tool developed by Southampton University, in United Kingdom (Jentsch, James, Bourikas, & Bahaj, 2013) was chosen to carry out this study. This tool works with the HadCM3 model output for the A2 Emissions Scenario (medium-high scenario), available from the IPCC data distribution centre. The *morphing* methodology for generating climate change weather data is based on the methods developed by Belcher et al.

Two future time periods are available in the CCworldWeatherGen tool, 2050 (supposing the 2041–2070 period) and 2080 (supposing the 2071–2100 period), and this study has been developed only with the first one. One of the main impacts of climate change in Southern Europe can be seen in the increase in mean temperatures. In Fig. 2.5, expected annual and monthly average temperatures for different locations are shown, calculated using CCWorldWeatherGen tool.

2.3 Conclusions

The Mediterranean Region defined by the European Environmental Agency, EEA, or the Southern Europe defined by the Intergovernmental Panel on Climate Change, IPCC, corresponds to a well-delimitated region with similar impacts and vulnerability, and recognized as a climate change hotspot.

Southern Europe has been traditionally considered within Mediterranean climates as temperate and dry, with hot to warm summers, and with coastal and continentalized ones mainly in the Iberian Peninsula. But it also includes other climates: temperate but without dry seasons and semi-arid ones. So, although the area has common and identified characteristics, there are differences mainly based for the scope of this book on the severity of winter and summer conditions that have impact directly on the indoor environment of buildings.

Observed and projected climate change confirms the trends in warming all throughout the year, and the increase in frequency, intensity and duration of heatwaves and warm spells, for the whole Mediterranean Region and especially over the Iberian Peninsula.

Acknowledgements The authors want to thank the assistance of María Dolores Juri and Ainhoa Arriazu in the elaboration of the maps provided in this chapter.

References

Barriopedro, D., Fischer, E. M., Luterbacher, J., Trigo, R. M., & García-Herrera, R. (2011). The hot summer of 2010: redrawing the temperature record map of Europe. *Science, 332*(6026), 220–224. https://doi.org/10.1126/science.1201224.

Cartalis, C. (2016). Climatic change in the built environment in temperate climates with emphasis on the Mediterranean Area. In S.-N. Boemi, O. Irulegi, & M. Santamouris (Eds.), *Energy performance of buildings. Energy efficiency and built environment in temperate climates* (pp. 19–36). Berlin: Springer.

EEA. (2016). Climate change, impacts and vulnerability in Europe 2016: An indicator-based report.

Giorgi, F. (2006). Climate change hot-spots. *Geophysical Research Letters, 33*(8), 1–4. https://doi.org/10.1029/2006GL025734.

Gualdi, S., Somot, S., May, W., Castellari, S., Déqué, M., Adani, M., … Xoplaki, E. (2013). Future climate projections. In A. Navarra & L. Tubiana (Eds.), *Regional assessment of climate change in the Mediterranean. Vol 1: Air, Sea and Precipitation and Water, Vol. 50* (pp. 53–118). Berlin: Springer Science + Business Media. http://doi.org/10.1007/978-94-007-5772-1.

Huang, J. (2011). ASHRAE research project 1477-RP development of 3,012 typical year weather files for international locations. Final Report.

Jentsch, M. F., James, P. A. B., Bourikas, L., & Bahaj, A. S. (2013). Transforming existing weather data for worldwide locations to enable energy and building performance simulation under future climates. *Renewable Energy, 55,* 514–524. https://doi.org/10.1016/j.renene.2012.12.049.

Kovats R. S., Valentini, R., Bouwer, L. M., Georgopoulou, E., Jacob, D., Martin, E., … Soussana, J. (2014). *Climate change 2014: Impacts, adaptation, and vulnerability. Part B: Regional aspects. Contribution of working group II to the fifth assessment report of the intergovernmental panel on climate change. Europe*, 1267–1326.

Metzger, M. J., Bunce, R. G. H., Jongman, R. H. G., Mücher, C. A., & Watkins, J. W. (2005). A climatic stratification of the environment of Europe. *Global Ecology and Biogeography, 14* (6), 549–563. https://doi.org/10.1111/j.1466-822X.2005.00190.x.

Metzger, M. J., et al. (2005b). Supplementary material. Table S1. Description of the location of the EnS strata. Retrieved May 10, 2017 from http://onlinelibrary.wiley.com/doi/10.1111/j.1466-822X.2005.00190.x/full.

Peel, M. C., Finlayson, B. L., & McMahon, T. A. (2007). Updated world map of the Köppen-Geiger climate classification. *Hydrology and Earth System Sciences, 15*(3), 259–263. https://doi.org/10.1127/0941-2948/2006/0130.

Ulbrich, U., Xoplaki, E., Dobricic, S., García-Herrera, R., Lionello, P., Adani, M., … Tsimplis, M. (2013). Past and current climate changes in the Mediterranean region. In A. Navarra & L. Tubiana (Eds.), *Regional assessment of climate change in the Mediterranean. Vol 1: Air, sea and precipitation and water* (pp. 9–51). Berlin: Springer Science + Business Media. http://doi.org/10.1007/978-94-007-5772-1.

Chapter 3
Vulnerable and Non-vulnerable Occupants in Residential Buildings

3.1 Introduction

The world population has grown from 3.2 billion in 1950 to 7.3 billion in 2014, and it is projected to be 9.6 billion in 2050. Urban population has grown from 746 million in 1950 to 3.9 billion in 2014 and is expected to be 6.4 billion in 2050, while rural population was 3.4 billion in 2014 and is projected to be 3.2 billion in 2050 (United Nations, 2014). So world population that lives in urban areas in 2014 is 54% of the total, while it was 30% in 1950 and is projected to be 65% in 2050, with an opposite trend in rural urbanization.

Following the same source, population in Europe has grown from 723 million in 1990 to 742 million in 2014 but is expected to decrease in 2050 to 709 million. In general, the trend is a decrease in population towards 2050, except in Western[1] and Northern Europe. The tendency in urban and rural population changes is the same as in the rest of the world, but with higher urban rates from 70.0% in 1990 to 73.4% in 2014, and being expected to be 82.0% in 2050. Finally, Southern Europe[2] has increased its population from 143 m in 1990 to 156 m in 2014, but it is expected to decrease to 151 m in 2050. Urban population constitutes 69.8% in 2014 and is expected to increase to 79.5% in 2050 (United Nations, 2014).

Alternatively, following the European Environment Agency, EEA, (n.d.), historical data from 1950 and projections for 2100 for the whole of the European Union and for some Mediterranean countries are shown in Fig. 3.1. Population in general will tend to decrease in the European Union as a whole and specifically in countries like Spain, Italy or Greece (although a continuous increase is estimated in France during all this century).

[1]Includes France.

[2]In this report, Southern Europe is Albania, Andorra, Bosnia and Herzegovina, Croatia, Gibraltar, Greece, Holy See (Vatican City State), Italy, Malta, Montenegro, Portugal, San Marino, Serbia (including Kosovo), Slovenia, Spain (including Canary Islands and Ceuta and Melilla) and TYR Macedonia (United Nations, 2014).

© Springer International Publishing AG 2018

A. Monge-Barrio and A. Sánchez-Ostiz Gutiérrez, *Passive Energy Strategies for Mediterranean Residential Buildings*, Green Energy and Technology,
https://doi.org/10.1007/978-3-319-69883-0_3

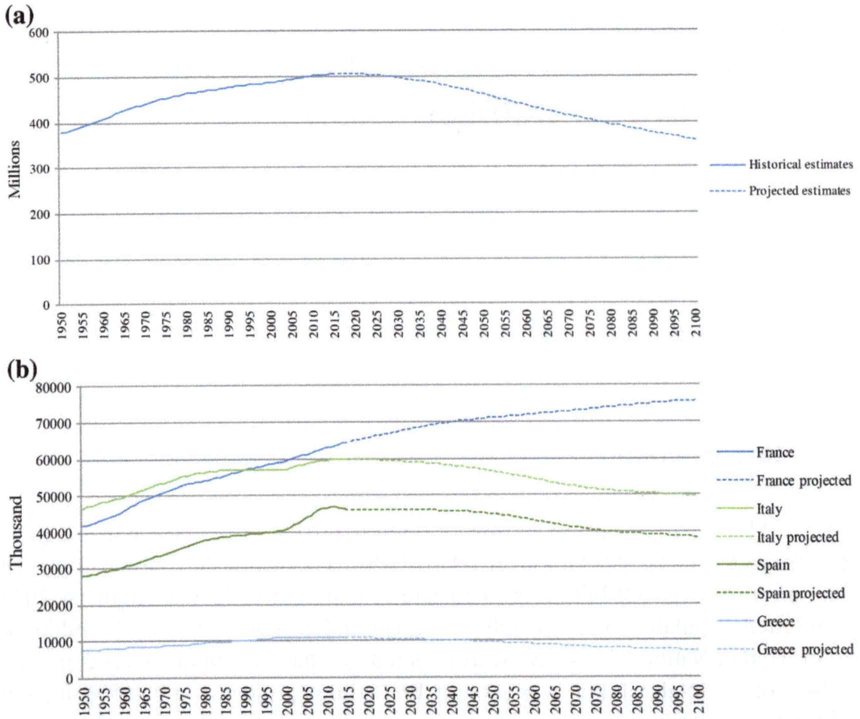

Fig. 3.1 Historical and projected population (1950–2100), in the European Union, EU-28 **a** and in France, Italy, Spain and Greece **b** Data retrieved from the European Environment Agency (EEA, n.d.)

3.2 Occupants as Main Actors in Residential Buildings

Buildings must endeavour to achieve adequate indoor environment conditions for people to be able to carry out their activities and rest properly. These conditionings are of hydrothermal, acoustic, light and air quality type, and are specifically considered in the comfort standards for the design of energy efficient buildings. Their achievement will depend on the design and characteristics of the building including its facilities: heating, ventilation and air conditioning systems.

In the search for reducing energy consumption of buildings and associated CO_2 emissions, aiming at adequate (or optimal) comfort conditions for users will be key in order to achieve the proposed objectives. If the users are not comfortable, they could use any type of energy to relieve or restore that discomfort situation (Nicol & Humphreys, 2002).

For this reason, although different research studies energy demand and consumption of *buildings*, it is worth considering that ultimately *people* use energy

(Janda, 2009), so occupants *decide to* switch on or switch off the devices. This issue is especially relevant in residential buildings where this use can be different and varied through time in each dwelling, and it is indispensable to know about this in order to properly evaluate energy reduction, always being associated to users' health and comfort.

Heating and cooling patterns of use of schedules and setpoints are related to the time spent in the dwelling and will depend on age and physical condition, on the user's situation (employed, unemployed, retired, at school), work or school times, and household's economic situation. It will also depend on the dwellings energy efficiency, expectative of users in relation to the building envelope and their services, which are closely related to the climate zones.

In centralized services (at building or district heating level), there are still old buildings where individual regulation and control systems have not been implemented, where interior conditions can get to be excessively hot and expensive in winter, and where the user has no deciding power (Tirado Herrero & Ürge-Vorsatz, 2012). In individual services or centralized services with individual regulation, the user can *choose* not to use heating or cooling systems due to not being able to afford them.

Lastly, every single person occupying residential buildings will be affected by the building's interior conditions if inadequate, whether warm or cold, but it will be the most vulnerable population the one that will be principally exposed to consequences on health.

3.3 Vulnerable Occupants

Vulnerable population is the one which is susceptible to be adversely affected by hazards, with high sensitiveness to harm and/or low capacity to adapt (IPCC, 2014). Although there are other groups of vulnerable population (see paragraph 1.2, in Chap. 1), we will mainly consider three groups in relation with residential buildings: ageing population, people in energy poverty and people with disabilities or long-term illnesses.

Each dwelling in residential buildings is characterized by the possibility of having an intergenerational population, from infants to the elderly (as opposed to the profile found in an office building, e.g.), the possibility of accommodating occupants with disabilities, or in situations defined as *transitory* related to health or socio-economic aspects.

The most vulnerable population is at the same time the one that is used to spending more time at the dwelling, either for being retired or for being unemployed, or for having disabilities that prevent or reduce their mobility. Vulnerable population can be specially exposed, even at home, to both severe winter and summer temperatures, precisely because of their condition. That is why dwellings, especially those built or rehabilitated with social means, should specifically respond to this target group due to the difficulties they can find in order to adapt themselves to these severe conditions.

Indoor temperature is mainly conditioned by the characteristics of the buildings, principally by the thermal envelope and the heating and cooling systems. For this reason, if thermal conditioning systems are not used, mostly due to socio-economic factors related to the occupants, the building in general and the envelope in particular will be essential in providing habitable conditions.

The dwelling is generally the most important expenditure or investment of a family and has a long timeline, so it will probably be the place where people will spend their old age. In Europe, 69.4% of dwellings are owned, being this percentage higher in the countries of Southern Europe, according to Eurostat (EUROSTAT, n.d.-b). The rental option in the European Union is 19.7% in standard market rentals and 10.9% in reduced or free rental accommodation (Fig. 3.2). In addition, the most vulnerable population presents a high percentage of dwellings under rental contracts. The oldest and most inefficient dwellings will be cheaper, and sometimes the only ones that can be accessed by a lower income population.

3.3.1 Ageing Population

The group of people over 65 years is experiencing a great rise in Europe fundamentally due to the improvement of life expectancy and the drop in birth rates. In Fig. 3.3, this trend is shown between 2005 and 2015 in the European Union (EU-28), as well as in Mediterranean countries, from the most aged countries like Italy to the youngest ones like Albania or Cyprus, according to Eurostat data (EUROSTAT, n.d.-a).

In projections for different groups of population in the European Union, according to the same source, although a drop in total population is previewed, the group that grows the most in relation to the present moment is the over 65 group (from 13.6 to 16.4%) and specially those over 80 (from 5.3 to 12.3%) years, as is shown in Fig. 3.4.

Although the vulnerability of ageing population can be reduced by the option of care homes, the present tendency is not to institutionalize them due to the greater benefits they achieve by staying in their own home. Older people maintain higher levels of autonomy for a longer time, and their level of social integration and satisfaction with the environment are higher. As an example, in Spain, 87.3% of elderly people prefer to live in their own houses even if they live alone (IMSERSO, 2012).

The percentage of dwellings in different countries in the south of Europe with occupants over 65 years is shown in Fig. 3.5, France having the highest percentage of homes occupied by ageing people and Cyprus the lowest, according to the Entranze project (ENTRANZE, n.d.).

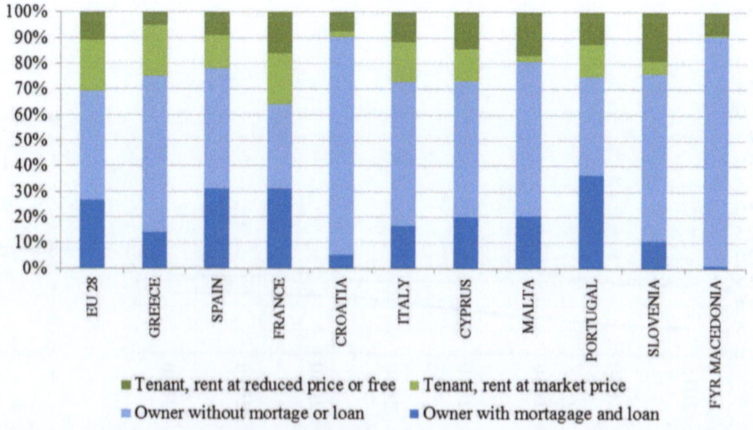

Fig. 3.2 Tenure status in the European Union and in countries of Southern Europe: owner with and without loan or mortgage, and tenants at market price, reduced or free. Data retrieved from Eurostat, Housing statistics, 2015 (EUROSTAT, n.d.-b)

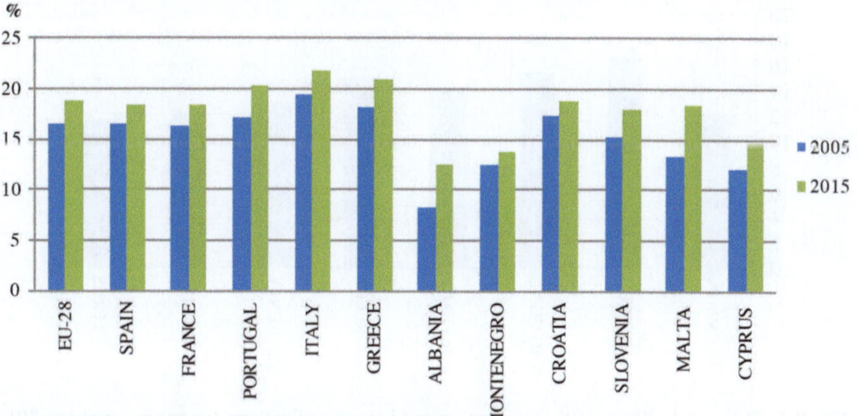

Fig. 3.3 Percentage of over 65s from total population, 2005–2015. Data retrieved from Eurostat (EUROSTAT, n.d.-a)

Dwellings for ageing people will need higher requirements than those for healthy adults precisely due to age conditions. However, it is more common for ageing people to live in homes acquired more than 40 years ago, when there were no regulations in relation to energy efficiency.

On the other hand, due to their condition as pensioners and to a lower life expectancy because of their age, rehabilitating the building is difficult for them, due to having high amortization periods in relation to their age, or for the inconvenience of refurbishment works in their daily routine. This is why, in general, the help of the administration both in the economic and in the social level is important for

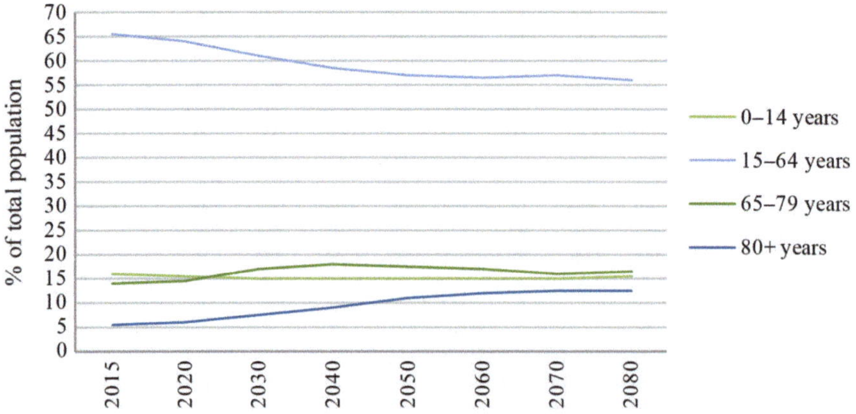

Fig. 3.4 Projections for different population age groups in the European Union, EU-28, 2015–2080. Percentage of the total population. Data retrieved from Eurostat (EUROSTAT, n.d.-a)

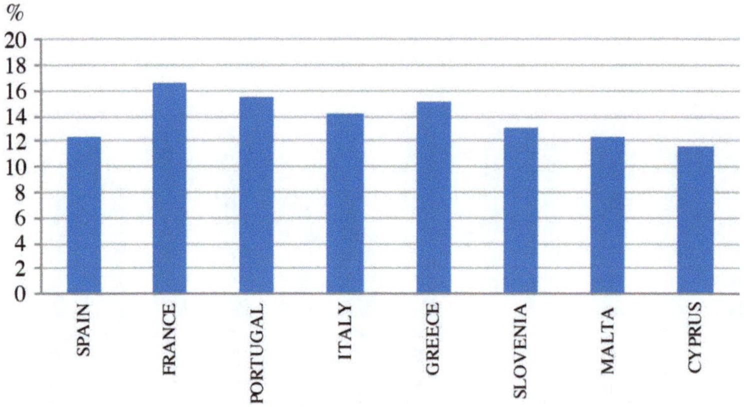

Fig. 3.5 Share of dwellings with occupants over 65 years in Southern European countries, 2014. Data retrieved from Entranze Project (ENTRANZE, n.d.)

achieving successful measures that allow the user to stay at home and live with adequate levels of comfort and health.

Finally, it would be desirable for housing to be flexible enough to give an answer to all necessary conditions in every stage of a person life. Dwellings should be able to be easily adapted instead of having to face strong costs of refurbishment or having to move. The rehabilitation or adaptation of ageing people's dwellings must guarantee all conditions: safety, accessibility, functionality and an adequate indoor environment.

3.3.2 *People in Energy Poverty*

Energy poverty (or fuel poverty) is understood to be the consequence of a more extended and diffuse condition namely energy vulnerability defined as the probability of a home of experimenting a situation where it does not get an adequate quantity of energy services (Bouzarovski & Petrova, 2015). It is therefore a global condition that affects vulnerable population both from the north and south of Europe, and it is understood as a temporary condition caused by structural and current conditions that could be even more important than the three factors traditionally considered: home income, energy prices and the dwelling's energy efficiency (Hills, 2011).

Some other factors related to the specific energy needs of a household, as size, gender, occupation or class, together with the housing tenure and heating system, have been proved not only as fuel poverty causes but also as obstacles or limitations to consider when dealing with energy efficiency measures (Bouzarovski, 2013).

The first cross-country analysis of fuel poverty using comparable data (1994–1997) proposes six objective and subjective social indicators (Healy, Clinch, Clinch, & Peter, 2002). Three of the indicators are subjective (self-reported): inability to afford adequate warmth in the home, inability to pay utility bills and lacking adequate heating facilities; and three objectives: presence of damp walls and/or floors, lacking central heating and rotten window frames, based on actual characteristics of the dwellings. A socio-demographic and socio-economic analysis was also conducted. The highest rates of fuel poverty were found in Southern Europe, specifically in Portugal, Greece, Spain and Italy, among the 14 European countries studied. In Southern Europe, single parents and lone pensioner households were identified as a great risk group declaring high and unacceptable rates of fuel poverty conditions (more than three-fourths, especially in the ageing group). Unemployed and tenant's households were other key risk groups.

More recently, reviewed evidence by Bouzarovsky indicates that the driving forces of energy poverty are embedded in locally specific social, political and environmental circumstances. So, although colder climates would be expected to exhibit higher rates of energy poverty than Southern European climates, the latter have the biggest, with higher rates of income poverty and poorly insulated homes, as well as inefficient heating systems (Bouzarovski, 2013).

The study compares among countries, indicators as a self-reported and subjective inability to keep the home warm, with three more objective ones: arrears on utility bills, inadequately insulated houses and disproportionately high housing expenditure, with Eurostat Data. Between 2003 and 2009, in the EU's Mediterranean countries, 16.6% of population have informed that they cannot maintain their home adequately warm, against 12.8% of EU average.

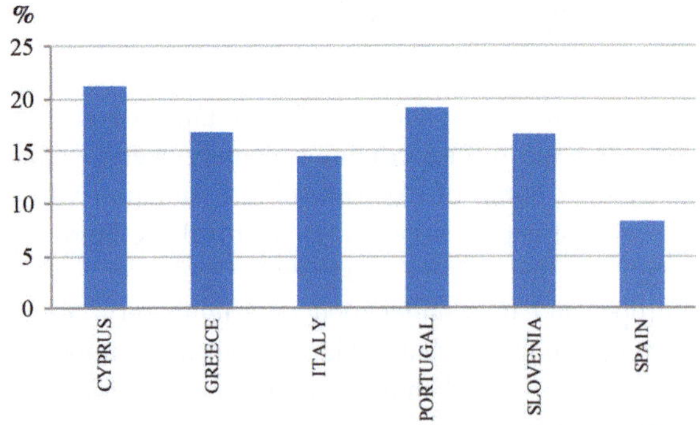

Fig. 3.6 Composite levels of fuel poverty in selected Southern European countries (% of households). Data extracted from Thomson's study across EU25 countries, supposing Scenario 4 (Thomson & Snell, 2013)

Thomson and Snell also found high levels of fuel poverty in Southern Europe in their study across the European Union, with nine indicators from the EU-SILC dataset and based on Healy and Clinch methodology exposed previously (Thomson & Snell, 2013). Composite level of fuel poverty in countries of the Southern Mediterranean, based on three indicators that hold an equal weight (households unable to pay to keep their home adequately warm, households in arrears on utility bills, and households in housing that has leaks, damp or rot), is shown in Fig. 3.6.

In Mediterranean countries with specific cooling needs, energy poverty in relation to the inability of keeping adequate temperatures in the home during summer is also studied. According to Bouzarovski and SILC data, 30% of the whole population in these countries is unable to keep their home cool enough in the summer (Bouzarovski, 2013). Much more systematic research will be needed in relation to energy poverty related to high temperatures in Southern Europe.

Although scenarios look at urban population's steady growth in the coming years as it has been until now (as seen in paragraph 3.1), rural population still deserves attention, due to the high rates of poverty that can be found there, together with poor energy infrastructure, although with differences among countries (Thomson & Snell, 2013; Geddes, Bloomer, Allen, & Goldblatt, 2011).

Lastly, as Bouzarovski summarizes, there is a consensus in academic literature about the prevalence of high rates of energy poverty among the elderly, families with children, households with disabilities and long-term illnesses, unemployed or low-income households, or households in rental homes (Bouzarovski, 2013). All these studies encourage the importance of improvements on the energy efficiency of domestic housing stock in order to contribute to eradicate energy poverty. At last, these measures have important synergies with climate change mitigation targets and are widely exposed and recognized (Mat Santamouris, 2016; Ürge-vorsatz & Herrero, 2012).

3.3.3 *People with Disabilities and Long-Term Illnesses*

People living with long-term illnesses or disabilities may be particularly vulnerable to experiencing fuel poverty, being more likely to have low incomes, and may have greater energy and thermal requirements associated with their condition, or may lead more sedentary lifestyles or spend more time in the home (Hills, 2011). They are also more affected in extreme heat-related events, mainly due to their thermoregulatory disorder and other problems. Therefore, the dwelling plays a main role due to the time these dwellers spend in it and the special requirements needed.

3.4 Comfort Conditions

The well-being parameters (related to energy consumption) that must be considered as the minimum threshold refer to thermal, hydrothermal, light, acoustic and indoor air quality characteristics. Indoor environment is closely linked to the productivity of employees in workplaces and the performance of students in an educational atmosphere, as well as resting and other everyday activities in dwellings.

Thermal comfort can be defined as the situation where the exchange of heat between a person and the environment has a neutral thermal balance and does not need internal or external regulatory mechanisms; but also as a desired thermal sensation that cannot coincide with the thermal neutrality (Humphreys & Hancock, 2007). The perception of comfort is variable and depends on factors like age and gender, economic and cultural aspects, and location and climatic conditions. There are also specific groups of people with more demanding needs like children and the elderly, the disabled and the sick.

As general criteria, most homes have an indoor temperature between 63 °F (17.2 °C) and 87 °F (30.5 °C) and people do not live comfortably outside this range, being individual tolerance lower in the edge of ageing and illness, in spite of their capacity to adapt (Kovats & Hajat, 2008). So, the acceptability range of conditions of an indoor environment is very narrow in comparison with the temperatures that are given in outdoor environments. On the other hand, there is a direct relationship between energy consumed for conditioning (mainly for thermal, ventilation and illumination purposes) and the type of building and services offered to the different population groups, so the regulation and evaluation of an adequate annual indoor environment quality are inherent in the European directive on buildings energy efficiency (EPDB, 2010) and regulated by the UNE-EN 15251 standard (UNE-EN 1525, 2008).

If users are not comfortable in the interior of a building, they will perform small actions trying to restore that comfort, trying to adapt themselves (Nicol & Humphreys, 2002), and further they could use any type of energy and system to alleviate that uneasy sensation, without possibly considering its energy efficiency or other architectural or urban implications of equipment's placement.

Different categories of indoor thermal environment are established depending on the users' expectations and needs. The most demanding category, with the highest degree of expectative, is applied to buildings inhabited by infants, disabled, ill or elderly people, that is, where people are most vulnerable. The description of these categories according to UNE-EN 15251 is attached in Table 3.1, having other names but keeping the same concept in other comfort standards, such as EN ISO 7730, with categories A, B and C (UNE-EN ISO 7730, 2006).

Design and evaluation of the indoor thermal environment of mechanically cooled and/or heated buildings will be done by the following criteria based on Predicted Mean Vote (PMV)[3] and Predicted Percentage of Dissatisfied (PPD)[4] indexes, which are described in detail in standard EN ISO 7730 (UNE-EN ISO 7730, 2006), and in parallel in American comfort standard ASHRAE-55 (2013).

As a summary, cooling and heating temperature ranges are attached in Table 3.2, according to categories recommended by the UNE-EN 15251, for mechanically conditioned residential buildings. In winter, low heating setpoints can oscillate between 18 and 21 °C, and in summer high cooling setpoints can oscillate between 25.5 and 27 °C, whether the level of expectation is high associated with vulnerable people, or moderate associated with existing buildings. These setpoints would be maximum or minimum in the given range, and take into account local traditions or the desire of saving energy, and occupants must have time and the chance of adapting to the modified design temperature (UNE-EN 15251, 2008).

The design and evaluation of thermal indoor environments that are naturally conditioned without mechanical cooling systems will follow criteria from chapter A.2 of standard UNE-EN 15251, admitting an expectative and adaptation level strongly related to the outdoor environment. ASHRAE-55 standard also incorporates the adaptive comfort approach, and although it is conceptually similar in both standards, there are some differences that do not allow their direct comparison: databases that come from different projects, classification for applicable buildings, derivation of neutral temperature and outdoor temperatures definition (Nicol & Humphreys, 2010).

ASHRAE-55 was the first regulation to introduce the adaptive comfort model in 2004 and is based on a world database elaborated by de Dear (ASHRAE RP-884, based on 21,000 measures, principally in offices around the world) (de Dear, 1998). The model presented in EN15251 is based on the EU Project Smart Controls and

[3]Predicted Mean Vote (PMV) is a statistic index based on the research developed by Povl O. Fanger, and widely recognized, used and implemented in comfort standards since 70s of the last century. Based on thermal balance of the human body, it will result from the combination of factors as metabolic rate, clothing, air temperature, mean radiant temperature, air velocity and relative humidity in the environment. PMV provides us information about the mean value of votes over thermal satisfaction that a group of people would emit in an indoor environment.

[4]Predicted Percentage of Dissatisfied (PPD) is directly related to PMV and establishes a quantitative prediction of the percentage of thermally dissatisfied people determined from PMV.

Table 3.1 Description of the applicability of indoor environment categories

Categories	Explanation
I	High level of expectation and is recommended for spaces occupied by very sensitive and fragile persons with special requirements like handicapped, sick, very young children and elderly persons
II	Normal level of expectation and should be used for new buildings and renovations
III	An acceptable, moderate level of expectation and may be used for existing buildings
IV	Values outside the criteria for the above categories. This category should only be accepted for a limited part of the year

Extracted from Table 3.1 of UNE-EN 15251 standard (UNE-EN 15251, 2008)

Table 3.2 Temperature ranges for hourly calculation of cooling and heating energy in three categories of indoor environment of residential buildings and living spaces

Categories	PPD	PMV	Temperature range for heating, °C Clothing—1.0 clo	Temperature range for cooling, °C Clothing—0.5 clo
I	<6	−0.2 < PMV <+0,2	21.0–25.0	23.5–25.5
II	<10	−0.5 < PMV <+0.5	20.0–25.0	23.0–26.0
III	<15	−0.7 < PMV <+0.7	18.0–25.0	22.0–27.0

Sedentary activity 1.2 met, 50% relative humidity and low air velocity. Extracted from Table A.3 (UNE-EN 15251, 2008)

Thermal Comfort (SCATs), elaborated by Nicol and McCartney, with data from 26 European offices in France, Greece, Portugal, Sweden and the UK. ASHRAE-55 is only applicable to naturally ventilated buildings, while EN15251 can be applied to any building in free oscillation, with or without conditioning installations. Both standards are applicable with metabolic rates ranging from 1 to 1.3 met, and persons must be free to adapt their clothing.

The concept of outdoor temperature from which ranges for indoor operational temperatures are derived from is different and is applicable within slightly different temperature limits, because of the database it is derived from. *Prevailing mean outdoor temperature* ($t_{pma(out)}$) from ASHRAE55 can vary between 10 and 33.5 °C, while *Running mean outdoor temperature* (Θ_{rm}) from EN15251 could vary between 10/15 and 30 °C, although over 25 °C the database is limited.

Running mean outdoor temperature (Θ_{rm}) is defined as an exponentially adjusted average of average daily outdoor air temperature (UNE-EN 15251, 2008). On the other hand, *Prevailing mean outdoor temperature* ($t_{pma(out)}$) is based on the arithmetic average of the mean daily outdoor temperatures, over some period of days (7–30 days), and its simplest form can be approximated by the climatically normal monthly mean air temperature from the most representative local meteorological station available. When used with dynamic thermal simulation, the preferred option is an exponentially weighted, running mean of a sequence of mean daily outdoor temperatures prior to the studied day (ASHRAE55, 2013), as *Running mean outdoor temperature,* Θ_{rm} is defined.

Although this approach is being widely investigated at a global level, especially in warmer climates and with higher relative humidity, a lot more research is needed in order to avoid overheating or excessive energy consumption, consequence of an inadequate environment. A low expectative, for example, does not mean that the situation is accepted or that there is a comfort sensation, especially if unable to change the situation. Adaptive margins are also in the limit of what is established in PMV as very hot/warm or very cold. On the other hand, users' productivity or school performance can be negatively compromised too, and possible impacts in the health of children and elderly must be taken into account (Roaf, Crichton, & Nicol, 2009).

Figures 3.7 and 3.8 show upper acceptability limits of operative temperature calculated for summer months (considered from June to September) with the ASHRAE55 adaptive approach, with acceptability limits of 80% for typical applications and 90% when a higher standard of thermal comfort is desired, according to the mentioned standards. They are calculated for different Southern European locations, based on data series 1983–2008 (IWEC2—ASHRAE). In general, these limits widely differ from the limits derived from PMV-PPD, including the most demanding category (90%, in Fig. 3.7). As a reference, graphics include the upper limit established for Category III (27 °C) and for Category II (26 °C) according to PMV-PPD (Table 3.2).

These indoor temperature limits can be used, according to adaptive approaches to design passive means to prevent overheating in summer conditions, for example,

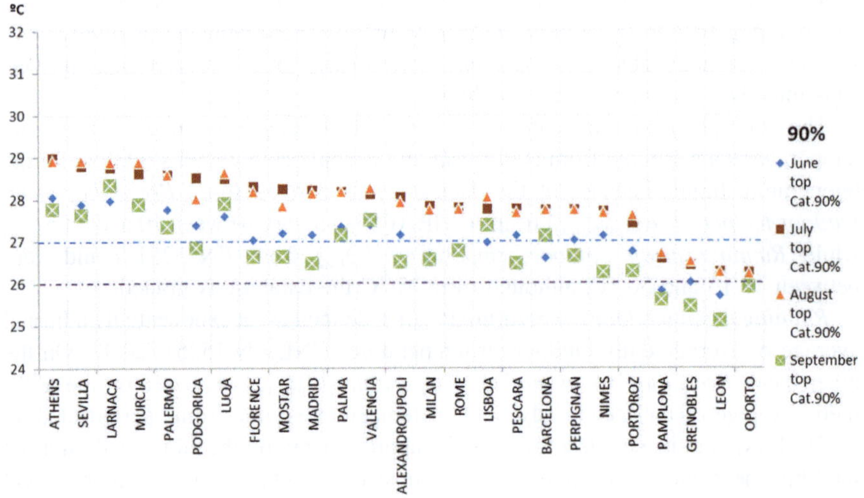

Fig. 3.7 90% upper acceptability limit of operative temperature calculated monthly, with ASHRAE55 adaptive approach (ASHRAE55-2013, 2013), calculated for different Southern European locations, based on data 1983–2008 (IWEC2—ASHRAE)

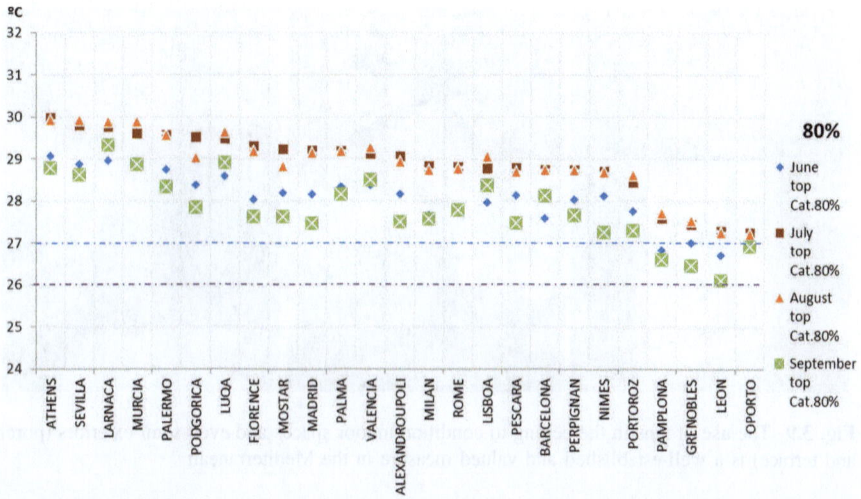

Fig. 3.8 80% upper acceptability limit of operative temperature calculated monthly, with ASHRAE55 adaptive approach (ASHRAE55-2013, 2013), calculated for different Southern European locations, based on data 1983–2008 (IWEC2—ASHRAE)

dimension and orientation of windows, shading devices or including thermal mass in constructive details. If these passive measures cannot guarantee temperature thresholds, mechanical cooling will be inevitable to ensure occupants' health and well-being (UNE-EN 15251, 2008).

On the other hand, the incorporation of airspeed, through fans or other means of personal adjustment (Fig. 3.9), like passive cooling strategy (or with very low energy demand) is becoming increasingly important in comfort standards in both approaches, from being merely informative to being regulated with a procedure. In any case, users must be able to control airspeed and it is applicable in summer conditions from an indoor operative temperature of 25 °C.

European regulation UNE-EN 15251 allows, in mechanically controlled buildings, that fans are used to compensate overheating in summer conditions. In this way and with a speed of 1 m/s, an increase of 1.5–2 °C, over the upper limits of acceptance for operative temperature from Table 3.2, could be accepted, that is, 27.5, 28 and 28.5 °C, for categories I, II and III (supposing 50% relative humidity and air temperature equal to operative temperature). ASHRAE55 (2013) has incorporated the *adjusted Predicted Mean Vote*, PMV_{adj}, that calculates the effect of airspeed in a range between 0.15 and 3 m/s (supposing a still air of 0.15 m/s).

Airspeed can also increase the upper limit of admissible ranges of adaptive approaches. According to UNE-EN 15251, it will be able to increase the upper limit of operative temperature for every category (Θ_i). ASHRAE55 also admits an increase in acceptable operative temperature limits as follows: 1.2 °C if average airspeed is 0.6 m/s; 1.8 °C if average airspeed is 0.9 m/s and 2.2 °C if average

Fig. 3.9 The use of fans in the ceiling to condition indoor spaces and even semi-exteriors (porch and terrace) is a well-established and valued measure in the Mediterranean

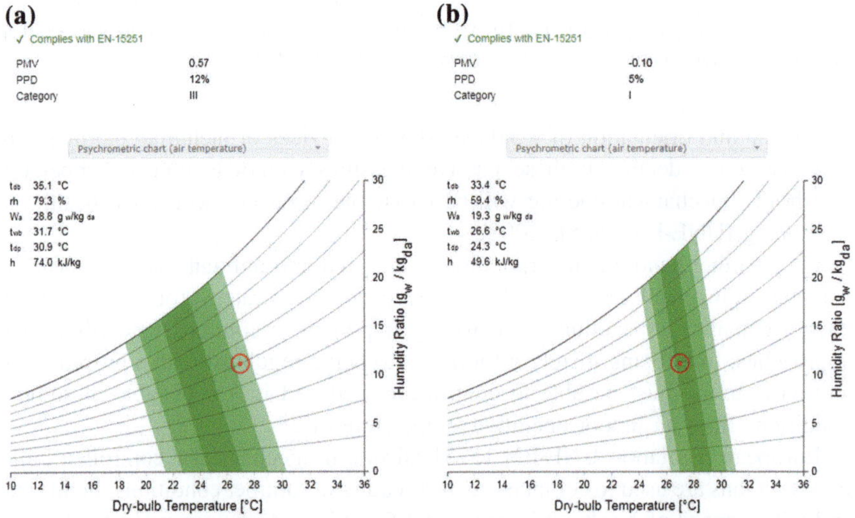

Fig. 3.10 Example of how environment category and thermal comfort change, if airspeed is modified from 0.15 to 1 m/s, considering 27 °C of operative temperature. Elaborated with CBE Thermal Comfort Tool (Hoyt et al., n.d.)

airspeed is 1.2 m/s. Therefore, these increases could be added to the limits given in Figs. 3.7 and 3.8 for Southern European locations for each summer month.

In order to assess the effect of airspeed, as well as the rest of the factors of indoor environment thermal quality, CBE's Thermal Comfort Tool, from the Centre for the Built Environment, University of California Berkeley website, is highly recommended (Hoyt, Schiavon, Piccioli, Moon, & Steinfeld, n.d.). In Fig. 3.10, as an example, we can see how environmental category and thermal comfort change following UNE-EN15251 in a mechanically cooled environment (that is, considering PMV-PPD), when airspeed is modified from 0.15 to 1 m/s and always considering that the user must be able to control airspeed (27 °C of operative temperature, 50% of relative humidity, 1.2 met and 0.5 clo).

3.5 Beyond Comfort: Talking About Health

Buildings must endeavour to give their occupants a comfortable indoor environment with a low energy demand, so they can pursue their activities guaranteeing users health under different assumptions. Residential buildings are considered very relevant due to the large number of hours that people spend in them, especially vulnerable people.

There are many studies on the increase of morbidity and mortality during summer and winter, as well as during cold and heatwaves and other climatic events. It is widely documented that effects on health are especially relevant in the most vulnerable population in both seasons.

The relation between energy poverty and health is especially relevant, being directly related to excess winter and summer mortalities, and also to the prevailing of certain sicknesses in vulnerable population like children and the elderly. According to research of Analitis et al. and Critchley et al., energy poverty derives into health problems such as breathing and coronary problems in adults, less calorie intake in babies, and higher chance of having respiratory problems in children (Analitis et al., 2008; Critchley, Gilbertson, Grimsley, & Green, 2007). Energy poverty is also associated with mental health problems. Adults present cases of anxiety and depression due to the amount of time spent in cold spaces, and also due to the connotations and social exclusion derived from this situation. Studies in children show a higher school absenteeism (13% compared to 3% for children in appropriately warmed homes), and in teenagers a bigger concern in bullying and mugging (27% compared to 15% in warmer homes) (Liddell & Morris, 2010).

Finally, climate change is likely to affect human health in Europe, and particularly in Southern Europe, with increases in morbidity and mortality rates related to heat (Kovats et al., 2014), considering the group of the elderly the most vulnerable (EEA, 2016).

3.5.1 Winter and Cold Conditions

It is generally established that with a temperature lower than 16 °C resistance to respiratory disease falls, and temperature lower than 12 °C results in raised blood pressure caused by the narrowing of blood vessels that can provoke cardiac failure and bad circulation, and the immunologic system is depressed so that there is a higher risk of suffering more infections (Wright, 2004) and also chronic problems related to cold temperatures like pneumonia, fever, asthma and arthritis (World Health Organization, 1987).

Excess Winter Mortality (EWM) has been widely studied during years and can be defined as the surplus number of deaths occurring during the winter season (December to March both inclusive) compared with the average of the non-winter seasons. With data 1988–1997 of 14 countries in Europe, Healy published a first key cross-country analysis, finding that Southern and Western Europe have the highest rates of EWM, while paradoxically having milder winters than in colder and northern regions (Healy, 2003). According to this study, analysing risk factors pertaining to climate, macroeconomy, health care, lifestyle, socio-economics and housing, a strong negative relation with the thermal efficiency of housing was found, so it was concluded that an improvement of energy efficiency in dwellings protecting people from cold indoors plays a fundamental role in the reduction of EWM, especially in Southern and Western Europe. Portugal (28%) and Spain (21%) presented higher rates than the average of 14 European countries studied (16%).

In 2008, Analitis et al. analysed cold-related mortality by age groups with data from 1990 to 2000 in 15 European cities, and the same trend was confirmed for the Southern Mediterranean (Athens, Barcelona, Valencia, Ljubljana, Milan, Rome and Turin), with clearly larger effects in the elderly (Analitis et al., 2008). The study also included a wide range of climatic, socio-demographic, cultural and health characteristics among the 15 European cities studied.

In a more recent study of 31 countries in Europe between 2002/3 and 2010/11, with over two million deaths attributed to Excess Winter Deaths (EWDI Index), Fowler et al. found an heterogeneity in the pattern, although again, southern countries like Malta (28.3%), Portugal (25.9%), Cyprus (19.4%) and Spain (18.6%) had EWDI significantly higher than the average (13.9%). Many deaths may be avoidable as environmental, social and personal factors are known to contribute to winter mortality, and as there was also substantial variation seen in countries with a similar climate (e.g. between Mediterranean countries), many winter deaths could be amenable to public health action (Fowler et al. 2015).

Therefore, although the focus has recently been on heatwave episodes and on future warming conditions across Europe, cold-related mortality continues being an important public health problem across Europe and should not be overlooked (Analitis et al., 2008), especially in countries of Southern Europe (Fig. 3.11).

3.5.2 *Summer and Hot Conditions: Heatwaves*

The World Health Organization (WHO) established in 1987 that summer conditions on indoor environments do not pose a risk when under 24 °C (World Health Organization, 1987), but this threshold could be higher, and various studies

Fig. 3.11 Architecture should aim to protect people's health specially the most vulnerable: ageing, disabled, young children, etc

demonstrate that it is strongly related to the climatic zone and to exterior temperatures.

In Baccini et al. study on 15 European cities, different thresholds for admissible temperatures related to mortality have been found: in the Mediterranean region it is 29.4 °C (Athens, Rome, Valencia, Turin, Milan and Ljubljana), and in the north continental region of Europe it is 23.3 °C, i.e. almost 6 °C difference (Baccini et al., 2008). Specifically by location, some of the thresholds found were Athens, 32.7 °C; Milan, 31.8 °C; Rome, 30.3 °C; Turin, 27.0 °C; and Valencia, 28.2 °C. Stronger associations were found between heat and mortality from respiratory diseases, and with mortality in the elderly. Finally, there is some suggestion of a higher effect of early season exposures, and there are differences in cities due to personal adaptation, among other aspects.

The main health effects of climate change in Europe are related to extreme weather events such as heatwaves, and human vulnerability is determined by a complex set of factors (EEA, 2016). Since the length, frequency and intensity of heatwaves are projected to increase, health effects will too in the absence of adaptation and physiological acclimatization. Heatwaves were the deadliest extreme weather event in Europe, and particularly in Southern Europe (Fig. 1.2, based on EEA 2016).

Heatwaves do not have a common definition, but they are generally understood as a prolonged period of excessive heat, climatic events being by nature unpredictable and uncomfortable. A comprehensive review of different definitions (and impacts and measures) was elaborated by Zuo et al. (2015). Definitions of heatwaves imply climatic variability, and they used to be based on statistical thresholds derived from daily maximum temperatures, while other researchers propose the use of mortality rate as a reference to define them.

Publications dealing with the severity and the incidence in the increase of mortality in different countries show that heatwaves affect high- and low-income countries. The incidence of heatwaves on health in Southern Europe, as Italy, France, Portugal or Spain, is summarized in different reviews (Haines, Kovats, Campbell-Lendrum, & Corvalan, 2006; Kovats & Hajat, 2008). The 2003 heatwave in Europe was probably the hottest since 1500, and climatologists consider that there is very likely an increasing risk of another similar heatwave caused by anthropogenic causes (Haines et al. 2006).

One of the major concerns about heatwaves is their impact on health, particularly on the most vulnerable population, the elderly, but also on persons with social and/or physical vulnerability such as very young children, homeless, disabled, etc. (Kovats & Hajat, 2008). Vulnerable populations and the way they cope with adaptability have important differences due to climate, culture, housing and other factors. Risk factors can be categorized as intrinsic (age or disability) and extrinsic (housing and behaviours), the latter differing according to location and adaptation to the local climate (Kovats & Hajat, 2008).

The main effects on ageing people are caused by changes in their thermoregulatory system. Babies and children also have limitations in their thermoregulatory system and can be more at risk of dehydration, although there is not, in general, an excess in children mortality due to heatwaves. In general, people with illnesses affecting their thermoregulatory system are at risk (Kovats & Hajat, 2008). As these same authors reviewed, the majority of European studies have also shown that women are more at risk, especially elderly women, being important not only physiological but also social factors.

Paris, with the highest percentage of elderly people in the country, got the worst from the 2003 heatwave in France, being a woman over 75 and living alone the highest risk factor according to the study by Cadot, Rodwin, & Spira, (2007). The study highlighted that the majority of deaths during heatwaves occurred at home (no homeless), that other populations of Asiatic and African origin with multigenerational families were less affected, and that a shift was detected from the poorest to the more wealthy neighbourhoods. Vandentorren et al. analysed risk factors that affected over 65 years mortality in the 2003 heatwave, considering the most important risk factor, the lack of mobility derived from pre-existing medical conditions (Vandentorren et al. 2006). A very interesting analysis of other variables that were considered included different factors such as personal (social, as a lower status or isolation), behavioural, housing (principally lack of insulation and living on the top floor) and environmental (urban temperature related to Urban Heat Island effect).

Spain experienced three heatwaves in 2003. Total associated excess deaths were 8%, and excess deaths were only observed in those aged 75 years and over (15% more deaths than expected for the age group 75–84 years, and 29% for those aged 85 or over), while among those 64 years and younger, mortality decreased during this period (Simón, Lopez-Abente, Ballester, & Martínez, 2005). Previously, Diaz analysed and quantified the effects of heatwaves in Madrid (Spain), between 1986 and 1997, and found an increase of mortality in the elderly (over 65), with a particular effect in women over the age of 75 (Díaz et al., 2002).

Other studies on Chicago heatwave in 1995 found a higher risk in men who lived alone, with the condition of *isolation* highlighted, as in the rest of studies, as one of the main risk factors (Kovats & Hajat, 2008).

Finally, it should be stressed that people who live in city centres are particularly affected by heatwaves, due to the urban heat island effect, UHI, together with a higher pollution concentration (Haines et al., 2006) (Wilkinson et al., 2009). In Madrid (Spain), Diaz et al. found that the effect of high temperatures with low humidity and a high level of pollution, specially ozone, contributed highly to the increase of mortality during heatwaves in Madrid (Díaz et al., 2002). Santamouris et al. have widely studied the effect of UHI and very high temperatures in Athens, especially relevant with low incomes or energy poverty in relation to heat, who cannot afford the cost of cooling energy demand (Santamouris, 2014; Sakka, Santamouris, Livada, Nicol, & Wilson, 2012).

3.6 Main Implications in Energy Efficiency of Residential Buildings

Energy efficient buildings or the improvement of efficiency in residential buildings contributes in securing adequate comfort conditions and protecting people's health, especially the most vulnerable, both in summer and winter conditions. It is directly related to an important drop in hot-related and cold-related morbidity and mortality rates, especially in the most vulnerable.

Energy efficiency contributes both to mitigation and to adaptation to climate change. Strategies that contribute to mitigation by reducing energy consumption and emissions in winter times may not be effective or enough in summer conditions, and while winter energy systems can have different types of fuel or renewable energy, summer ones mostly require electricity.

Beside an increase in heatwaves, warming conditions are expected especially in Southern Europe, as has been widely exposed. Heatwaves affect the health of people of all ages and social classes although they particularly affect ageing people living in city centres with a lower capability of adaptation. The quickest measure seems to be the installation of air conditioning in dwellings to protect people from excessively hot summer conditions. However, the protection of the most vulnerable should not be based only on a measure that depends on the economic capacity of the household, that can have energy supply issues during peak times and that implies a consumption of energy and associated emissions which contribute to the aggravation of global warming.

This is the reason why the studies reviewed stress the need of boosting passive measures inherent to architecture itself, in order to reduce overheating to acceptable health limits and to minimize energy demand for cooling services. The efficiency of measures will also depend on the climatic severity of the zone together with extreme heat-related events as heatwaves or warm spells.

Users' behaviour and housing conditions are much better adapted in the Mediterranean zone, as can be deduced from the adaptive approaches of comfort (see Figs. 3.7 and 3.8), or from the thresholds considers safe for health (Baccini et al. 2008), compared to other European locations. Consequently, the latter could learn from Southern European traditional strategies implemented in buildings (Ward, Lauf, Kleinschmit, & Endlicher, 2016) (Fig. 3.12).

The exponential growth that has taken place these last years in Mediterranean countries in air conditioning systems installation in residential buildings, due to a reduction in energy costs and in cooling systems, must be stated too. Many of them have been installed in buildings without pre-installation and with inefficient systems that often respond to inadequate designs for the location's climate (e.g. apartments with a single west orientation in residential high-rise buildings in city centres) or without a preview for future warming conditions (Fig. 3.13).

Lastly, there is a consideration to be made in relation to indoor environmental categories in relation to regulations. In existing buildings, and specially in the ones with worse characteristics, as can be deduced from the increased threshold in the level of expectations (Category III), is where we can find a higher percentage of

Fig. 3.12 Illustrative images of Southern European buildings with passive strategies. Nice and Montpellier (France)

Fig. 3.13 Installation of air conditioning systems anywhere on the façade and integrated. Examples in Barcelona and Madrid (Spain)

vulnerable population. However, in this group of vulnerable population, the elderly or the sick are given a high level of expectative (Category I), which highly reduces admissible ranges of comfort. In individual or in collective buildings where population is intergenerational and with a very varied socio-economic profile, it is a paradox to consider on acceptable levels (Category III) existing buildings that precisely leave the most vulnerable population unattended. Therefore, a more ambitious consideration for the indoor environment would be advisable in the construction and rehabilitation of residential buildings and dwellings, and the administration should favour it and promote it in the most sensitive and vulnerable neighbourhoods.

3.7 Conclusions

Residential buildings must give an answer to occupants' health and comfort conditions, bearing in mind the specific characteristics of vulnerable population, since they have a diverse and changing profile. Together with an ageing population, warmer future conditions with more extreme events are predicted, which buildings will have to deal with, prioritizing passive strategies in winter and summer conditions that do not imply additional energy consumption that contributes to aggravate global warming.

The consideration of indoor environmental category in relation to the objectives of energy efficiency should be considered according to the varied and intergenerational occupants of residential buildings and not to the state of the building, therefore prioritizing very ambitious actuations and projects in residential buildings.

Acknowledgements The authors want to thank the assistance of Ainhoa Arriazu in the collection of information and in the elaboration of some graphs provided in this chapter.

References

Analitis, A., Katsouyanni, K., Biggeri, A., Baccini, M., Forsberg, B., Bisanti, L., … Michelozzi, P. (2008). Effects of cold weather on mortality: Results from 15 European cities within the PHEWE project. *American Journal of Epidemiology, 168*(12), 1397–1408. https://doi.org/10.1093/aje/kwn266.

ASHRAE55-2013. (2013). Thermal Environmental Conditions for Human Occupancy.

Baccini, M., Biggeri, A., Accetta, G., Kosatsky, T., Katsouyanni, K., Analitis, A., … Michelozzi, P. (2008). Heat effects on mortality in 15 European cities. *Epidemiology (Cambridge, Mass.), 19*(5), 711–719. http://doi.org/10.1097/EDE.0b013e318176bfcd.

Bouzarovski, S. (2013). Energy poverty in the European Union : landscapes of vulnerability, 276–289. http://doi.org/10.1002/wene.89.

Bouzarovski, S., & Petrova, S. (2015). A global perspective on domestic energy deprivation: Overcoming the energy poverty-fuel poverty binary. *Energy Research and Social Science, 10,* 31–40. https://doi.org/10.1016/j.erss.2015.06.007.

Cadot, E., Rodwin, V. G., & Spira, A. (2007). In the heat of the summer: Lessons from the heat waves in Paris. *Journal of Urban Health, 84*(4), 466–468. https://doi.org/10.1007/s11524-007-9161-y.

Critchley, R., Gilbertson, J., Grimsley, M., & Green, G. (2007). Living in cold homes after heating improvements: Evidence from Warm-Front, England's Home Energy Efficiency Scheme. *Applied Energy, 84*(2), 147–158. https://doi.org/10.1016/j.apenergy.2006.06.001.

De Dear, R. J. (1998). A global database of thermal comfort field experiments. *ASHRAE Transactions 104b*, (January 1998), 1141–1152.

Díaz, J., Jordán, A., García, R., López, C., Alberdi, J. C., Hernández, E., et al. (2002). Heat waves in Madrid 1986-1997: Effects on the health of the elderly. *International Archives of Occupational and Environmental Health, 75*(3), 163–170. https://doi.org/10.1007/s00420-001-0290-4.

EEA. (n.d.). Agencia Europea de Medio Ambiente. Retrieved April 7, 2017, from http://www.eea.europa.eu/es.

EEA. (2016). *Climate change, impacts and vulnerability in Europe 2016: An indicator-based report.*

ENTRANZE. (n.d.). ENTRANZE: Enforze the transition to Nearly Zero Energy buildings in the EU-27. Retrieved April 7, 2017, from http://www.entranze.enerdata.eu/.

EPDB. (2010). Directive 2010/31/EU of the European Parliament and of the Council of 19 May 2010 on the energy performance of buildings (recast). *Official Journal of the European Union*, 13–35. http://doi.org/doi:10.3000/17252555.L_2010.153.eng.

EUROSTAT. (n.d.-a). Eurostat. Retrieved April 7, 2017, from http:/ec.europa.eu/eurostat EUROSTAT. (n.d.-b). Eurostat. Housing statistics. Retrieved September 1, 2017, from http://ec.europa.eu/eurostat/statistics-explained/index.php/Housing_statistics.

Fowler, T., Southgate, R. J., Waite, T., Harrell, R., Kovats, S., Bone, A., … Murray, V. (2015). Excess winter deaths in Europe: A multi-country descriptive analysis. *European Journal of Public Health, 25*(2), 339–345. http://doi.org/10.1093/eurpub/cku073.

Geddes, I., Bloomer, E., Allen, J., & Goldblatt, P. (2011). *The health impacts of cold homes and fuel poverty*. London.

Haines, A., Kovats, R. S., Campbell-Lendrum, D., & Corvalan, C. (2006). Climate change and human health: Impacts, vulnerability and public health. *Public Health, 120*(7), 585–596. https://doi.org/10.1016/j.puhe.2006.01.002.

Healy, J. D. (2003). Excess winter mortality in Europe: a cross country analysis identifying key risk factors. *Journal of Epidemiology and Community Health, 57*(10), 784–789. https://doi.org/10.1136/jech.57.10.784.

Healy, J. D., Clinch, J. P., Clinch, H. D., & Peter, J. (2002). Fuel Poverty in Europe A Cross-Country Analysis Using a New Composite Measurement. *Environmental Studies Research Series Working Papers 2002.*

Hills, J. (2011). Fuel poverty: the problem and its measurement. Retrieved from http://www.decc.gov.uk/en/content/cms/funding/fuel_poverty/hills_review/hills_review.aspx.

Hoyt, T., Schiavon, S., Piccioli, A., Moon, D., & Steinfeld, K. (n.d.). CBE Thermal Comfort Tool. Center for the Built Environment, University of California Berkeley. Retrieved from http://cbe.berkeley.edu/comforttool/.

Humphreys, M. A., & Hancock, M. (2007). Do people like to feel "neutral"?. Exploring the variation of the desired thermal sensation on the ASHRAE scale. *Energy and Buildings, 39*(7), 867–874. https://doi.org/10.1016/j.enbuild.2007.02.014.

IMSERSO. (2012). *22023/ Informe 2010. Las personas mayores en España. Datos estadísticos estatales y por Comunidades Autónomas*. Madrid.

IPCC. (2014). Climate Change 2014: Impacts, Adaptation, and Vulnerability. *Climate Change 2014: Impacts, Adaptation, and Vulnerability—Contributions of the Working Group II to the Fifth Assessment Report*, 1–32. http://doi.org/10.1016/j.renene.2009.11.012.

Janda, K. B. (2009). Buildings don't use energy - People do! In *PLEA2009—26th Conference on Passive and Low Energy Architecture, Quebec City, Canada, 22-24 June 2009*. http://doi.org/10.3763/asre.2009.0050.

Kovats, R. S., & Hajat, S. (2008). Heat stress and public health: A critical review. *Annual Reviews of Public Health, 29,* 41–55. https://doi.org/10.1146/annurev.publhealth.29.020907.090843.

Liddell, C., & Morris, C. (2010). Fuel poverty and human health: A review of recent evidence. *Energy Policy, 38*(6), 2987–2997. https://doi.org/10.1016/j.enpol.2010.01.037.

Nicol, F., & Humphreys, M. (2010). Derivation of the adaptive equations for thermal comfort in free-running buildings in European standard EN15251. *Building and Environment, 45*(1), 11–17. https://doi.org/10.1016/j.buildenv.2008.12.013.

Nicol, J. F., & Humphreys, M. A. (2002). Adaptive thermal comfort and sustainable thermal standards for buildings. *Energy and Buildings, 34*(6), 563–572. https://doi.org/10.1016/S0378-7788(02)00006-3.

R. S. Kovats, Valentini, R., Bouwer, L. M., Georgopoulou, E., Jacob, D., Martin, E., … Soussana, J. (2014). Climate Change 2014: Impacts, Adaptation, and Vulnerability. Part B: Regional Aspects. Contribution of Working Group II to the Fifth Assessment Report of the Intergovernmental Panel on Climate Change. Europe. *Climate Change 2014: Impacts, Adaptation, and Vulnerability. Part B: Regional Aspects. Contribution of Working Group II to the Fifth Assessment Report of the Intergovernmental Panel on Climate Change*, 1267–1326.

Roaf, S., Crichton, D., & Nicol, F. (2009). *Adapting Buildings and Cities for Climate Change : A 21st Century Survival Guide* (Vol. 2nd. ed). Amsterdam: Routledge. Retrieved from http://ezproxy.si.unav.es:2048/login?url=http://search.ebscohost.com/login.aspx?direct=true&AuthType=ip,url&db=edsebk&AN=297204&lang=es&site=eds-live&scope=site.

Sakka, A., Santamouris, M., Livada, I., Nicol, F., & Wilson, M. (2012). On the thermal performance of low income housing during heat waves. *Energy and Buildings, 49,* 69–77. https://doi.org/10.1016/j.enbuild.2012.01.023.

Santamouris, M. (2014). On the energy impact of urban heat island and global warming on buildings. *Energy and Buildings, 82,* 100–113. https://doi.org/10.1016/j.enbuild.2014.07.022.

Santamouris, M. (2016). Innovating to zero the building sector in Europe: Minimising the energy consumption, eradication of the energy poverty and mitigating the local climate change. *Solar Energy, 128,* 61–94. https://doi.org/10.1016/j.solener.2016.01.021.

Simón, F., Lopez-Abente, G., Ballester, E., & Martínez, F. (2005). Mortality in Spain during the heat waves of summer 2003. *Euro Surveillance : Bulletin Européen Sur Les Maladies Transmissibles = European Communicable Disease Bulletin., 10*(7), 156–161. Retrieved from http://ezproxy.si.unav.es:2048/login?url=http://search.ebscohost.com/login.aspx?direct=true&AuthType=ip,url&db=edselc&AN=edselc.2-52.0-33745184363&lang=es&site=eds-live& scope=site.

Thomson, H., & Snell, C. (2013). Quantifying the prevalence of fuel poverty across the European Union. *Energy Policy, 52,* 563–572. https://doi.org/10.1016/j.enpol.2012.10.009.

Tirado Herrero, S., & Ürge-Vorsatz, D. (2012). Trapped in the heat: A post-communist type of fuel poverty. *Energy Policy, 49,* 60–68. https://doi.org/10.1016/j.enpol.2011.08.067.

UNE-EN 15251. (2008). Parámetros del ambiente interior a considerar para el diseño y la evaluación de la eficiencia energética de edificios incluyendo la calidad del aire interior, condiciones térmicas, iluminación y ruido.

UNE-EN ISO 7730. (2006). Ergonomía del ambiente térmico. Determinación analítica e interpretación del bienestar térmico mediante el cálculo de los índices PMV y PPD y los criterios de bienestar térmico local.

United Nations. (2014). *World Urbanization Prospects: The 2014 Revision, Highlights (ST/ESA/SER.A/352). New York, United.* http://doi.org/10.4054/DemRes.2005.12.9.

Ürge-vorsatz, D., & Herrero, S. T. (2012). Building synergies between climate change mitigation and energy poverty alleviation, (October). http://doi.org/10.1016/j.enpol.2011.11.093.

Vandentorren, S., Bretin, P., Zeghnoun, A., Mandereau-Bruno, L., Croisier, A., Cochet, C., … Ledrans, M. (2006). August 2003 Heat wave in France: Risk factors for death of elderly people living at home. *European Journal of Public Health, 16*(6), 583–591. http://doi.org/10.1093/eurpub/ckl063.

Ward, K., Lauf, S., Kleinschmit, B., & Endlicher, W. (2016). Heat waves and urban heat islands in Europe: A review of relevant drivers. *Science of the Total Environment, 569–570,* 527–539. https://doi.org/10.1016/j.scitotenv.2016.06.119.

Wilkinson, P., Smith, K. R., Davies, M., Adair, H., Armstrong, B. G., Barrett, M., … Chalabi, Z. (2009). Public health benefits of strategies to reduce greenhouse-gas emissions: Household energy. *The Lancet, 374*(9705), 1917–1929. http://doi.org/10.1016/S0140-6736(09)61713-X.

World Health Organization. (1987). Health impact of low indoor temperatures. Report on a WHO meeting.

Wright, F. (2004). Old and cold: Older people and policies failing to address fuel poverty. *Social Policy and Administration, 38*(5), 488–503. https://doi.org/10.1111/j.1467-9515.2004.00403.x.

Zuo, J., Pullen, S., Palmer, J., Bennetts, H., Chileshe, N., & Ma, T. (2015). Impacts of heat waves and corresponding measures: A review. *Journal of Cleaner Production, 92,* 1–12. https://doi.org/10.1016/j.jclepro.2014.12.078.

Chapter 4
Residential Architecture in Mediterranean Climates. Towards Optimized Passive Solutions for the Whole Year

4.1 Residential Buildings' Energy Consumption in Southern Europe

The reduction of energy consumption in buildings is a major issue in securing the European environmental goals to contribute to the mitigation of climate change by reducing CO_2 emissions to the atmosphere. European buildings suppose a 40% of total energy consumption and around 55% of the EU's electricity consumption in 2012, building sector being the one that consumes the most energy, ahead of transport (32%), industry (26%) and agriculture (2%) (Gynther, Lappillone, & Pollier, 2015). From that 40%, approximately two-thirds correspond to the residential sector, although with differences among countries.

European Directive 2010/31/EU on the Energy Performance of Buildings, EPBD (2010) establishes specific objectives in reduction of greenhouse gas emissions, share of renewable energy and improvement in energy efficiency for the 2020 horizon, on the way to a transformation towards a low-carbon economy for 2050, reducing CO_2 emissions to an 80% from 1990 levels.

EPBD has defined a *nearly zero-energy building* (nZEB) as a building that has a very high energy performance, and the nearly zero or very low amount of energy required should be covered to a very significant extent by energy from renewable sources. European objectives stress the need for high energy efficiency in buildings taking into account outdoor climatic and local conditions, as well as indoor climate requirements and cost-effectiveness, covering the annual energy performance of a building.

Emphasis is also placed on the significant increase in conditioning systems in Europe, and the related problems at peak load times, increasing the cost of electricity and disrupting the energy balance. From here on, the EPBD emphasizes that there should be a focus on strategies which enhance the thermal performance of buildings during the summer period avoiding overheating and improving indoor climatic conditions but also microclimate around buildings, namely shading, thermal capacity in building construction and other passive cooling techniques.

© Springer International Publishing AG 2018

A. Monge-Barrio and A. Sánchez-Ostiz Gutiérrez, *Passive Energy Strategies for Mediterranean Residential Buildings*, Green Energy and Technology,
https://doi.org/10.1007/978-3-319-69883-0_4

According to data from Odyssee-Mure, household energy consumption has been decreasing regularly in most countries since 2000, household energy efficiency has improved thanks to the boost in efficiency of heating systems and the efficiency of electrical appliances (Gynther et al., 2015). Household energy efficiency has improved by 18% at European Union level since 2000, and the greatest enhancements have taken place in space heating (20%), water heating and large appliances (15%). There has been an increase in the number of dwellings and appliances[1] (especially in those with small appliances) that contribute to rise the household's energy consumption, although it is counterbalanced due to energy efficiency improvements (Gynther et al., 2015). Energy Efficiency Gains in Southern Europe and EU total, since 2000, are shown in Fig. 4.1.

The most important end use of energy in the UE household sector is heating consumption, being on average 67% of total energy consumption, but having less incidence in Mediterranean countries, in Spain for example being just below 50%, and in Malta, Cyprus and Portugal, below 30% (Gynther et al., 2015). Figure 4.2 shows the evolution of heating consumption (kWh/m^2 a), for the EU28 average, and for Southern European countries (Croatia, France[2], Italy, Malta, Spain, Greece, Cyprus and Portugal) with climatic corrections (ODYSSEE-MURE, n.d.). The graph shows average data per country, although it should be highlighted that relevant differences can be found in each country due to the different climate zones (e.g. in Spain, in cold climatic zones, non-rehabilitated buildings consumption can reach 200 kWh/m^2 a).

The heating system is also a key factor in order to ensure an adequate thermal comfort, together with an efficient use of energy. Countries with milder winters will present a higher percentage of room heating systems. Figure 4.3 shows data on the number of these dwellings, together with individual, collective central or district heating, according to data from the Tabula project (TABULA, n.d.).

Data from this source are also attached in Fig. 4.4, which shows the housing stock permanently occupied (habitual housing) by selected countries of the Mediterranean Region. The European Union has a total stock of 212,512 dwellings in 2014. According to these figures, France, Italy and Spain have the higher number of dwellings in Southern Europe.

In the whole European Union, air conditioning represents only 0.5%, although in the Mediterranean countries, an increasing number of cooling systems have been installed during these last years. As an example, in Spain, dwellings with air conditioning systems constitute 35.5% (INE, 2008), although with different percentages according to the climate classification. The incidence of climate is also related to the type of cooling system, as can be seen in Fig. 4.5.

Lastly, building typology plays an important role in housing energy consumption, as is widely presented in Chap. 5. In fact, energy and thermal studies

[1]Large appliances include, according to Odyssee-Mure data, cold appliances (refrigerators and freezers), washing appliances (washing machines, dish washers and dryers) and TVs.

[2]This book has only included the part of France with a Mediterranean climate as can be appreciated in the maps, although this section shows statistic data per all the country.

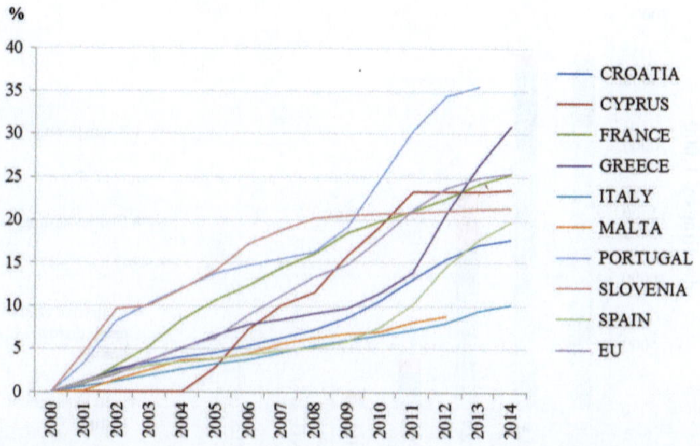

Fig. 4.1 Energy efficiency gains in residential sector in the European Union (EU) and in Southern European Countries, 2000–2014 (%). Data retrieved from ODYSSEE-MURE (n.d.)

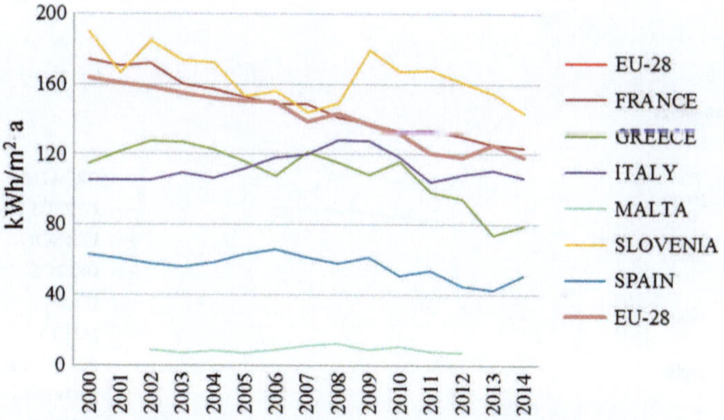

Fig. 4.2 Housing heating consumption (kWh/m^2 a), for the EU28 average for Southern European selected countries, 2000–2014. Data extracted from ODYSSEE-MURE (n.d.)

performed in this research (Chaps. 5, 6, 7 and 8) are based on representative typologies that help us understand the scope of this problematic and the efficiency of the implemented measures.

The TABULA project developed national building typologies representing the residential building stock of 13 European countries ("Episcope," n.d.). Different individual residential buildings were distinguished, such as *Single-Family House* or *Terraced House*, and collective housing buildings such as *Multifamily House* (less than 10 apartments) and *Apartment Blocks* (more than 9 apartments). There are some differences in the percentage of typologies among Southern European countries, which can be seen in Chap. 5.

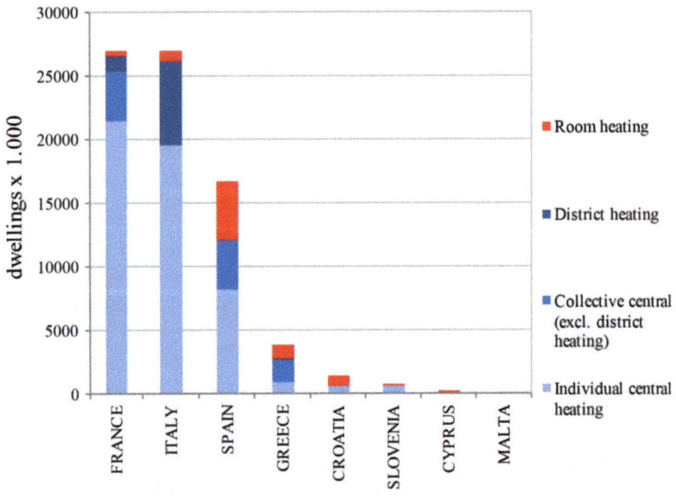

Fig. 4.3 Heating systems in selected Southern European countries. Data extracted from the Tabula project (TABULA, n.d.)

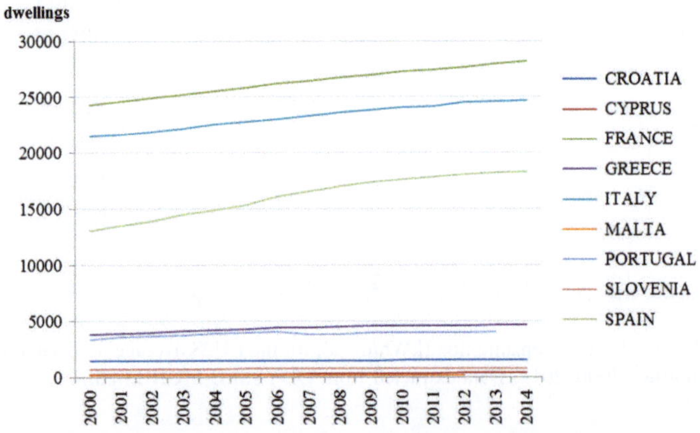

Fig. 4.4 Housing stock permanently occupied in Southern European countries, 2000–2014. Data retrieved from ODYSSEE-MURE (n.d.)

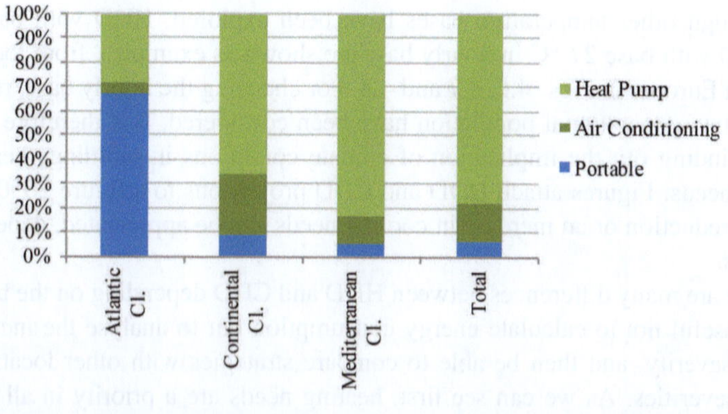

Fig. 4.5 Type of cooling systems in Spain (2012), from IDAE2012, and retrieved from Tabula project (TABULA, n.d.)

4.2 Different Scenarios and Implications on Energy in Residential Buildings, Based on HDD and CDD

A first approach on the implications of climate and climate change in the impact of energy on the built environment can be taken through heating degree days (HDD) and cooling degree days (CDD) calculations. They are defined in relation to a base temperature below or over which a building is assumed to need heating or cooling. They can be calculated in hourly or daily base, and this base temperature or balance point differs from country to country in the codes (e.g. in Spain[3] is 20 °C for HDD and CDD) (CTE-HE, 2013). This balance point depends largely on the characteristics of the building.

According to the European Environmental Agency, EEA, in its study of energy households based on trends of change of HDD (Base 15.5 °C) and CDD (Base 22 °C), from 1951 to now there has been a decrease on HDD and an increase on CDD in Europe (EEA, 2016). According to EEA, between the periods 1951–1980 and 1981–2014, HDD showed a decrease of 8.2%, mainly in Northern and Northwestern Europe, while CDD presented an increase of 49.1%, mainly due to the increase given in Southern Europe. Future projections show the same trend in HDD and CDD, and although the projected decrease in HDD is estimated to be higher than the projected increase in CDD in absolute terms, the EEA highlights that, in economic terms, these two effects have the same weight because in Europe, cooling consumption is generally more expensive than heating consumption.

[3]HDD and CDD are used in Spanish Technical Code to define the severity of the climate zones, beside the amount of sun hours, with which maximum heating and cooling demand and consumption are established.

Although other temperature bases have been explored, HDD with base 18 °C and CDD with base 27 °C in hourly base are shown as examples, from locations in Southern Europe, in Figs. 4.6, 4.7 and 4.8. For choosing the hourly base, residential use and intergenerational population have been considered. It is therefore an initial way of finding out the implication of climate conditions in buildings heating and cooling needs. Figures attach HDD and CDD projections to a future 2050 scenario where a reduction or an increase in cooling needs can be appreciated, depending on locations.

There are many differences between HDD and CDD depending on the base used, so it is useful not to calculate energy consumption but to analyse the incidence of climate severity, and then be able to compare strategies with other locations with similar severities. As we can see first, heating needs are a priority in all Southern European locations, before cooling needs.

In Mediterranean climates, differences in latitude are not significant (from Milan 45° to Larnaca 34°), but winter and summer severities do change, as do the different climate combinations that can be found. According to the figures, locations with a higher climate severity in Southern Europe are Leon or Grenobles with more than 2800HDD, and there are some locations with more than 2000HDD, like Madrid, Mostar, Portoroz or Milan. It is useful to understand the complexity of passive measures in these kinds of climates by comparing the different CDD of each

Location	Country	Koppen-Geiger	Altitude	Latitude	HDD (18°C) EA	HDD (18°C) 2050	CDD (27°C) EA	CDD (27°C) 2050
Sevilla	ESP	Csa	31	37°25'	1029	732	260	539
Larnaca	CYP	Csa	2	34°52'	799	581	180	370
Murcia	ESP	BSk	62	38°0'	990	694	179	440
Athens	GRC	Csa	15	37°54'	1100	791	161	388
Madrid	ESP	Csa	582	40°27'	2104	1641	153	405
Podgorica	MNE	Cfa	33	42°22'	1911	1472	128	406
Palermo	ITA	Csa	117	38°5'	940	692	109	228
Mostar	BIH	Cfb	108	43°19'	2027	1560	105	341
Florence	ITA	Cfb	38	43°47'	1797	1355	98	279
Lisbon	PRT	Csa	114	38°46'	985	686	94	198
Luqa	MLT	Csa	91	35°50'	713	481	92	198
Palma	ESP	Csa	7	39°32'	1341	1026	83	205
Alexandroupoli	GRC	Csa	3	40°50'	2024	1534	70	281
Valencia	ESP	BSk	62	39°30'	1142	835	66	272
Nimes	FRA	Csa	62	43°52'	1808	1406	62	191
Pescara	ITA	Cfa	11	42°25'	1805	1396	49	214
Portoroz	SVN	Cfb	95	45°31'	2115	1582	49	229
Perpignan	FRA	Cfa	47	42°43'	1523	1176	44	178
Pamplona	ESP	Cfb	453	42°46'	2489	2041	44	139
Milan	ITA	Cfa	103	45°25'	2264	1766	41	209
Leon	ESP	Csb	914	42°34'	3131	2725	33	98
Rome	ITA	Csa	3	41°47'	1445	1085	26	129
Grenobles	FRA	Cfb	386	45°20'	2845	2287	17	121
Barcelona	ESP	Csa	6	41°16'	1332	989	14	131
Oporto	PRT	Csb	77	41°13'	1566	1146	10	44

Fig. 4.6 Selected locations in Southern Europe. HDD and CDD, for current and 2050 scenario

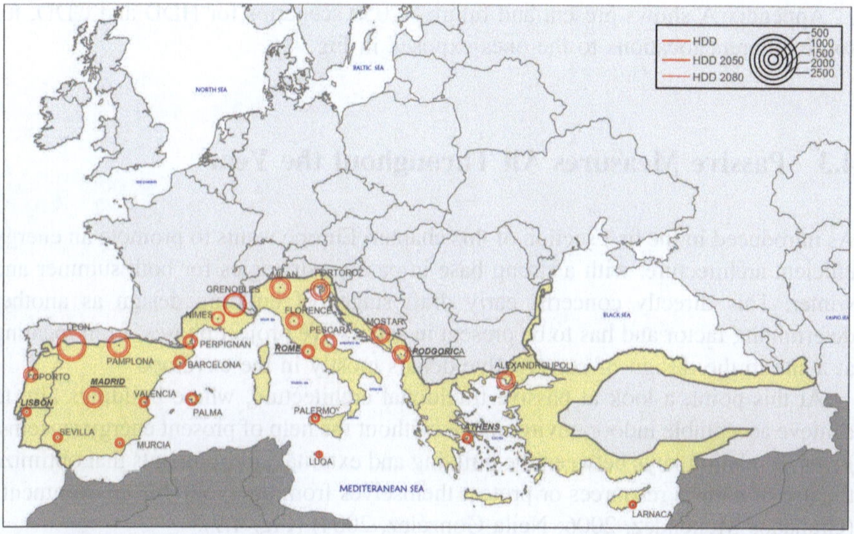

Fig. 4.7 Map of Southern Europe with selected locations and their heating degree days, HDD (18 °C) for current scenario and H2050 future scenario, over EEA delimitation

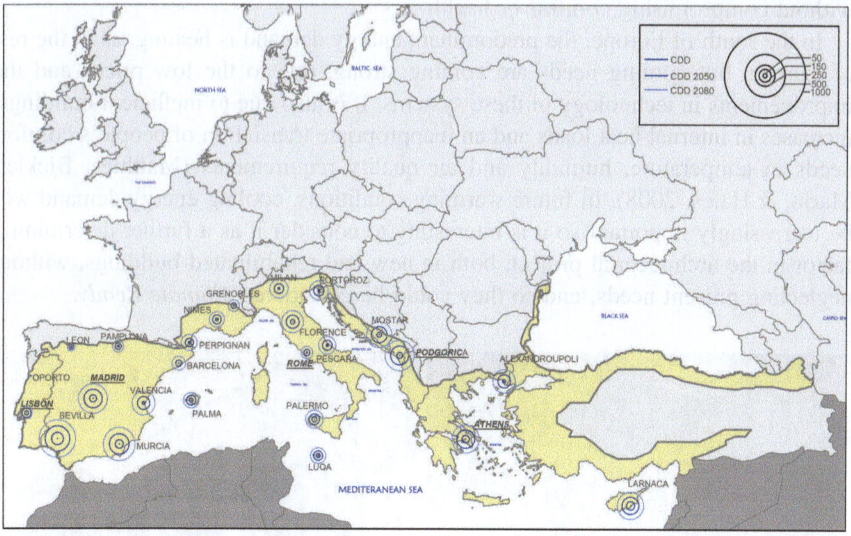

Fig. 4.8 Map of Southern Europe with selected locations and their cooling degree days, CDD (27 °C), for current scenario and H2050 future scenario, over EEA delimitation

location. On the other hand, locations with the mildest winters as Larnaca and Luqa have very different severities in summer, especially when comparing them to a future scenario of warming conditions.

Appendix A shows present and future H2050 scenarios for HDD and CDD, for some different locations to the ones exposed in Fig. 4.6.

4.3 Passive Measures All Throughout the Year

As introduced in the first section of this chapter, Europe wants to promote an energy efficient architecture, with a strong base in passive strategies for both summer and winter. This directly concerns early draft stages of building design as another determining factor and has to be present in successive project phases, from locating at a site to the design of constructive details mostly in the envelope.

At this point, a look at passive traditional architecture, where buildings aim to achieve acceptable indoor environments without the help of present energy systems, gives us relation keys between the building and external environments that optimize the use of natural resources or protect themselves from unfavourable environments (Granados Menéndez, 2006; Neila González, 2004) (Fig. 4.9).

Buildings in warm climates of Southern Europe meet the difficulty of having to respond to summer and winter conditions which require different, even opposed designs, so an optimized design that responds in the best way to its annual total energy will be necessary, aiming for an indoor environment adequate for occupiers, without compromising comfort or health.

In the south of Europe, the predominant energy demand is heating, as in the rest of Europe, but cooling needs are coming strong, due to the low prices and the improvements in technology of these systems. It is also due to inefficient buildings, increases in internal heat loads and an inappropriate translation of people's comfort needs in temperature, humidity and air quality requirements (Matthies, Bickler, Marin, & Hales, 2008). In future warming conditions, cooling energy demand will be increasingly important so it is interesting to consider it as a further determining factor in the architectural project, both in new and rehabilitated buildings, without neglecting present needs, and so they could be considered *Climate Ready*.

Fig. 4.9 Passive architecture in the Mediterranean Region. Montpellier (France) and Cuenca (Spain)

The way to deal with new or rehabilitated building design is totally different from the moment that rehabilitation starts at a pre-existing building, generally with a limited field of action. On the other hand, in new buildings projects, the possibilities offered by location, form, volume and orientation will be key in an adequate passive design.

Rehabilitations from building inadequate for their climate conditions will need solutions that will involve a higher energy consumption to reach comfort conditions. There are however examples of very efficient rehabilitations, with very inadequate pre-existing conditions and also with socioeconomic limitations (de Luxán García de Diego & Gómez Muñoz, 2006; Giancola, Soutullo, Olmedo, & Heras, 2014).

The building's thermal envelope plays a key role in the limitation of energy demand both in summer and in winter conditions, although the insulation and airtightness required may vary between both seasons, the most optimal solution will be sought. The control of ventilation to ensure the quality of indoor air, together with the optimization of solar gains (passive solar heating), has been traditionally considered in bioclimatic architecture (Mazria & Serra Florensa, 1983; Behling, Behling, Schindler, & Foster, 2002).

Strategies in summer are different than in winter and although insulation characteristics in envelopes improve heat transfer also in summer conditions, other strategies will be required. The first and most evident refers to solar control, together with measures to evacuate and dissipate heat inside the building, from ventilation, thermal mass and evaporative cooling to simple systems such as overhangs or a fan, to more sophisticated ones (Szokolay, 2008).

The bioclimatic chart based on the one proposed by Givoni is still very useful in the early phases of a project. Based on a psychrometric chart, it shows the exterior climatic conditions and displays the comfort zone with the different passive strategies of winter and summer (Givoni, 1969). The Climate Consultant Tool, developed by the Energy Design Tools Group of the University of California, USA, is easy to use and very effective in understanding local climate (UCLA, 2017). Figures 4.10 and 4.11 show the psychrometric chart for Athens and Oporto, to graphically understand the differences in climatic severities of winter and summer, and the ranges of temperature and humidity.

4.4 The Envelope as a Key Factor: Façade and Roofs

Residential buildings in Southern Europe present similar characteristics with respect to typologies and building systems, with variations depending on climatic, social and cultural factors. The thermal envelope plays a main role in the energy efficiency of buildings and we define it mainly as the integrated elements of a building which separate its interior from the outdoor environment, namely façade and roof. Elements that separate residential spaces from other non-conditioned spaces must be also taken into consideration, such as slabs between first floor and ground floor in residential collective buildings, or ground slabs in single houses.

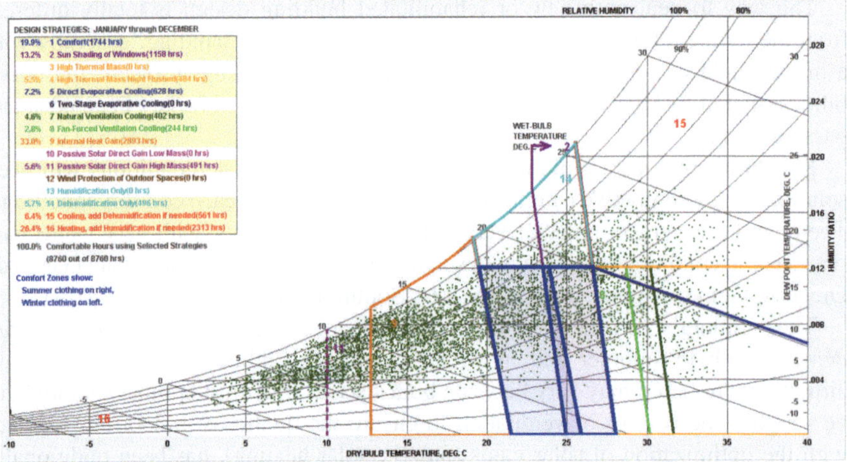

Fig. 4.10 Psychrometric chart of Athens, Greece (IWEC2 ASHRAE, 1983–2008), elaborated with climate consultant tool (UCLA, 2017)

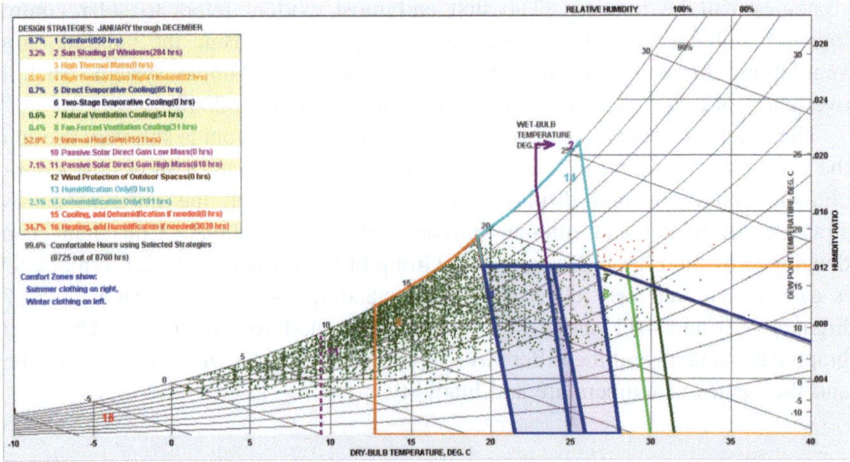

Fig. 4.11 Psychrometric chart of Oporto, Portugal (IWEC2 ASHRAE, 1983–2008), elaborated with climate consultant tool (UCLA, 2017)

In façade requirements, insulating characteristics and their situation inside the wall are key factors, such as thermal bridging and airtightness control, and their relevance will depend on the climatic severity of winter and objectives set for reducing energy consumption (Sánchez-Ostiz Gutiérrez & Campo Baeza, 2011). Its thermal mass characteristics will also be of interest both in winter and summer conditions in Southern Europe, beside its materials and configuration as, for example, ventilated façades or green façades (Fig. 4.12).

Fig. 4.12 Examples of green façades

The window characteristics, the frame and the glass as well as the reception in the wall, also play an important role. In the Mediterranean zone, shading systems are of special importance when they are part of the envelope, and in some cases can be the origin of important infiltrations and thermal bridges. Shading systems should be designed mainly according to the orientation and used for summer conditions but without interfering with solar gains in winter. For this reason, movable and adaptable designs will be prefered.

The roof is a key element of the envelope in summer and winter energy consumption, and as it is getting the highest solar radiation in summer, overheating or important cooling consumptions in the building could take place if it is inadequately designed and constructed (Sánchez-Ostiz Gutiérrez, 2007). Figure 4.13 shows the values for irradiation in Pamplona, Spain (latitude 42°46′), where the roof in summer conditions doubles the south façade. In a roof, the level of insulations, ventilated constructive details, colour and kind of materials or the consideration of green roofs, all constitute part of the roof design (Fig. 4.14).

Finally, the main characteristics of envelopes and systems, building typology, age and even climatic zone, can be found by Southern European country in the Tabula Project (TABULA, n.d.).

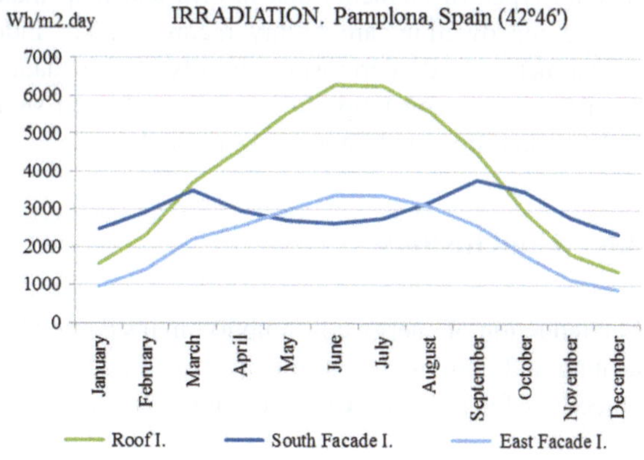

Fig. 4.13 Irradiation Graph in Pamplona, Spain (Latitude 42°46′) in façades and roof, based on PVGIS data ("PVGIS," n.d.)

Fig. 4.14 "Examples of green roofs" in the Mediterranean zone

4.5 Active Occupants in Passive Buildings

Residential buildings and passive strategies implemented in them require the active use of occupiers to ensure a low-energy demand and comfortable and healthy indoor conditions. Domotics and automatized systems, which are having a great development, can be used but probably sometime will pass before they reach the most vulnerable population, whether for economic causes, distrust in the systems, etc.

Active occupants act according to their own experience, therefore elderly people who have lived without heating and cooling systems may have more resources to act on buildings (Loughnan, Carroll, & Tapper, 2014). Specific actions, like wearing thicker clothes in the winter, night ventilation and use of shading devices in the summer, are normally done.

However, dealing with an increasingly warming climate and extreme heatwave events, there may be a time when users do not know how to act or even act in a counterproductive way. Then, even if conscious about the need for actions, they certainly need more information on the elements of the building's envelope, and on the incidence of their actions in the dwelling, that could affect indoor conditions. This is especially relevant in those locations where buildings do not normally have air conditioning, and where wrong actions could provoke continuous overheating in dwellings making it hard to return to a comfortable and healthy situation. This is why all types of elements that inform and help make decisions related to conditioning the dwelling are highly recommended: indoor and if possible, outdoor thermometers and thermostats, hourly online data from near weather stations, detailed energy consumption invoices and even smartphone-controlled electronic gadgets that are slowly appearing in the home.

4.6 Conclusions

The reduction of energy consumption in residential buildings, particularly related to heating and cooling, has a principal role in achieving the European environmental objectives for reducing CO_2 emissions which are contributing to global warming.

Energy efficiency improvements are responsible for the significant drop in heating consumption currently taking place in the European Union.

However, cooling consumption is seriously increasing together with the installation of air conditioning systems, especially in the Mediterranean zone, due to the hardening of summer climatic conditions and the improvement of life conditions. Percentages in the whole of the European Union are small (0.5% of global consumption in residential buildings), but the trend is important looking at future scenarios where climatic conditions will be warmer.

A first approach on heating degree days and cooling degree days in Southern Europe is presented in this chapter, both for the present and for a future scenario of warming conditions, allowing us to see the different climatic severities and their associated cooling consumption.

Therefore, design and construction of new and rehabilitated buildings must have the objective of being *Climate Ready*, without forgetting they must also respond to actual conditions, and focus on optimized solutions throughout the year. Implemented passive strategies in architecture itself, both for summer and winter conditions, come a long way in Southern European traditional architecture, from location and orientation aspects to form and volume, to materials and systems, etc. Therefore, higher latitude locations used to having milder summers but now experiencing warming conditions can benefit from Southern European experience in traditional passive architecture.

Finally, thermal envelopes are highlighted as key elements, as well as the active role that occupants should have in passive residential buildings.

Acknowledgements We would like to thank Germán Campos (www.ecoeficiente.es) for his assistance in the heating degree days and cooling degree days calculation for any hourly base. The authors also want to recognize the assistance of Ainhoa Arriazu and María Dolores Juri in the elaboration of maps provided in this chapter.

References

Behling, S., Behling, S., Schindler, B., & Foster, N. (2002). *Sol power : la evolución de la arquitectura sostenible*. Barcelona : Gustavo Gili, 2002. Retrieved from http://ezproxy.si.unav.es:2048/login?url=, http://search.ebscohost.com/login.aspx?direct=true&AuthType=ip,url&db=cat00378a&AN=bnav.b1703370&lang=es&site=eds-live&scope=site.

CTE-HE. (2013). Documento Básico CTE-HE Ahorro de energía. *Septiembre*. Retrieved from http://www.codigotecnico.org/images/stories/pdf/ahorroEnergia/DBHE.pdf.

de Luxán García de Diego, M., & Gómez Muñoz, G. (2006). Dos bloques de viviendas y locales comerciales en San Cristobal de los Angeles, Madrid. *Informes de La Construcción*, *58*(502), 5–16.

EEA. (2016). *Climate change, impacts and vulnerability in Europe 2016: An indicator-based report*.

Episcope. (n.d.). Retrieved from http://episcope.eu/building-typology/

EPBD. (2010). Directive 2010/31/EU of the European Parliament and of the Council of 19 May 2010 on the energy performance of buildings (recast). *Official Journal of the European Union*, 13–35. http://doi.org/10.3000/17252555.L_2010.153.eng.

Giancola, E., Soutullo, S., Olmedo, R., & Heras, M. R. (2014). Evaluating rehabilitation of the social housing envelope: Experimental assessment of thermal indoor improvements during actual operating conditions in dry hot climate, a case study. *Energy and Buildings, 75,* 264–271. https://doi.org/10.1016/j.enbuild.2014.02.010.

Givoni, B. (1969). *Man, climate, and architecture.* Amsterdam [etc.]: Elsevier, 1969. Retrieved from http://ezproxy.si.unav.es:2048/login?url=, http://search.ebscohost.com/login.aspx?direct=true&AuthType=ip,url&db=cat00378a&AN=bnav.b1350591&lang=es&site=eds-live&scope=site.

Granados Menéndez, H. (2006). *Principios y estrategias del diseño bioclimático en la arquitectura y el urbanismo: eficiencia energética.* [Madrid]: Consejo Superior de los Colegios de Arquitectos de España, [2006].

Gynther, L., Lappillone, B., & Pollier, K. (2015). Energy efficiency trends and policies in the household and tertiary sectors. An analysis based on the ODYSSEE and MURE databases, (June), 97. Retrieved from http://www.odyssee-mure.eu/publications/br/energy-efficiency-trends-policies-buildings.pdf.

INE. (2008). Encuesta de hogares y medio ambiente. Retrieved from http://www.ine.es/dyngs/INEbase/es/operacion.htm?c=Estadistica_C&cid=1254736176950&menu=resultados&idp=1254735576508.

Loughnan, M. E., Carroll, M., & Tapper, N. (2014). Learning from our older people: Pilot study findings on responding to heat. *Australasian Journal on Ageing, 33*(4), 271–277. https://doi.org/10.1111/ajag.12050.

Matthies, F., Bickler, G., Marin, N., & Hales, S. (2008). Heat–health action plans. WHO Europe, 45. Retrieved from http://www.euro.who.int/__data/assets/pdf_file/0006/95919/E91347.pdf.

Mazria, E., & Serra Florensa, R. (1983). *El libro de la energía solar pasiva.* México: Gili, 1983. Retrieved from http://ezproxy.si.unav.es:2048/login?url=, http://search.ebscohost.com/login.aspx?direct=true&AuthType=ip,url&db=cat00378a&AN=bnav.b1028610&lang=es&site=eds-live&scope=site.

Neila González, J. (2004). *Arquitectura bioclimática en un entorno sostenible.* Madrid: Munilla-Lería, 2004.

ODYSSEE-MURE. (n.d.). Energy efficiency policies | ODYSSEE-MURE. Retrieved April 7, 2017, from http://www.odyssee-mure.eu/.

PVGIS. (n.d.). Retrieved from http://re.jrc.ec.europa.eu/pvgis/apps4/pvest.php#.

Sánchez-Ostiz Gutiérrez, A. (2007). *Cerramientos de edificios: cubiertas.* Madrid: CIE Inversiones Editoriales Dossat 2000, 2007. Retrieved from http://ezproxy.si.unav.es:2048/login?url=, http://search.ebscohost.com/login.aspx?direct=true&AuthType=ip,url&db=cat00378a&AN=bnav.b1840806&lang=es&site=eds-live&scope=site.

Sánchez-Ostiz Gutiérrez, A., & Campo Baeza, A. (2011). *Fachadas: cerramientos de edificios.* Madrid: CIE Inversiones Editoriales Dossat-2000, 2011. Retrieved from http://ezproxy.si.unav.es:2048/login?url=, http://search.ebscohost.com/login.aspx?direct=true&AuthType=ip,url&db=cat00378a&AN=bnav.b2256922&lang=es&site=eds-live&scope=site.

Szokolay, S. V. (2008). *Introduction to architectural science: The basis of sustainable design/Steven V. Szokolay.* Oxford: Architectural Press is an imprint of Elsevier. Retrieved from http://ezproxy.si.unav.es:2048/login?url=, http://search.ebscohost.com/login.aspx?direct=true& AuthType=ip,url&db=edsnuk&AN=edsnuk.vtls001348063&lang=es&site=eds-live&scope=site.

TABULA. (n.d.). IEE Project Tabula (2009–2012). Retrieved from http://episcope.eu/index.php?id=169.

UCLA. (2017). Climate consultant. Retrieved from http://www.energy-design-tools.aud.ucla.edu/.

Ward, K., Lauf, S., Kleinschmit, B., & Endlicher, W. (2016). Heat waves and urban heat islands in Europe: A review of relevant drivers. *Science of the Total Environment, 569–570,* 527–539. https://doi.org/10.1016/j.scitotenv.2016.06.119.

Chapter 5
Retrofitting Focus on Vulnerable Residential Buildings in Winter

5.1 Tackling Inefficiently Constructed Buildings to Improve People's Well-Being and Health

5.1.1 Value of Retrofitting. Benefits and Co-benefits of Retrofitting

On different fronts, a search for greater energy efficiency in construction is being carried out, due to the increase in energy consumption which has occurred over the last two decades in all Western countries. As a result of this increase, natural resources are being exhausted, and the CO_2 emissions into the atmosphere, associated with the use of fossil fuels, are growing exponentially. Different international agreements such as the Kyoto Protocol (Naciones Unidas, 1998) and the Paris Agreement (UNFCCC, 2015), European directives like that of energy efficiency (Comisión Europea, 2010) and national regulations (in Spain the Código Técnico de la Edificación, CTE) (CTE-HE, 2013), stress the adoption of measures aimed at reducing these associated environmental impacts. However, due to the age of the building stock and the lack of maintenance, the European objectives of reaching the goal of nearly zero energy buildings are far from being achieved, because of the low energy efficiency of the buildings already constructed.

The struggle against climate change in the next few decades will be directed towards the retrofitting of existing buildings in order to improve their thermal performance. With this in mind, we must be aware of what their critical points are, what the different levels of intervention may be and their costs and where investment should be made in order to achieve the best results.

In the year 2014, Spain developed a strategy (Ministero de Fomento, 2014) which, within the energy objectives, includes ten measures regarding construction and fitting, the first four are great interest for our subject: energy retrofitting of the thermal envelope of existing buildings; improvement of the energy efficiency of the thermal systems of existing buildings; improvement of the energy efficiency of the

© Springer International Publishing AG 2018

A. Monge-Barrio and A. Sánchez-Ostiz Gutiérrez, *Passive Energy Strategies for Mediterranean Residential Buildings*, Green Energy and Technology,
https://doi.org/10.1007/978-3-319-69883-0_5

indoor lighting of existing buildings; and construction of new buildings and retrofitting of existing ones with high energy rating.

In line with the first measure established, the thermal envelope retrofitting offers direct benefits such as a reduction in energy demands and in the CO_2 emissions associated with the consumption of energy from fossil fuels. Apart from considering these advantages, we should take into account other indirect benefits such as:

- Improvement of the indoor comfort conditions. Placing thermal insulation on the outer envelope not only reduces heat loss through same but also increases the radiant temperature, and so, the operative temperature and comfort.
- Positive effects on the health of the residents because the indoor temperature is maintained at a higher level. There is a drop in the mortality rate and winter morbidity, and in the negative psychological effects resulting from unsuitable indoor conditions. This is particularly important in the cases of energy poverty. In addition, the quality of air indoors increases due to the occasional use of heaters which generate CO_2, or of braziers, which may also cause fires in the home. This is of greater concern in the case of elderly people.
- Increase in the acoustic insulation from airborne noise, for example, by replacing old windows and improving their airtightness and their placement.
- Improvement of the outer appearance of the building when the insulation is placed on the outside, and the windows and the shading devices are uniformly replaced.
- A new appreciation in the building's value caused by the positive perception of the user, who rents or buys at a higher price because the building is more efficient and consumes less energy.
- Reuse of the building. Dwellings that are uninhabited due to bad interior habitability are improved by retrofitting the thermal envelope. Besides, the retrofitting of the envelope becomes an opportunity to improve other aspects of the building such as, for example, its accessibility.
- Renovation is more sustainable than demolition and new construction as there is a reduction in waste generation, consumption of building materials and the energy involved in their manufacture.
- Reduction of land use for the construction of new dwellings in new urban developments, which causes urban sprawl and greater demands for infrastructure and transport.
- The creation of jobs in the construction sector which the envelope retrofitting market generates. This sector is still in crisis in many countries including Spain, and retrofitting is at present the driving force for this activity.

5.1.2 Old Residential Buildings with Greater Retrofitting Potential. Residential Vulnerability

Both in Europe and in Spain, refurbishment focuses mainly on dwellings built before the approval of the first regulations on thermal insulation (around 1980). Table 5.1 shows the data on this type of dwellings in some Mediterranean countries, distinguishing

Table 5.1 Number of dwellings constructed in European Mediterranean countries in different periods (TABULA Episcope, n.d.)

		Before		1941–80 Spain 1949–81 France 1946–80 Italy 1945–80 Slovenia		After and till 2000		Total dwellings	
		%	N° Dwellings	%	N° Dwellings	%	N° Dwellings	%	N° Dwellings
Spain	Single-family house	7	1,527,806	14	2,961,166	11	2,193,619	32	6,682,591
	Collective building	7	1,421,978	42	8,653,920	20	4,064,966	68	14,140,864
France	Single-family house	21	5,080,000	22	5,208,000	14	3,273,000	57	13,561,000
	Collective building	12	2,743,000	24	5,664,000	8	1,794,000	43	10,201,000
Italy	Single-family house	7	2,171,496	10	3,178,142	4	1,280,153	21	6,629,791
	Collective building	16	5,054,957	45	1,4042,338	17	5,221,797	79	24,319,092
Slovenia	Single-family house	21	165,183	26	208,771	16	128,048	62	502,002
	Collective building	13	104,214	18	141,581	7	57,282	38	303,077

between three construction periods, the period with the greatest volume of construction of suburban dwellings: between 1940 and 1980 (in other European countries, this may be between 1945 and 1980), before and after this period.

In Spain, dwellings constructed between 1940 and 1980 have the greatest potential for renovation as they make up 56% of the total of constructed dwellings.

Most of the dwellings from this period are located in suburban areas around big cities and are the result of the ongoing city construction processes, where some residential and urban complexes were constructed in response to the pressing and immediate need to supply housing for sectors of the immigrant population who had moved there to satisfy the demand of the incipient industrial development of the 1950s; this continued and spread until the late 1970s. These dwellings have shown to be unsuitable for present-day demand, as they were constructed before the basic regulations, with criteria that prioritized quantity over quality. Over time, they have fallen into a gradual physical decline in parallel with their occupation by sectors of the population with low economic resources, little cultural and professional education, who also have social problems, and in many cases are marginalized.

In general, we are not referring to particularly old building stock. Its main characteristic is that it was originally very low-quality construction, and moreover, very little investment has been made in it since its construction. These buildings have very repetitive construction types and very common shortcomings due to the poor construction of the thermal envelope. Windows with no thermal insulation, uncontrolled

Fig. 5.1 Examples of social housing buildings constructed between 1940 and 1980 in Spain

air infiltration and thermal bridging are some of the most common problems which result in high energy consumption in an attempt to achieve comfort indoors. The poor insulating capacity of the façades and roofs in turn results in pathologies of condensation damp, fundamentally in rooms facing north or in shady areas.

Because of what we have said above, **residential vulnerability** may be identified in older buildings, buildings without a lift or with architectural barriers, with individual heating or none, with thermal envelopes without thermal insulation and high air infiltration. In addition, one vulnerability leads to another, that is, social vulnerability, and people with the least resources end up living in these buildings.

Nowadays, in Spain and in Europe, the renovation of social housing complexes must consider the retrofitting of the thermal envelope in order to reduce energy consumption and to increase thermal comfort, together with improvement of the heating systems. This would permit the establishment of real retrofitting strategies with a major impact in order to achieve the environmental objectives of reduction of energy consumption and of emissions. Moreover, this retrofitting of buildings would also lead social and economic regeneration (Fig. 5.1).

5.1.3 Retrofitting Experiences in Europe and Spain

Different renovation actions on buildings of these characteristics are an example and a starting point for proposed interventions in the constructed building stock, with the intention of adapting and mitigating climate change.

First, since the 1990s the **European Union** has promoted urban regeneration. The main strategy was the development of the URBAN Community Initiative (IC URBAN), which begun in 1994 as part of the EU cohesion policy. Its objective was the economic and social revitalization of cities and neighbourhoods in crisis. Using European structural funds, it subsidized the rehabilitation of buildings and actions intended to improve the quality of life in cities and neighbourhoods (Gutiérrez Palomero, 2010). Various publications (Ma, Cooper, Daly, & Ledo, 2012; Morandi, Pessina, & Scavuzzo, 2010; Deponds, 2010) refer to renovation interventions in suburbs in Europe, in order to achieve an urban regeneration which takes into account social and economic aspects besides energy savings.

Additionally, there has been an important parallel concern regarding the retrofitting of apartment blocks. We cannot here mention all the different experiences, but through case studies, we can see different proposals with results that are poles apart. For example, the cases presented in the working group in Annex 56 (IEA Annex 56, 2016; Domingo-Irigoyen, Sánchez-Ostiz, & San Miguel-Bellod, 2015) show successful building renovation interventions in various European countries such as Austria, Denmark, Netherlands, Portugal, Switzerland, Sweden and Spain. These interventions were intended to affect the thermal envelope and the heating, ventilation and lighting systems, with the aim of saving energy, reducing CO_2 emissions and life cycle costs. Likewise, these cases describe the retrofitting that were found and the potential solutions. One of the most habitual problems is financial. Increasing the suitability for construction, adding one or two floors to the building or enlarging its surface area may be a means of making the necessary investment or the retrofitting of the whole building profitable. In countries such as Germany, Austria, France and England, upward extensions are a strategy to reactivate the construction industry, to make the existing city more compact and to make the renewal profitable. To do so, each of these countries has approved regulations for this type of intervention.

Second, in **Spain** since 1990, diverse retrofitting interventions have been carried out on buildings constructed between 1940 and 1980 (Prat Navarro & Wadel, 2010; Rubio del Val & Molina Costa, 2010). Suburbs on the outskirts of big cities such as, for example, Madrid and Barcelona, and of smaller cities such as Zaragoza, Vitoria, Pamplona and Tudela are some examples which, moreover, have received funding from Europe, Spain or local governments to tackle renovation.

Pioneering examples in Spain were the renovation of the San Cristóbal de los Ángeles suburb in Madrid and the Mina neighbourhood in Barcelona. The renovation programme of the first one was led by the Empresa Municipal de Suelo y Vivienda of the Madrid City Council. It is a social housing neighbourhood to the south of the capital with a little over 4.000 dwellings whose current state is typical of this kind of neighbourhood: mainly residential use, residents who are older than the mean, a higher density of immigrants, deterioration of the physical space, very high social vulnerability, etc. The intervention taken consists of designing a Special Plan intended to detect buildings in need of re-structuring or remodelling, and the designation of new arrangements or build areas for the new blocks to permit more suitability for building (greater funding) or the possibility of placing exterior lifts. Over the last two decades, renovation of different buildings and new vulnerable neighbourhoods has occurred (Rubio del Val, 2011).

In the last 25 years, ADIGSA, the company that manages social housing in the Government of the Autonomous Community of Catalonia, has promoted and carried out the refurbishment of the social housing stock in Catalonia. Approximately, 60.000 dwellings underwent a systemized diagnostic process of their condition. Since then, maintenance and renovation projects were carried out directed mainly to their appropriateness in terms of urban services, outdoor urbanization and improvement of the thermal insulation of the whole envelope (roofs and façades). A very well-known example was the Proyecto de Transformación Urbana del Barrio de La Mina (Sant Adrià de Besòs, Barcelona). The initial conditions in the area for intervention were very complex, and a particular concern was the process of deterioration, principally in the social area (López de Lucio, 2008; Rubio del Val & Molina Costa, 2010).

As an example of the strategy followed in Zaragoza, a systemized programme for maintenance or refurbishment of social housing was used. Rubio del Val (2011) underlines a proposal for the retrofitting of different urban complexes, led and coordinated by the Sociedad Municipal Zaragoza Vivienda, with the economic support of the Ministry for Housing and the Government of Aragón. This initiative intends to encourage building retrofitting policies for over 8.000 dwellings belonging to various urban complexes of interest, which are over 40 years old in the following suburbs: Arrabal and Picarral, Fuentes and San Agustín, San José, Torrero, Delicias, Oliver and Casetas. The complexes studied have been ordered into categories depending on the decade of construction and their construction-architectural characteristics.

In Navarre also, the Government of Navarra and NASUVINSA (Navarra de Suelo y Vivienda/Navarre Land and Housing) has initiated several retrofitting interventions in apartment blocks. Two pilot cases stand out: Integral Energy Retrofitting of the Lourdes neighbourhood in Tudela (2005–2012) and the project Efidistrict in Pamplona (ongoing since 2014). In the first case, the project Lourdes Renove, which has received several national awards, is part of the Concerto Program, an EU initiative that supports local communities and the reduction of CO_2 emissions by means of improving energy efficiency and inclusion of renewable energies. The improvement actions have focused on public space, the renovation of district heating for a group of 486 dwellings and the retrofitting of three social housing buildings in order to improve accessibility, the thermal envelope and the fittings. In Pamplona, the objective of the Efidistrict project is the energy retrofitting of the social housing neighbourhoods constructed between 1950 and 1980, the first phase of which is being developed in Chantrea (See Fig. 5.9) and will later be extended to other suburbs of Pamplona and further Navarrese localities. The main interventions planned are the following: retrofitting of the thermal envelopes, renovation of the fittings of the buildings with efficiency criteria, the inclusion of renewable energies and solution to accessibility problems.

Finally, this chapter will show the results of the **prestaRener** research project, 'Protocol for action in the retrofitting of the envelope of buildings, with performance-based design', [1] developed by the SAVIArquitectura research group at the University of Navarra, Spain, in collaboration with the Worcester Polytechnic

[1]Project funded by MINECO, Ministerio de Economía y Competitividad de España, BIA 2012-38666

Institute, USA. Carried out between 2013 and 2016, its objective was to establish an intervention procedure for the retrofitting of the thermal envelope of buildings constructed between 1940 and 1980 in social housing areas in Pamplona, Spain. Based on the identification of the building typologies and simulation of the energy demands of the current building and different retrofitting levels, it detected the most cost-effective measures for the retrofitting of the thermal envelope.

The methodology used is as follows. First, establishing the most representative typologies of these buildings, recognizing the constructive characteristics of the different parts of the envelope and typifying the retrofitting measures. Second, monitoring case studies to find the true performance of the envelope, followed by computer simulation of different levels of intervention to assess how much energy could be saved. Finally, assessment of the cost-effectiveness of the different interventional levels.

The interventions proposed for the retrofitting and the savings produced in the energy demands allow the establishment of interventions which are translatable to the Spanish and European residential building stock, taking into account the climate differences. This chapter presents some of the results obtained in the project, which may be consulted at http://www.unav.edu/centro/saviarquitectura/prestarener/index.html.

5.2 Diagnosis of the Current Status of Existing Vulnerable Residential Buildings

The retrofitting interventions must be based on a correct diagnosis of the actual thermal performance of the existing buildings. In this way, the retrofitting measures adopted will be the most suitable for solving the problems.

The multiplicity of criteria which define the thermal performance of a building makes proposing standard intervention solutions complicated, unless they are based on the identification of the most common typologies. Therefore, it is indispensable to establish these typologies beforehand and later make the diagnosis of the actual performance and its critical points.

5.2.1 Representative Typologies of Residential Buildings Built 1940–1980

Juan Rubio del Val analyses urban renovation strategies in Spain between 1989 and 2010 (Rubio del Val, 2011) with the intention of diagnosing problems which impede advances in integral urban regeneration and renovation. As part of his final conclusions on the analysis, he states that it is 'necessary to carry out systematic studies depending on typologies'.

Identification of these building typologies permits the establishment of intervention areas with the same pathologies and needs for improvement. These may be renovated jointly, so as to improve the living standards of a broader population group and to profit from the synergy of intervening on a broader scale: lower investment costs, the possibility of installing more efficient joint heating systems and greater ease of administrative procedure management.

Dascalaki et al. (2011) define the 'building typology' as a classification of buildings according to certain specific characteristics, which for this case refers to the building energy performance. The energy consumption in buildings depends on a number of factors including the envelope construction, age distribution of the existing building stock, outdoor weather conditions, building size, type, age and efficiency of the existing systems. Successful strategies towards minimizing the energy consumption and greenhouse gas emissions attributed to the building sector require knowledge of the energy-related characteristics of the existing building stock.

A revision of typology studies of examples built before the first insulation regulations in Spain and Europe was carried out. The most significant was the research project TABULA (TABULA Episcope, n.d.), which established different residential typologies for each European country. On a large scale, and quite accurately, the impact of different rehabilitation scenarios can be established for each typology. In the case of Spain, three different climate zones and four different building typologies were identified in TABULA in accordance with their construction date: single-family house, terraced house, multifamily house and apartment block.

Nevertheless, after this analysis, to make an in-depth study of the differences in Pamplona and the Navarre region, a specific study was carried out. A higher proportion of multifamily buildings was found with more typologies which showed differences in thermal performance. Within the **prestaRener** project, the classification of building typologies constructed in Navarre between 1940 and 1980 was carried out together with their potential for renovation. The appeal of this classification is the possibility of developing solutions which are applicable to each one of the typologies.

Given the focus on existing buildings of social housing in suburban towns, five typologies have been defined: T1: linear block, T2: H-shaped block, T3: tower, T4: other types of blocks and T5: single-family house. These are illustrated in Table 5.2. T4 latter has been excluded from this study because its repercussion on the global analysis is less significant. The percentage of each typology of buildings is given in Chaps. 6 of this book (See Sect. 6.4.3).

Each building typology has more sub-typologies, based primarily on the year of construction, height, number of dwellings per floor, size of dwellings, main orientation and construction characteristics of envelopes. All these parameters have an influence on their energy performance. The thermal envelope is composed of façade (F), roof (R), windows (W) and partitions with non-habitable spaces like ground floor (L) and stairs (C). The differences between typologies and sub-typologies are summarized in Table 6.1 in Chap. 6. But in this chapter, in order to simplify the sub-typologies for the simulation, the two most representative in each typology have been selected and are shown in Table 5.3, where the codes of characteristics of envelopes are shown in Table 5.4. These characteristics have been defined based on the study of original projects and visual inspection of the buildings.

Table 5.2 Typologies of social housing 1940–1980 identified in Pamplona (prestaRener project)

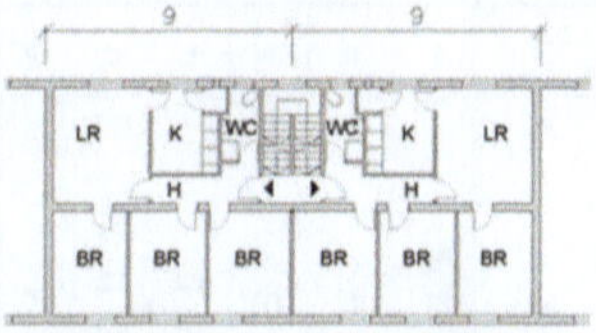		
Typology T1. Linear block		
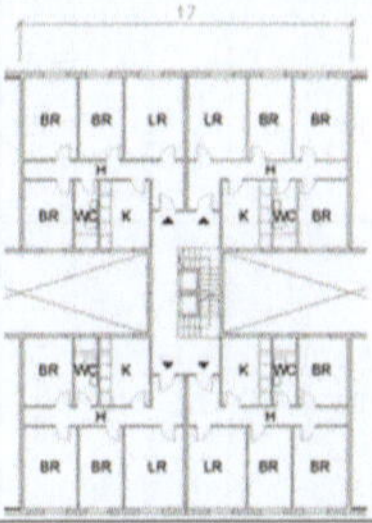		
Typology T2. H-shaped block with linear agrupation		

Typology T3. Tower		

Typology T4. Other types of blocks		
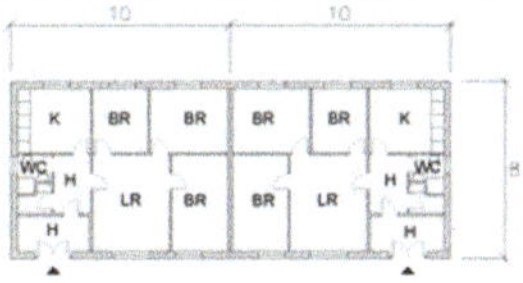		
Typology T5. Single-family house		

Legend *LR* Living room; *BR* bedroom; *K* kitchen; *WC* bathroom; *H* entrance hall; *D* hall

Table 5.3 Characteristics of sub-typologies and most representative sub-typologies

Scheme	Typology Subtypology	Description	Year of construction	Height	Dwelling per floor	M2/ dwelling	F	R	W	L	C	P
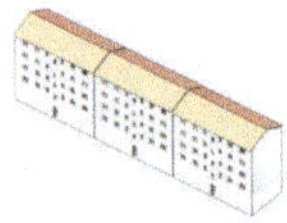	T1 **T1A**	Compact Linear block	1940-1960	GF+4	2	62,4	F1 F2 F3	R2	G1	L2 L4	C2 C3	P2
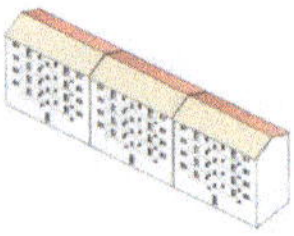	T1 **T1B**	Linear block with balconies	19460-1980	GF+4	2	77,5	F1 F2 F3	R3	G1	L2 L3	C2	P2
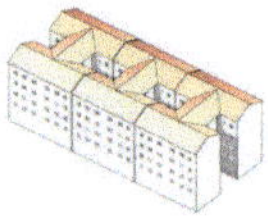	T2 **T2A**	Compact H-shaped block with linear agrupation	1950-1960	GF+4	4	60,4	F1 F2	R2	G1	L2	C2	P1 P2
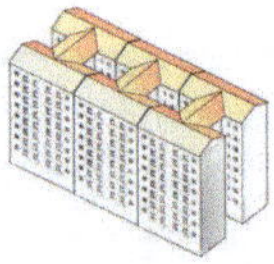	T2 **T2B**	H-shaped block with balconies and linear agrupa-tion	1960-1980	GF+8	4	87	F3 F4	R3	G1	L3	C2	P1 P2
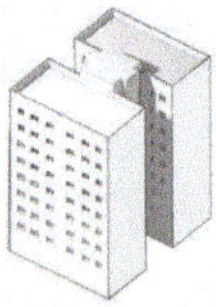	T3 **T3A**	Compact Tower	1960-19 80	GF+8	4	60,5	F2 F3	R1	G1	L2	C2	-
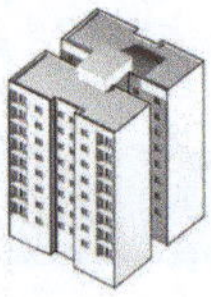	T3 **T3B**	Tower with balconies	1960-1980	GF+8	4	80,5	F2 F3	R1	G1	L3	C2	-
	T5 **T5A**	Compact Terraced Houses	1940-1960	GF+1	1	119	F2	R2	G1	L5	C2 C3	P1
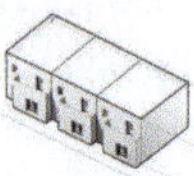	T5 **T5B**	Terraced Houses with balconies	1960-1980	GF+1	1	124	F4	R1	G1	L5	C2 C3	P1

Table 5.4 Building envelope characteristics

Façade, **F**	F1—Façade of one wythe, solid brick, 24 cm, face view or to cover U = 3.77 W/m² K	F2—Façade of one wythe, hollow bricks, 24 cm to cover U = 2.46 W/m² K	F3—Façade of cavity wall, one sheet brick face view and another hollow brick U = 1.95 W/m² K	F4—Façade of cavity wall, double sheet hollow bricks U = 1.37 W/m² K	
Roof, **R**	R1—Flat roof U = 2.65 W/m² K	R2—Pitched roof, unheated U = 1.74 W/m² K	R3—Pitched roof, heated		
Windows, **W**	G1—Wood frame + single glass U = 5.8 W/m² K	G2—Other material frame + single glass			
First floor separation, **L**	L1—Ground floor open 100%	L2—Ground floor enclosed unheated U = 2.22 W/m² K	L3—Ground floor enclosed heated (local)	L4—Ground floor enclosed (house)	L5—Ground slab U = 1.18 W/m² K
Stair shaft, **C**	C1—Simple hollow brick (12 cm) U = 2.56 W/m² K	C2—Double hollow brick (7 cm) U = 2.17 W/m² K	C3—Double hollow brick (12 cm) U = 1.78 W/m² K	C4—Perforated brick (12 cm) U = 2.04 W/m² K	
Walls within other apartments or other building, **P**	P1—Solid brick (12 cm) U = 2.43 W/m² K	P2—Double hollow brick (12 cm) U = 1.78 W/m² K	P3—Double hollow brick (24 cm) U = 1.18 W/m² K	P4—Double hollow brick (7 cm) U = 2.17 W/m² K	

Legend: *U* thermal transmittance (W/m² K) considering thermal bridges

5.2.2 Diagnostic Monitoring

To obtain reliable and consistent results, it is essential to clearly identify the characteristics of the thermal envelope and its critical points. These aspects can be evaluated by monitoring. The monitoring is a key phase in the study of the real conditions found in the buildings to be renovated or which have been renovated. In this way, various parameters of the buildings are observed, to verify their performance and detect the most relevant problems or parameters regarding the energy efficiency of the buildings and the comfort of the users. These data permit correct evaluation of the priorities for intervention and the later verification that the planned improvement objectives have been reached.

Therefore, the monitoring includes gathering, in winter and in summer, 10-minutes data reviews of the room, surface, radiant temperatures, the relative humidity, concentration of CO_2 ppm, etc. In addition, the following tests were carried out: a heat flow metre test (International Organization for Standardization, 2014) to calculate the transmittance of the façades, a comfort test to obtain the radiant and operative temperatures, blower-door tests (International Organization for Standardization, 2000) to calculate the airtightness of the dwellings and to detect the origin of the infiltrations by combining them with indoor thermography, and outdoor infrared thermography inspection (International Organization for Standardization, 2002) of the façades to assess the thermal bridges. All these assays were complemented with surveys on the thermal and usage satisfaction of the users of the dwellings together with the energy consumption data for each dwelling.

The prestaRener Project was used to monitor 103 dwellings in 33 buildings with different typologies. Some of these buildings were the same, but some still had the original thermal envelope (which we will now call SE) and in others, it had been retrofitted (called CE). In this way, the characteristics of the envelope and the interior temperatures reached could be verified. In each building, dwellings were selected because of their different performance: dwellings facing mainly south (living rooms) but with some rooms in other directions, dwellings on different floors within the building, at mid-level and others at more unfavourable levels which imply greater energy consumption, such as, for example, top floor under-roof dwellings and on the first floor above the ground floor premises.

The surveys on the use and thermal satisfaction by dwelling, together with the energy consumptions, are fundamental to correctly analyse the results obtained in the hygrothermal monitoring. The simply structured surveys open the door to semistructured interviews with the resident, which are used to detect their problems, needs or perceptions regarding the effectiveness of the measures. The survey model used includes:

- General information: postal address, members in family group, ages, level of education, work situation, unemployment benefits, income, health problems, disability or special needs, tenancy regime.
- Winter conditions: heating system, fuel, radiators, additional heating systems, heating timetables and setpoints, ventilation timetables, use of shutters at night, thermal sensation, cosiness and problems detected in the heating, etc.
- Summer conditions: the air conditioning system and solar protection systems, ventilation timetables, thermal sensation and problems detected.
- Expenses: housing expenses (mortgage or rent), other expenses and energy expenses, whether there have been delays in payment of bills.
- Retrofitting and investment: the user is asked if he/she would want to retrofit the envelope and control and regulate the heating, and how much he/she would be prepared to pay in certain conditions (in monthly instalments).

The following sections will graphically show some of the most relevant results obtained in the prestaRener project.

5.2.3 *Critical Elements of the Thermal Envelope*

The data obtained by monitoring allow adjusting the computer simulation model, together with the climate data registered and the energy consumption in heating. In this way, we obtain the current demands of the building and the heating demand associated with the different parts of the thermal envelope, detecting those of greater impact which will receive priority attention in the retrofitting measures in order to achieve an important reduction in the energy demands (Fig. 5.2). The thermal envelope, according to the CTE-HE1 (2013), is composed of all roofs, walls, subfloor, exterior doors, windows which separate the habitable spaces from the outside air, the plot or another building, and of all interior partitions which separate the habitable spaces from non-habitable spaces in contact with the exterior. That is, the façades, windows, roofs, separation from the ground floor when it is not used as a dwelling, separation from the slab when the ground floor is a dwelling, interior separations in contact with the stairwell and with other dwellings or adjoining buildings.

In these graphs, we see that the most important thermal losses are due to the façades (because of the lack thermal insulation), to the windows (generally they are single glazed or have high thermal transmittance) and the uncontrolled air infiltrations. These draughts occur through the window seals, the shutter encasements and/or where the windows join the façade.

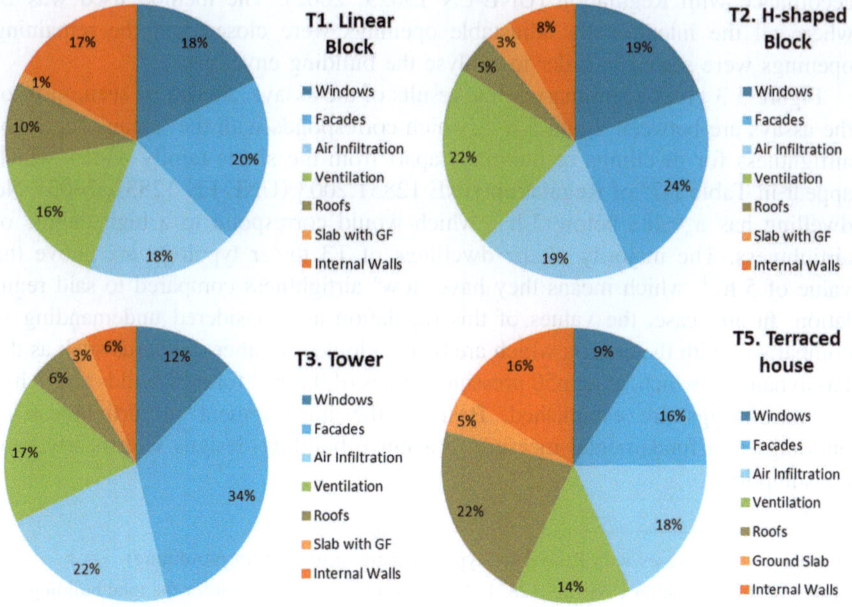

Fig. 5.2 Distribution of the heating demand associated with the envelope components. Typologies T1, T2, T3 and T5

In T1, the loss through the façades (which include thermal bridging), windows and air infiltrations is 56%, in T2 it is 62%, in T3 it is 58%. In T5 is where the percentage associated with these components drops to 33% as the losses through the roof rise to 22%, whereas in the other typologies the roof has less repercussion. Regarding ventilation, in all the types the percentage is also high and varies between 14 and 22%. During monitoring, we found a great variety of timetables and length of time which complicated the establishment of a pattern of use by typology. For this reason, as the ventilation patterns depend on the user, we have used the value of 0.63 h^{-1} as established in the Spanish regulations CTE-HS3 (CTE-HE, 2013), in order to ensure indoor air quality.

5.2.4 *Effect of Air Infiltration*

As regards air infiltration, we must emphasize the major effect it has on the thermal losses through the envelope, and therefore the increase in energy demands. This is a point which is not yet being tackled systematically[2] in Spain and is fundamental for the reduction of energy consumption associated with heating.

In the prestaRener project, a total of 42 blower-door assays were carried out, in 35 dwellings, covering all the typologies studied. In addition, some of these tests were done in buildings with the original envelope (SE) and others in buildings where the envelope had been renovated (CE). These tests have been carried out in accordance with Regulation (UNE-EN 13829, 2002). The method used was B, where all the intentionally adjustable openings were closed and the remaining openings were sealed in order to analyse the building envelope.

Figure 5.3 gives a summary of the results of the assays. As can be seen, most of the assays are between 2 and 5 h^{-1}, which corresponds with the 'mean' degree of airtightness for dwellings or buildings apart from the single-family homes which appear in Table D7[3] of Regulation UNE 12831:2003 (UNE-EN 12831, 2003). No dwelling has a value below 2 h^{-1} which would correspond to a high degree of airtightness. The majority of the dwellings of T3 tower typology are above the value of 5 h^{-1}, which means they have 'low' airtightness compared to said regulation. In any case, the values of this regulation are considered undemanding in comparison with the criteria which are being adopted in other standards such as the Passivhaus, in which, at an n50 pressure, values of 0.6 h^{-1} for new builds and 1 h^{-1} for retrofittings are established. Besides, the improvement in airtightness of buildings is a fundamental measure to achieve building designs with nearly zero consumption.

[2]Except in certain cases with Passivhaus-type standards (Enerphit for retrofitting)

[3]Table D7 of Regulation UNE-EN 12831:2003, index for air renovation for the total building at a pressure of n50, for single-family homes, establishes a high degree of airtightness when the value is below 4 h^{-1}, average when it is between 4 and 10 h^{-1}, and low for over 10 h^{-1}. For other dwellings or buildings: high (below 2 h^{-1}), average (between 2 and 5 h^{-1}), low (below 5 h^{-1}).

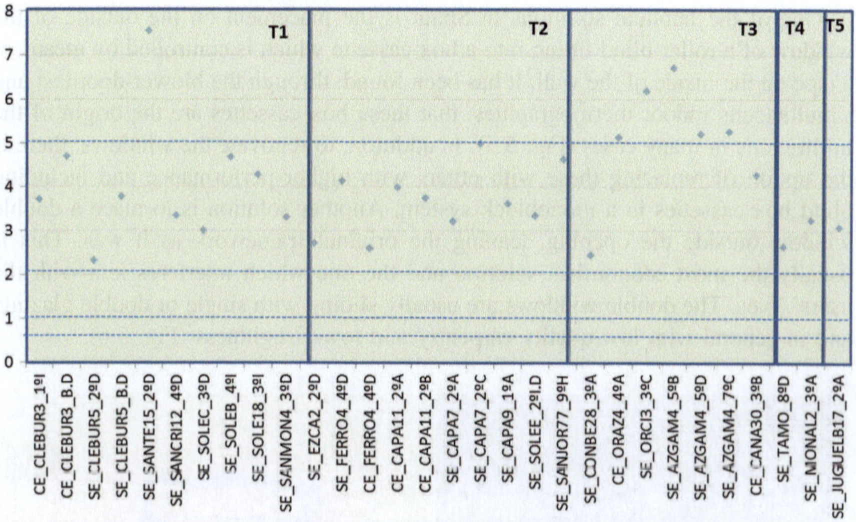

Fig. 5.3 Summary table of results of blower-door tests by typologies. Graph units n50 (h^{-1})

Figure 5.4 shows the results differentiating between buildings with (CE) and without (SE) renovation of the thermal envelope. As can be seen, the improvement in airtightness is an objective which has not been effectively reached as similar values are obtained, between 2 and 5 h^{-1}. Perhaps lack of knowledge or the complexity of carrying out this assay, because the results obtained after the retrofitting were not verified, is the reason why this point shows no improvement, as this is related to the quality of the work carried out and the placement of the exterior carpentry.

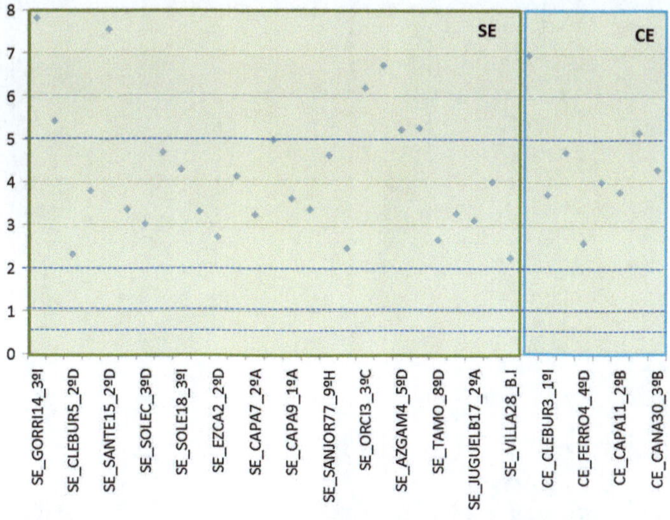

Fig. 5.4 Summary table of results of blower-door assays ordered by CE or SE buildings. Graph units n50 (h^{-1})

One of the habitual solutions in Spain is the placement on the outside of the window of a roller-blind fitting into a box cassette which is controlled by means of a tape on the inside of the wall. It has been found, through the blower-door test and simultaneous indoor thermographies, that these box cassettes are the origin of the infiltrations in many cases (Fig. 5.5). In addition, to renovate the windows, there is the option of replacing these with others with higher performance and including blind box cassettes in a monoblock system. Another solution is to place a double window outside the opening, leaving the original framework as it was. This is usually the most economical solution and the one which interferes less with the users' lives. The double windows are usually sliding with single or double glazing, and in general with low-quality carpentry and low airtightness (Fig. 5.6).

Fig. 5.5 Origin of infiltrations through the blind box cassette. Detection by means of blower door and thermography

Fig. 5.6 Intervention in windows: placement of a new window outside the existing one

It has been found that airtightness values of 2 h^{-1} may be reached when the windows and blind box cassettes are substituted by monoblock types (Fig. 5.23), due to both the characteristics of the carpentry itself and its placement, and the airtightness of the box cassettes of the blinds. The double window solution offers very variable values, which depend fundamentally on the characteristics of the inner window and the blind box cassette, together with the proper placement of the outer window with correct sealing with their façade.

5.2.5 Key Factors Regarding Monitored Indoor Temperatures

By monitoring the indoor temperatures reached in the dwellings, we can know the comfort level of the residents and the use they make of heating and ventilation. Besides, monitoring permits the detection of families which may be at risk of energy poverty as they do not heat, and the temperatures in their homes are below levels which may be dangerous for health.

The winter monitoring actions of the prestaRener project were carried out mainly in Pamplona (Spain) during the winters of 2013, 2014, 2015 and 2016, generally during the months of December, January, February and March.[4] Pamplona is a city in the north of Spain, and it has a Cfb climate, a temperate oceanic climate, with a mean annual temperature of 12.9 °C, varying between 5.2 °C as the monthly mean for January and 21.4 °C as the monthly mean for August, according to the values of the AEMET airport weather station from climate series 1981–2010. Figure 5.7 shows the mean monthly data for the city.

By monitoring the selected dwellings, the profiles of the indoor winter temperatures have been collected. The obtained results show differences depending on:

- Orientation of the dwellings and/or rooms, generally north and south.
- Location of the block: block on the corner and block in the middle between other buildings. The buildings selected were generally in the latter location.
- Position in the building: intermediate floor, sub-roof, first floor over commercial premises or over the ground slab.
- Type of heating installation, hours of use and setpoint. We find different situations: dwellings with district heating (with or without individualized controls), central heating for the building (with or without individualized controls), individual heating in each dwelling, dwellings without heating installations or with auxiliary installations such as butane gas or electric heaters.
- Use of the dwelling and socio-economic aspects of the occupier.

[4]The typical heating campaign in Pamplona is from October to May

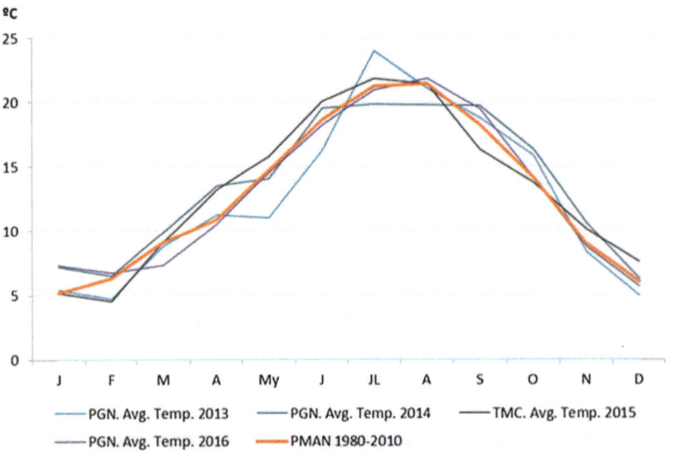

Fig. 5.7 Graphs of mean temperatures in Pamplona (PMAN & PGN) (Pamplona weather stations: data from the PMAN weather station of the climate series 1980–2010, and from the Pamplona Gobierno de Navarra PGN weather station from the years 2013–2016)

Monitoring graphs are presented illustrating these aspects in the most representative typologies of buildings from the years 1940–80 (T1, T2, T3), with different heating systems, given the noteworthy effect they have on the indexes of indoor temperatures. Three cases are dealt with:

- Grupo ORVINA, located in the Chantrea suburb, one of the social housing neighbourhoods of Pamplona with a high percentage of buildings constructed between 1940 and 1970 (Figs. 5.8 and 5.9). The building selected is a T3 tower. The heating type is district heating without individual regulation. The population is ageing but has a middle to low socio-economic level, and in general has savings at their disposal.
- Grupo FERRO, located in the Santa Engracia suburb, which is considered one of the most vulnerable neighbourhoods due to the construction typology and the low social economic profile of the residents (Figs. 5.10 and 5.11). The building is a T2 H-shaped block, with individual heating.
- Edificio GORRI in Pamplona, located in the II Ensanche, an area which is currently appreciated and in great demand (Figs. 5.12 and 5.13). Type T1 linear block, with individual heating and middle-class socio-economic profile.

First, we practically see that in all cases the **dwellings located on intermediate floors** are those with the most appropriate thermal performance because of their height placement between other dwellings. However, the dwellings located on the floor under the roof and on the first floor have the most shortcomings. This has been detected both in buildings with a retrofitted envelope (CE) and those without (SE), with central heating (see Fig. 5.8) or individual (taking into account the energy consumptions in addition to the indoor temperatures). The difference in temperature

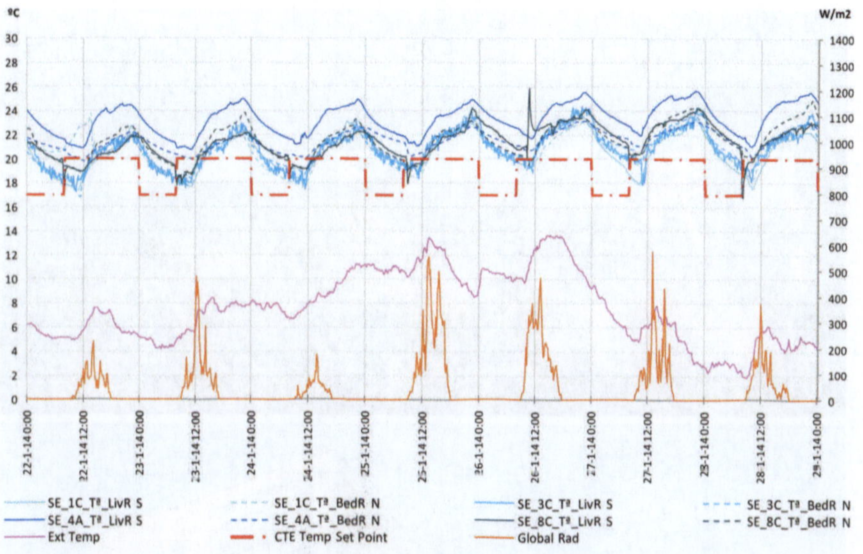

Fig. 5.8 Examples of monitored indoor temperatures in winter in a non-renovated building. ORVINA (SE). Typology T3 Tower (*Note* In the graphs, the first SE code (in the title) means non-renovated envelope, not the orientation which is indicated underneath)

between the ground floor and the top floor compared to intermediate floors depends on the construction characteristics of each building (insulation, separation from the ground floor or the slab, roof type), and the use made of the ground floor (whether it is in use or not, if it is a car park, a commercial premise or a dwelling, and its timetable of use). In general, the dwelling with fewest advantages is at the top, located under the roof, either because of the heating system (in centralized systems, the load losses produce lower temperatures on the floors which are furthest away) or because of the lack of insulation, as this is the dwelling that has the greatest surface of the thermal envelope in contact the outdoor weather. For this reason, the indoor temperatures are lower on this floor than on the others and/or require a greater consumption of energy for heating.

On the aspect of **orientation**, the dwellings which have a greater surface of spaces facing south, south-west and south-east usually have a better performance than when the spaces face in another direction. For example, in the case of ORVINA (Fig. 5.8), we can see that the living room of the dwelling SE_4A (south-west) stands out for its higher mean temperatures (3.1 °C on average compared with the coldest, a living room of SE_1C to the northeast), and a higher minimum temperature (3.7 °C higher than the coldest, living room SE_3C to the northeast). This dwelling is the only one whose living room faces south-west, and there is a notable solar gain in the afternoons. Some days it reaches temperatures of over 25 °C. The temperatures reached are much higher than those set out in the Spanish regulations (20 °C during the day and 17 °C at night) and the standards of comfort. This building belongs to an unregulated heating group, so the energy

Fig. 5.9 Photographs of ORVINA group in Chantrea, **a** panoramic view, **b** building renovation example and **c** ORVINA case study

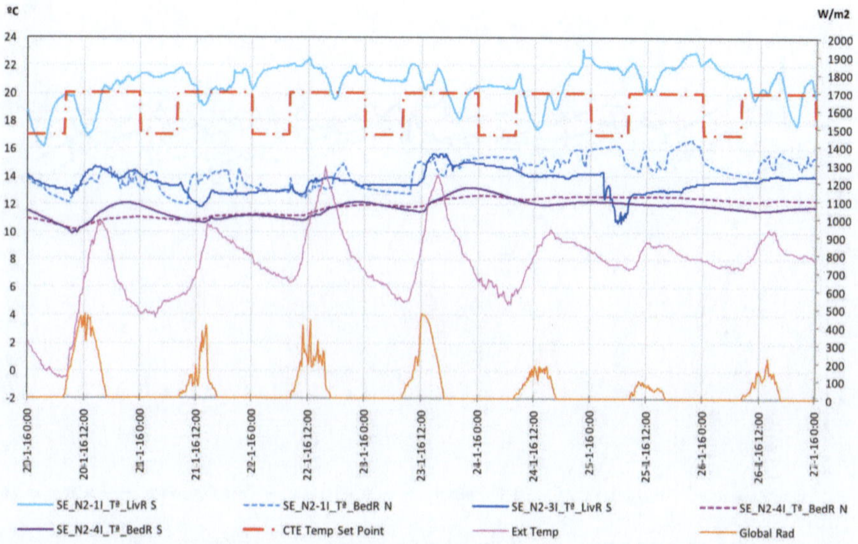

Fig. 5.10 Examples of monitored indoor temperatures in winter in a non-renovated building. FERRO (SE), Typology T2 H-shaped block

Fig. 5.11 Photograph of FERRO case study

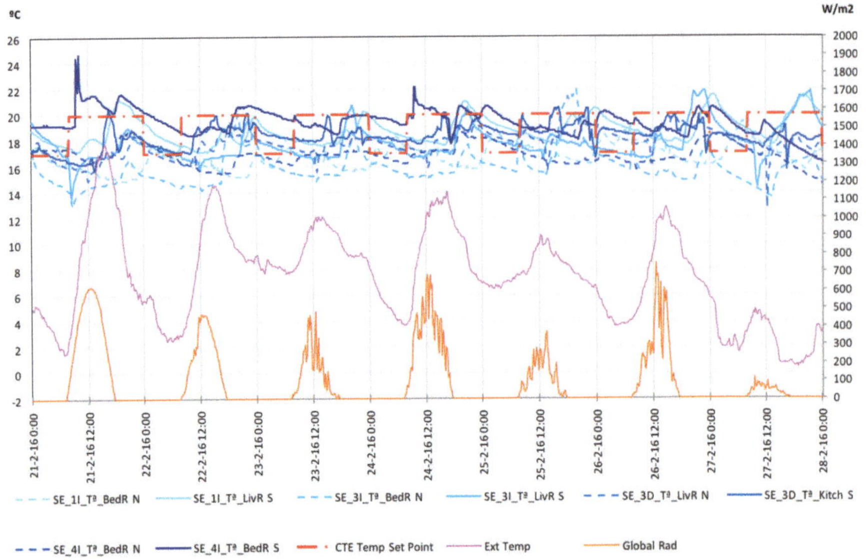

Fig. 5.12 Examples of monitored indoor temperatures in winter in a non-renovated building. GORRI (SE), Typology T1 Linear Block

Fig. 5.13 Photograph of GORRI case study

consumption of each of the dwellings is estimated a priori as 107 kWh/m^2 yr, and therefore the bill is the same for all the dwellings. The temperature pattern is quite uniform, following the heating use pattern which has been seen to be very inefficient because of the lack of regulation, although there are differences due to the effect of the orientation and height, as has been commented.

Figure 5.10 deals with the case of FERRO, a building with individual heating in which we observe important differences in indoor temperatures between dwellings, because of their orientation, height and/or use of heating. It shows the monitoring of three dwellings, with different rooms, located on the first floor (with living room facing south and bedroom north), third floor (living room facing south) and fourth floor (one bedroom to the south and another to the north). In general, we found that the residents had a low socio-economic profile as the temperatures were very low, between 11 and 15 °C, with barely any use of heating on the third floor, and with no use of heating in the bedrooms on the fourth floor. Comparing by heights, we can state that the worst conditions were found on the fourth floor under the roof for the reasons mentioned above: greater surfaces of the envelope without insulation and no use of heating. The first floor does use the heating to maintain the living room (southern orientation) at a level close to the profile established by the Spanish CTE regulations (indicated with a red line), although the pattern of use is chaotic, and the bedroom in the same dwelling, facing north, is at around 14 °C.

We must state that both the occupier of the fourth floor and that of the third floor may be at risk of energy poverty/vulnerability. We must not forget the threat to health caused by such low temperatures (see Chap. 3, Point 3.5.1). Particularly, we must highlight the dwelling on the fourth floor which is kept within the range of 10 to 12 °C, below the thresholds which endanger health.

The third case studied, GORRI (Fig. 5.12), is also a building with individual heating, but with an occupier socio-economic profile which is higher than in the previous case. We found that the temperature profiles are also higher, although all the rooms which face north (dotted lines) are below the CTE temperature profile (red line), and are between 14 and 18 °C. In the rooms facing south, the temperature is higher, between 17 and 20 °C due to solar radiation and the occasional use of heating. We also observed that when the heating is switched off, the temperature drops quite quickly because the walls and windows have no thermal insulation.

In short, the differences between the buildings by typologies are as follows: height difference in all the typologies: T1, T2 and T3; differences of orientation (south-north) in typologies T2 and T3, and in T1 as it has double orientation between the spaces facing south and that facing north. Likewise, very different temperature patterns are found depending on the type of heating used in the dwellings and the cases of energy poverty or energy and social vulnerability.

5.3 Retrofitting Can Provide Better Conditions for a Vulnerable Population

As we have seen, the population group considered as most vulnerable because of age (old people, children) or due to disability and/or a low socio-economic profile which makes it difficult for them to afford energy expenses. Also, they end up living in the most vulnerable buildings: non-renovated pre-1980 buildings with no thermal insulation, with individual heating systems or with no installation in order to spend as little as possible on energy; in addition, these may be buildings with no lifts to reduce the property management fees. A study conducted in France found that vulnerable older people living in top floor flats and poorly insulated houses were most at risk (Vandentorren et al., 2006).

On the basis of these conditions, we can assess how the retrofitting of the thermal envelope of the building may offer better indoor temperature conditions, even when the heating is not used, moving away from the range of temperatures which are considered critical for health.

We can compare the monitoring of identical buildings, of which some have a retrofitted thermal envelope and others not, and assess the improvement produced by retrofitting. In the 1940–1980 period, large groups of social housing were constructed as a single project, and so we have been able to find cases of dwellings which were exactly the same as when they were constructed, some in their original state (SE) and others in which the thermal envelope was renovated later (CE).

The results of the graphs are clear: in the dwellings with a retrofitted envelope (CE), the mean indoor temperatures are improved, rising to comfortable temperatures of between 18 and 20 °C; the daily thermal fluctuations are reduced and the minimum temperatures rise. In addition, user satisfaction increases and is reflected in the surveys carried out where they state that the thermal sensation is much better than before the retrofitting.

As examples, we present the cases mentioned in the previous section (ORVINA, FERRO and GORRI) comparing the earlier graphs with those monitored simultaneously in identical buildings which now have retrofitted envelopes. Two new cases (SOLE and CP) are also added. The characteristics of the thermal envelope of these cases are given in Table 5.5.

Figure 5.14 is the graph of the ORVINA building comparing the performances of SE and CE buildings. In this case, there has been intervention on the retrofitting of the thermal envelope, but not on the regulation and control of the district heating. As there is no regulation or control of the heating system per dwelling, the central heating has to offer a minimum temperature regime to the most badly affected dwelling in a group of buildings with a CE retrofitted envelope and without renovated SE. The temperatures are found to be too high, over 20 °C, rising to 26 °C. The CE dwellings (green lines) have higher temperatures than the SE (blue lines). In fact, some of the dwellings have the windows open all day during the winter because of the heat. In addition, the CE homes have less thermal fluctuation. This

Table 5.5 Characteristics of the thermal envelope of case studies

Case Study	Thermal envelope	Façades	Roofs	Windows
ORVINA	SE	F2	R1	G1
	CE	F2+rFei08	R1+rRei08	r2W
FERRO	SE	F2	R2	G1
	CE	F2+rFei08	R2+rRei08	rGLoE6.16.8B
GORRI	SE	F4	R3	G1
	CE	F1+rFei08	R3+rRei08	rGLoE6.16.8B
SOLE	SE	F1	R3	G1
	CE	F1+rFei08	R3+rRei08	rGLoE6.16.8B
CP	SE	F1	R3	G1
	CE	F1+rFei06	R3+rRei06	r2W

Legend: *SE* original thermal envelope no renovated, *CE* renovated thermal envelope. See codes in Table 5.4 for SE and Table 5.6 for CE

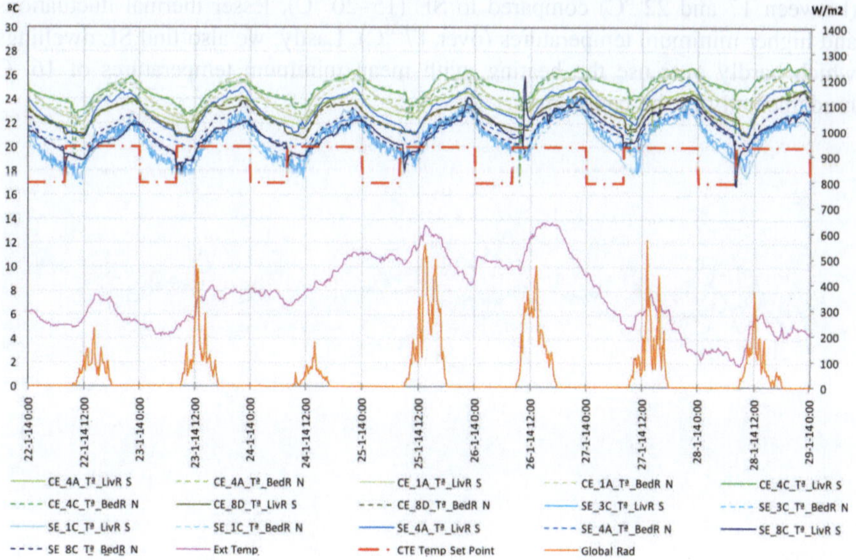

Fig. 5.14 Comparison of the monitored indoor temperatures in winter in the ORVINA building (CE green lines/SE blue lines) Typology T3, Tower

does not occur in the SE which have brusque drops due to the high thermal transmittance of the non-insulated windows.

We have also been able to monitor two apartment blocks, SOLEC and FERRO, before and after the retrofitting. Both are located in vulnerable neighbourhoods or areas of Pamplona (see Chap. 6) and are apartment blocks with individual heating.

In the case of FERRO, Typology T2, the retrofitting of two of the buildings was carried out in the year 2016 and can be compared the block which has not been renovated. Figure 5.15 shows the comparative graphs of the indoor temperatures. If we compare dwellings CE_N6_1C y CE_N4_4th with those mentioned in the point above 5.2.5 SE_N2_4I y SE_N2_3I, we find that none of the dwellings use heating or use it very occasionally. There is a clear difference of 2 and 4 °C between the buildings with the CE retrofitted envelope and those which have not been retrofitted. While the two SE apartments are between 12 and 14 °C, the CE are at 16–17 °C, in very tight temperature ranges, but they are above the threshold which is considered dangerous for health.

In SOLE (Fig. 5.16) similar differences between SE and CE can also be observed. However, the most significant point, in this case, is that despite having a retrofitted envelope, there are some homes which do not use the heating and their temperatures are between 12 and 15 °C, with clear risks for the residents' health.

In the case of GORRI (Fig. 5.17), the dwellings also have individual heating systems. The residents have a socio-economic profile higher than those in FERRO and SOLE. Likewise, the temperature ranges are higher, in this case, responding to greater use of the heating. We also observed that CE has higher temperatures (between 17 and 22 °C) compared to SE (15–20 °C), lesser thermal fluctuations and higher minimum temperatures (over 17 °C). Lastly, we also find SE dwellings which hardly ever use the heating, with mean minimum temperatures of 16 °C fundamentally in the bedrooms.

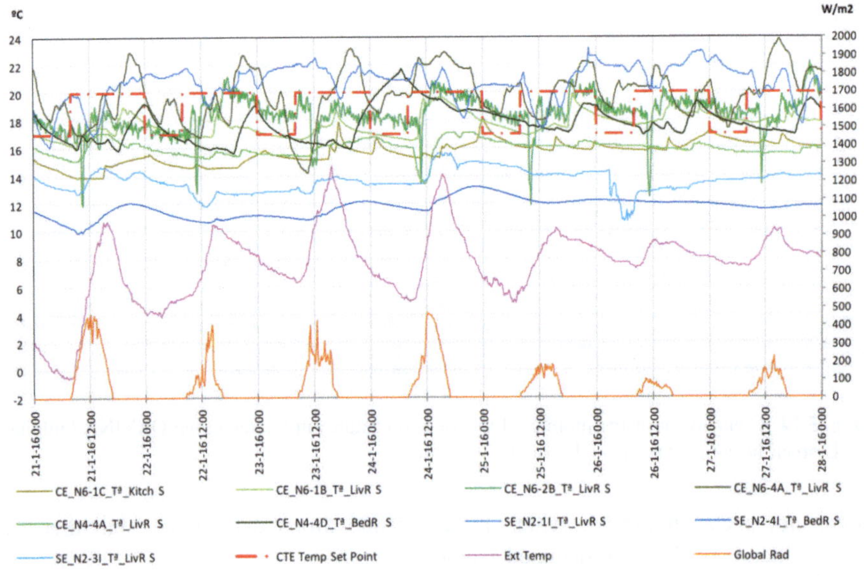

Fig. 5.15 Comparison of the monitored indoor temperatures in winter in the FERRO building (CE green lines/SE blue lines) Typology T2

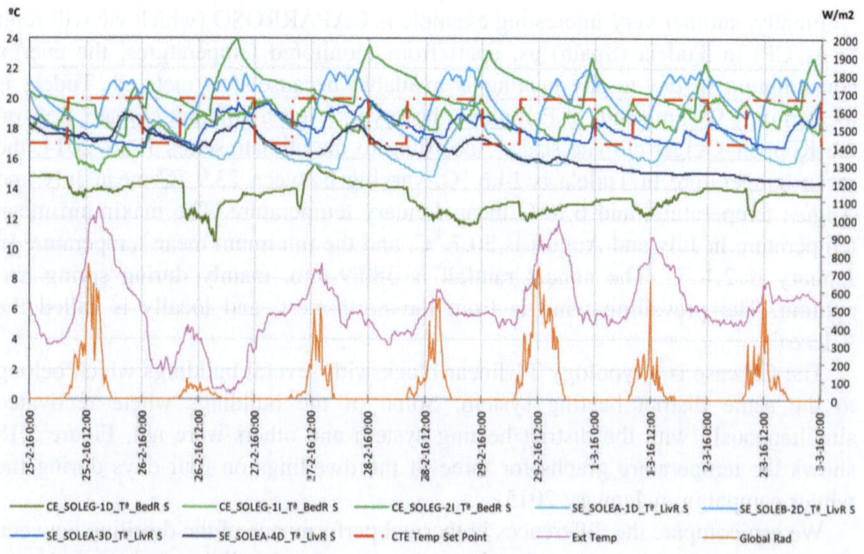

Fig. 5.16 Comparison of the monitored indoor temperatures in winter in the SOLE building (CE green lines/SE blue lines) Typology T1

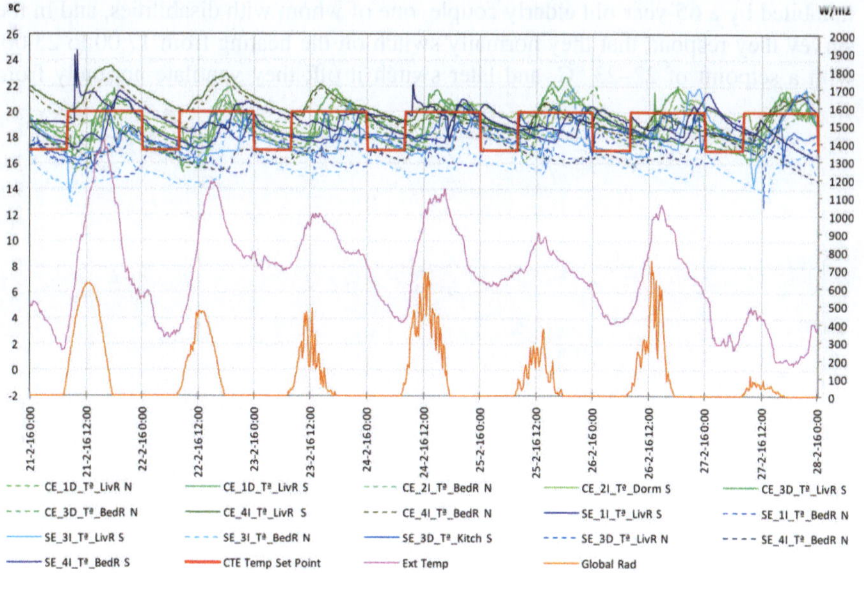

Fig. 5.17 Comparison of the monitored indoor temperatures in winter in the GORRI building (CE green lines/SE blue lines) Typology T1

Finally, another very interesting example is CAPARROSO (which we will refer to as CP) in Tudela (Spain) as, apart from monitored temperatures, the energy consumption in day-to-day heating is available because it is metered. Tudela is located only 95 km south of Pamplona; however, it has a different climate, BSk on the Köppen–Geiger climate scale. According to the climate series 1986–2014, the mean temperature in Tudela is 14.6 °C, varying between 23.9 °C mean July and August temperature, and 6.1 °C mean January temperature. The maximum mean temperature in July and August is 30.7 °C, and the minimum mean temperature in January is 2.1 °C. The annual rainfall is 380,9 mm, mainly during spring and autumn. The prevailing wind is from the north-west, and locally is called the 'Cierzo'.

The CP case is a typology T1 linear block with several buildings which belong to the same district heating system. Some of the buildings where renovated simultaneously with the district heating system and others were not. Figure 5.18 shows the temperature graphs for three of the dwellings on four days during the winter campaign in January 2015.

We can compare the differences in thermal performance of the dwelling between the living room of an intermediate home in a retrofitted building facing south-east (CE CP_2B), and the living room of an intermediate home in a non-retrofitted building facing northeast (SE CP_2D), together with a dwelling on the top floor facing mainly south-east (SE CP 4B). The retrofitted apartment CE CP_2B is inhabited by a 65-year-old elderly couple, one of whom with disabilities, and in the survey they respond that they normally switch on the heating from 17.00 to 23.00, with a setpoint of 22–23 °C, and later switch it off; they ventilate normally from

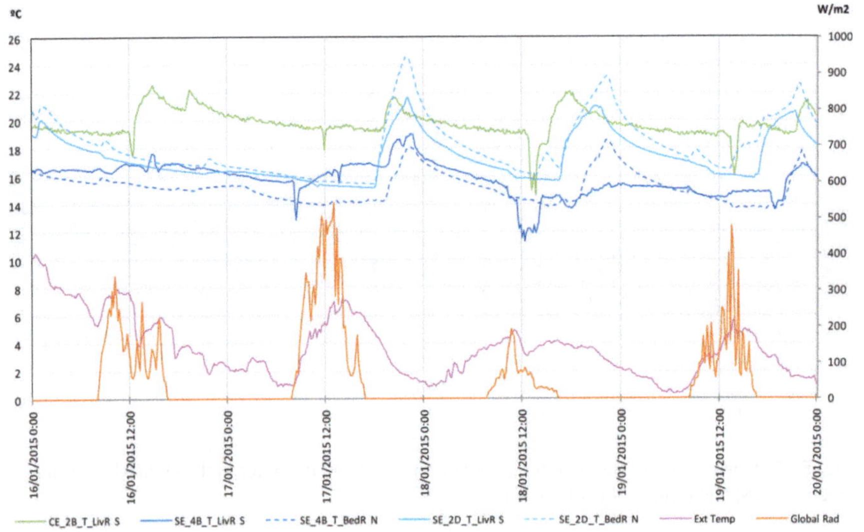

Fig. 5.18 Comparison of the monitored indoor temperatures in winter in the CP (CE green lines/ SE blue lines) Typology T1. Renovated District Heating

13.00 to 14.00 and have a typical heat sensation. The SE CP_2D apartment is inhabited by a 65-year-old (woman), and in the survey she answers that she switches on the heating only from 19.00 to 22.00 with a setpoint of 25 °C and later switch is it off; that she only ventilate for half an hour every three days, and typically feels cold. While the first of these had an energy consumption in the 2013–14 campaign of 14.32 kWh/m^2 yr, the second had a consumption of 26.12 kWh/m^2 yr, that is, its consumption in heating was 80% higher than the former, and its temperature pattern was very uncomfortable. Likewise, we can see the difference in the thermal performance of the dwelling due to the insulation and thermal inertia when the heating is switched off at night. The retrofitted dwelling CE CP_2B, with the heating is switched off, maintains a temperature of approximately 20 °C, while in the SE CP_2D non-retrofitted dwelling it drops to 16 °C, with much greater thermal fluctuations in this dwelling. In the retrofitted dwelling CE CP_2B, we can see the daily ventilation patterns, and how, once the windows are closed, the temperature in the dwelling returns to normal. Dwelling SE CP_4B maintains a temperature of approximately 16 °C, with a very occasional use of heating.

5.4 Towards Efficient Retrofitting

Once determined which parts of the thermal envelope affect more winter energy demands (See Sect. 5.2.3), then the retrofitting measures to apply for demand reduction may be studied.

The first measure to improve the performance of the thermal envelope is to instal insulation in order to decrease the thermal transmittance, reduce heat loss and, in most cases, avoid mould caused by condensation (Fig. 5.19). Insulation can be placed either on the interior, in the cavity, if there is one, or on the exterior of the façade (rF) (Figs. 5.20, 5.21 and 5.22) or roof (rR).

Fig. 5.19 Condensation dampness from façades (left) and roofs (right)

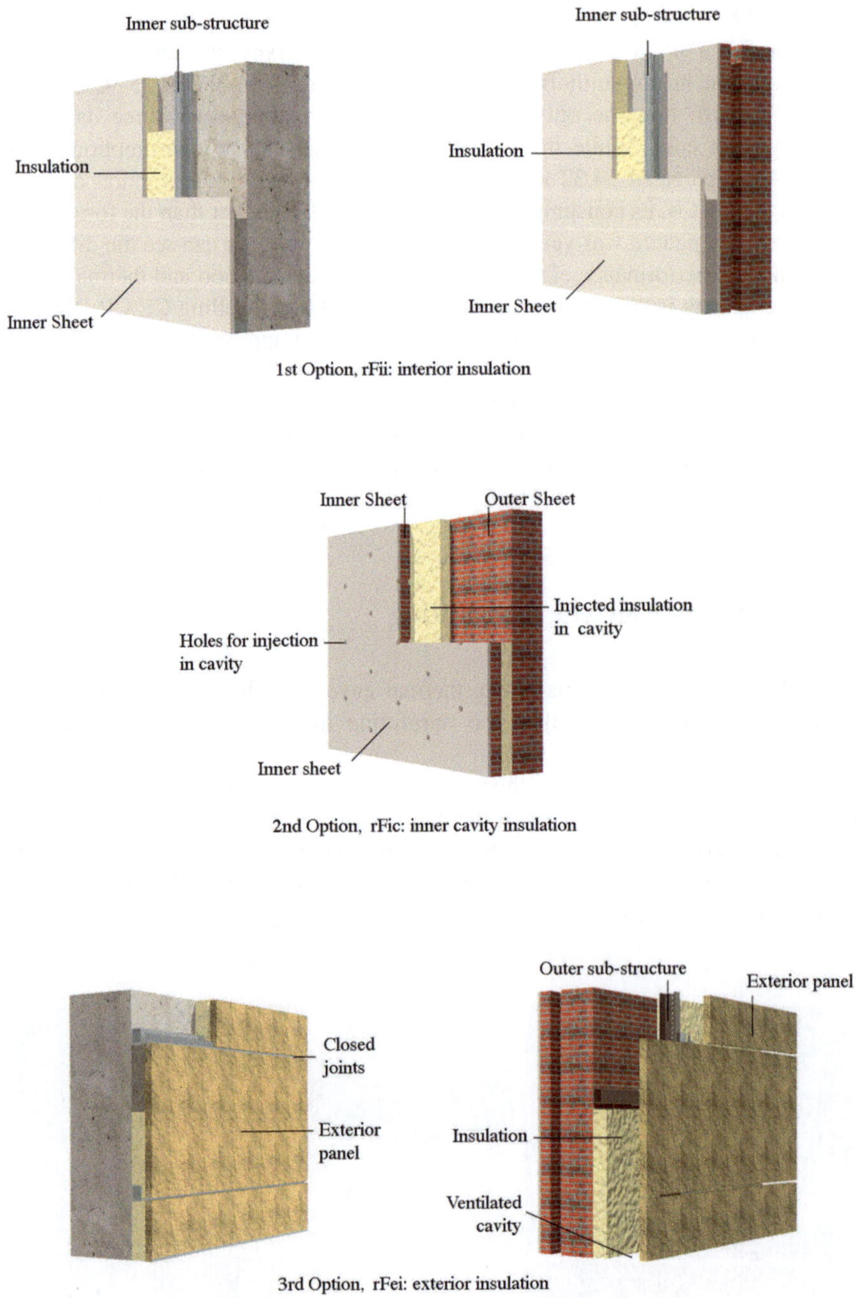

Fig. 5.20 Façade retrofitting systems: interior insulation (rFii), inner cavity insulation (rFic) and exterior insulation (rFei)

Fig. 5.21 Interior insulation, rFii (left) and in cavity insulation, rFic (right)

Fig. 5.22 Exterior insulation rFei

The two first insulation options allow for individual interventions by the owners, but the third, with major advantages over the others (for instance, thermal bridging is solved, thermal inertia is achieved, the aesthetics of the façade may be renovated, the usable space in the dwellings is not reduced, there is no bother for the residents and the provisional rehousing of the residents is avoided, etc.), represents an integral intervention on the façade or roof, and therefore requires the agreement of all the proprietors. In addition, other measures must be taken such as the replacement of windows (rW) (Fig. 5.23) to resolve problems with the original windows which are usually in poor condition (Fig. 5.24) and the improvement of insulation in the separation of floors with the ground floor (rL), and walls with stairwells (rC) and other buildings (rP).

The energy savings associated with the two last separations (rLC y rP) were identified as not being significant, between 1 and 5% depending on the typology. In addition, the installation of thermal insulation in these two separations results in

Fig. 5.23 Replacement of window in process (left) and replacement of windows with blind box cassettes by monoblock system (right)

Fig. 5.24 Condition of windows found on inspection of dwellings

some complications: given the inner distribution of the dwellings, if insulation is placed on the inside of the wall separating the dwelling from the stairwell, the width of the corridor and possibly of some rooms is reduced, thereby losing habitability; if it is installed on the outside, the width of the staircase of the building is reduced resulting in inferior evacuation conditions in case of fire. These situations may impede the insulation of this type of walls.

The retrofitting measures identified from these analyses are presented in Table 5.6. Whenever we refer to retrofitting measures, they are indicated with an 'r'

Table 5.6 Winter retrofitting measures adopted in thermal envelope

Façade, **rF** (included thermal bridges)	Exterior insulation, rFei Thickness: 4, 8, 12, 16 cm rFei04, rFei08, rFei12, rFei16	Interior insulation, rFii Thickness: 4.8 cm rFii04, rFii08	Insulation in cavity, rFic Thickness: 4 cm rFic04
Roof, **rR**	Exterior insulation, rRei Thickness: 8, 12, 20 cm rRei08, rRei12, rRei20	Interior insulation, rRii Thickness: 8, 12, 20 cm rRii08, rRii12, rRii20	Insulation in cavity, rRic Thickness: 8, 12, 20 cm rRic08, rRic12, rRic20
Windows, **rW** If replacing window and roller-blind box with a 'monoblock' solution, rWb	Replacing window with a new frame low transmittance + double glazing, rG = rG4.12.6	Replacing window with a new frame low transmittance + Low-E double glazing, rGLoE = rGLoE 6.16.8	Incorporation of double window outside of the existing, r2W
Stair Shaft, rC	Exterior insulation, rCei Thickness: 4 cm, rCei04	Interior insulation, rCii Thickness: 4 cm, rCii04	
Walls with other buildings, rP	Exterior insulation, rPei Thickness: 4 cm, rPei04	Interior insulation, rPii Thickness: 4 cm, rPii04	
First floor separation, rL	Exterior insulation, rLei Thickness: 4, 8, 12 cm, rPei04, rPei08, rPei12		

**Note* the 'r' in front indicates a retrofitting option, e.g. rF: retrofitting of façade

plus the code of the retrofitted element. For example, rF is retrofitted façade; rFei08 indicates exterior insulation 8 cm thick; rFii04 indicates interior insulation 4 cm thick; rFic04 is insulation in cavity of the façade 4 cm thick, etc.

By combining the different measures, different levels of intervention may be reached, together with measures for reduction of air infiltration and taking into account the thermal bridges in the equivalent thermal transmittance of the façade and roof. In the prestaRener Project, a parametric analysis was carried out to measure the reduction of demands on applying different levels of intervention in all the typologies identified. 159.744 simulations were carried out which allowed us to detect the key factors for energy demands and which interventional levels are more cost-effective (http://www.unav.edu/centro/saviarquitectura/prestarener/index.html).

The objective of the next sections is, on the one hand, to compare the simulation results for Pamplona with the patterns of use established by the Spanish regulations (CTE-HE1) and the values of the patterns of use found in the monitored cases

(Sect. 5.5). On the other hand, to compare the current demands and that of the different levels of intervention, in the current scenario and in the climate change scenario for the year 2050, by typologies and in different cities in the south of Europe with different climates (Sect. 5.6).

To do so, two levels of intervention have been chosen that will be called M1 and M2 and consist of the following:

M1: rFei08+rR12+rW+rLei08 means the rehabilitation of the façade (rF) with 8 cm exterior insulation (ei); 12 cm insulation on the roof (rRic12), generally in attic space, except for flat roofs where it would be placed on the outside (rRei12); replacement of windows by monoblocks with blinds and Low-E double glazing (rGLoE6.16.8B). Finally, 8 cm insulation under the separation of the first floor with the ground floor (rLei08), except in single-family dwellings where 4 cm would be placed on the ground slab; likewise, a general improvement of the airtightness the dwellings is contemplated (varying according to typologies, building height, etc.).

M2: rFei16+rR20+rW+rLei08 means the rehabilitation of the façade (rF) with 16 cm exterior insulation (ei); 20 cm insulation on the roof (rRic12), generally in attic space, except for flat roofs where it would be placed on the outside (rRei12); replacement of windows by monoblocks with blinds and Low-E double glazing (rGLoE6.16.8B). Finally, 8 cm insulation under the separation of the first floor with the ground floor (rLei08), except in single-family dwellings where 4 cm would be placed on the ground slab; likewise, a general improvement of the airtightness the dwellings is contemplated (varying according to typologies, building height, etc.).

On the other hand, as has been commented, assessment must be made of the effects that the retrofitting measures of the thermal envelope will have on the other requirements that the buildings must fulfil, such as fire protection (Meacham, Poole, Echeverria, & Cheng, 2012), sound insulation, protection against damp, etc. (Sánchez-Ostiz, Meacham, Domingo-irigoyen, Echeverria, & González, 2014). Different possibilities of intervention for each measure have been analysed in Table 5.7 from the point of view of different requirements, evaluating their effect (1, best; 2, medium; 3, worst). For example, the exterior insulation of the façade is the best from the point of view of minimizing thermal bridges, taking advantage of thermal inertia and increasing airtightness, but the execution process is more complicated, as scaffolding is needed, and it is more complex and far more expensive. In Table 5.7, ei is exterior insulation, ic is insulation in cavity and ii means internal insulation. In addition, rG refers to replacing windows with a new frame with low transmittance and double glazing, rLoE is replacing window with a new frame with low transmittance and Low-Emissivity double glazing; r2W is placing a double window outside the existing one.

Table 5.7 Assessment of the different measures adapted (Sánchez-Ostiz et al., 2014)

Requirements	rFaçades			rRoof			rC		rL		rW		
	ei	ii	ic	ei	ii	ic	ei	ii	ei	ii	rG	rLoE	r2 W
1. Thermal behaviour													
1.1. Insulation thickness	1	2	3	1	2	1	3	3	1	3	3	1	2
1.2. Thermal bridges	1	2	3	1	2	3	–	–	1	3	–	–	–
1.3. Thermal mass	1	3	2	1	3	2	–	–	1	3	–	–	–
1.4. Airtightness	1	2	3	1	2	2	–	–	–	–	1	1	1
1.5. Solar gains (winter)	2	–	–	–	–	–	–	–	–	–	2	3	3
1.6. Solar protection (summer)	2	–	–	–	–	–	–	–	–	–	2	1	1
2. Humidity protection													
2.1. Condensation	1	2	2	1	2	2	–	–	1	2	1	1	2
2.2. Solve exterior infiltration	1	3	3	1	3	3	–	–	–	–	1	1	1
3. Fire protection													
3.1. Exit width	3	–	–	–	–	–	3	2	–	–	–	–	–
3.2. Exterior safe space	–	–	–	–	–	–	–	–	–	–	1 (a)	1(a)	3
3.3. Separation distance	3	–	–	2	–	–	2	–	–	–	3	3	3
3.4. Fire reaction materials	3 (b)	1	1	2	1	1	3	2	2 (c)	–	–	–	–
4. Other aspects													
4.1. Acoustic	3	2	3	2	2	2	2	2	–	–	3	2	1
4.2. Natural light	2	–	–	–	–	–	–	–	–	–	2	3	3
4.3. Natural ventilation	2	–	–	–	–	–	–	–	–	–	1	1	2
4.4. Fall protection	–	–	–	–	–	–	–	–	–	–	2	2	2
4.5. Usable area	1	3	1	1	3	1	1	3	1	3	–	–	–
4.6. Habitability	1	3	1	1	3	1	1	3	1	3	–	–	–
4.6. Envelope's thermal movements	1	3	3	1	3	3	–	–	–	–	–	–	–
4.7. Aesthetic aspects	1	3	3	1	3	3	2	2	–	–	–	–	–
5. Execution process													
5.1. Need of scaffolds	3	1	1	3	1	1	1	1	1	1	1	1	2
5.2. Occupants' disturbance	1	3	2	1	3	2	1	3	1	3	1	1	1
5.3. Complexity	3	1	1	2	1	1	1	1	1	1	1	1	1
5.4. Singular points	3	2	1	3	1	1	1	1	1	2	1	1	1
6. Cost													
6.1. Materials	3	3	1	3	2	1	1	2	2	3	2	3	1
6.2. Execution	3	2	1	3	2	1	2	1	1	3	1	1	1

Legend = **1** the best, **2** medium, **3** the worst, (–) not applicable, **ei** = exterior insulation, **ii** = interior insulation and **ic** = insulation in cavity

Notes (a): If there are balconies; (b): depends on the material of the exterior skin or the insulation material in ventilated cavity wall (combustible or non-combustible); and (c) when it is the ceiling of evacuation exit

5.5 Influence of Patterns of Use in Achieving the Objectives of Reduction of Energy Demands

The energy simulations are generally carried out using the pattern of use for heating established by the regulations (in our case Spanish Regulation CTE-DB-HE1), thereby considering that the demands for comfort are fulfilled.

However, in the consecutive monitoring campaigns carried out in buildings in social housing neighbourhoods constructed between 1940 and 1980, with and without retrofitted thermal envelopes, the reality of the patterns of use of the dwellings has become clear and is far from what is stipulated in the above-mentioned regulations. This is so both in non-renovated dwellings and even in renovated ones. The most important parameters to be considered are the hours of heating and the temperature setpoints for its use. The result is that the energy savings are less than predicted in the simulation, based on an actual demand which is less than the simulated theory.

On the one hand, we find patterns of heating use which are strictly fixed to certain hours of the day, and even patterns which suggest family groups that may suffer from energy poverty, where the heating system is individual and is not used, or even where there is no heating system and use is made of movable auxiliary butane or electric heaters, etc. On the other hand, we also find dwellings where excessive use is made of heating, with temperatures as high as 24 °C, which corresponds mainly to district heating installations with no possibility of individual regulation.

For this reason, in this section, we wish to compare the effect of the different heating patterns of use on the energy demands and to find the percentage of deviation produced between the demands resulting from the different patterns compared to the pattern of use in the regulations.

5.5.1 *Estimating Patterns of Use*

Using the monitoring of the indoor temperatures, the user surveys and the thermal sensation drawn from the surveys, the following patterns of use of heating with the temperature setpoint used (Table 5.8):

- Use 0. CTE type
- Use 1. Intensive,
- Use 2. Aware and active,
- Use 3. During lunch and dinner times,
- Use 4. Occasional and very economic and
- Use 5. Without heating.

Next, a description and justification of each of the estimated patterns of use is given, adjusted to examples of dwellings where the typical use corresponds to the

Table 5.8 Used Patterns of heating hours and temperature setpoints

	1	2	3	4	5	6	7	8	9	10	11	12	13	14	15	16	17	18	19	20	21	22	23	24
U0.	17	17	17	17	17	17	17	20	20	20	20	20	20	20	20	20	20	20	20	20	20	20	20	17
U1.	20	20	20	20	20	20	20	20	20	20	20	20	20	20	20	20	20	20	23	23	23	23	23	23
U2.	OFF	OFF	OFF	OFF	OFF	OFF	OFF	OFF	OFF	OFF	OFF	OFF	OFF	OFF	OFF	OFF	OFF	OFF	20	20	20	20	20	OFF
U3.	OFF	OFF	OFF	OFF	OFF	OFF	OFF	OFF	OFF	OFF	OFF	OFF	OFF	OFF	20	20	OFF	OFF	OFF	OFF	20	20	20	OFF
U4.	OFF	OFF	OFF	OFF	OFF	OFF	OFF	OFF	OFF	OFF	OFF	OFF	OFF	OFF	OFF	OFF	OFF	OFF	OFF	OFF	21	21	OFF	OFF
U5.	OFF	OFF	OFF	OFF	OFF	OFF	OFF	OFF	OFF	OFF	OFF	(ill	OFF	(ill	OFF	OFF	OFF	OFF	OFF	OFF	OFF	OFF	OFF	OFF

studied pattern. In this way, we can estimate the deviation percentage of each pattern of use compared to the CTE.

USE 0. USE OF HEATING CTE REGULATIONS TYPE
The heating is used all day, with a low (17 °C) setpoint during the night, and a 20 °C setpoint during the day, in accordance with Appendix C 'Profiles for use' for residential use, of the CTE-HE-1 'Energy-saving. Limitation of demand'. This use is not very habitually detected in the monitoring carried out in social housing neighbourhoods. Figure 5.25 gives two examples of dwellings that may have a similar use to Use 0.

USE 1. INTENSIVE
This type of heating pattern use corresponds to dwellings in buildings with central or district heating where there is no regulation system or where it is very inefficient. The heating may be working all day and all night, or only during the day (e.g. from 12.00 to 24.00), although on the coldest days and during holidays (e.g. Christmas) the heating works 24 h a day. Unwarranted temperatures of over 24 °C are reached, frequently because they must offer a minimum temperature (e.g. 18 °C) to the dwellings which are furthest away or located on a high floor in the building, and because the system does not allow for proper regulation. It is very common to find dwellings with the windows open, due to 'excessive heat' in midwinter with the heating on. In general, the owners have high incomes, as the vulnerable population who cannot pay the expenses of central heating live in dwellings with poor characteristics and individual heating or none. Besides, as these patterns of use generally coincide with people over the age of 65 who spend long periods at home because they do not go out to work, it is not frequent to hear complaints about the thermal sensation although there is a growing 'awareness' of the energy waste involved.

Heating working 24 h a day with setpoints between 20 and 23 °C have been established (the latter varying from 18 to 24 °C). Figure 5.26 shows an example of a dwelling which is similar to Use 1.

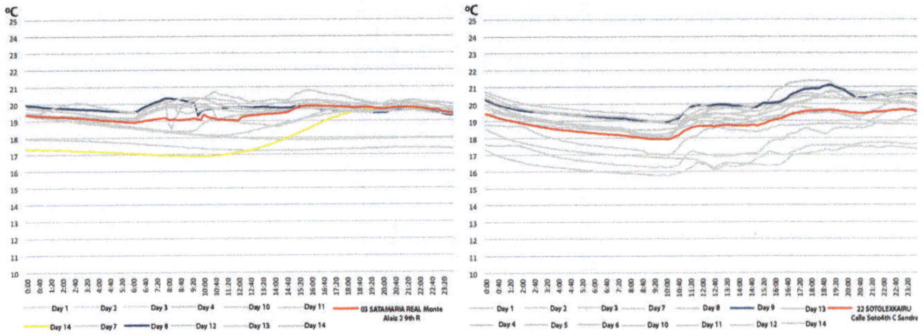

Fig. 5.25 Examples of monitored dwellings with a pattern similar to Use 0. Type CTE

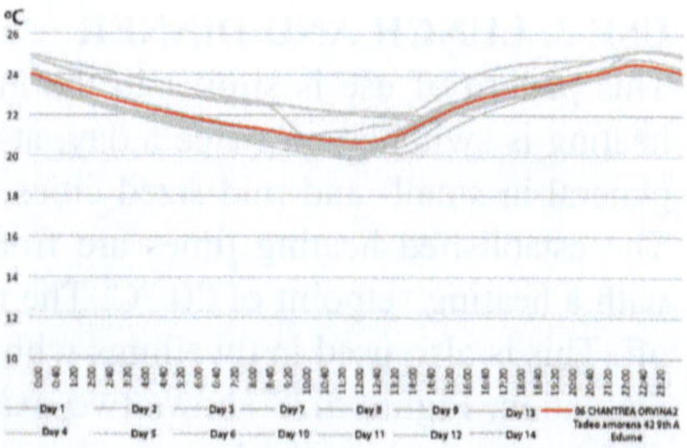

Fig. 5.26 Examples of monitored dwellings with a pattern similar to Use 1. INTENSIVE

USE 2. AWARE AND ACTIVE

The fact is that family habits and uses have changed very much over the last few years. It is very common to find families or people who live alone, where people work all day outside the home and the children are at school; thus, the heating is on only when the users are at home or is programmed to begin working 30 min before they come home so that the house is comfortable when they arrive. This corresponds to buildings with individual heating in each dwelling or central heating with individual regulation. The setpoints are very close to 20 °C, a setpoint that is deeply rooted in society and considered within the usual ranges for comfort (PMV −0.5, PPD 10%). In general, there already is clear awareness of how important it is both economically and ecologically to adjust the timetables and heating setpoints.

The thermal sensation of comfort and really having a comfortable atmosphere are clearly linked to the constructive parameters of the dwellings although this use is less efficient and less comfortable in apartment blocks from between 1940 and 1980 with no retrofitting, and to the original thermal envelope without insulation.

Heating from 18.00 to 23.00 (a total of five hours heating), with the heating off during the remaining time, has been established as the pattern of use. Figure 5.27 shows two examples of dwellings which are similar to Use 2.

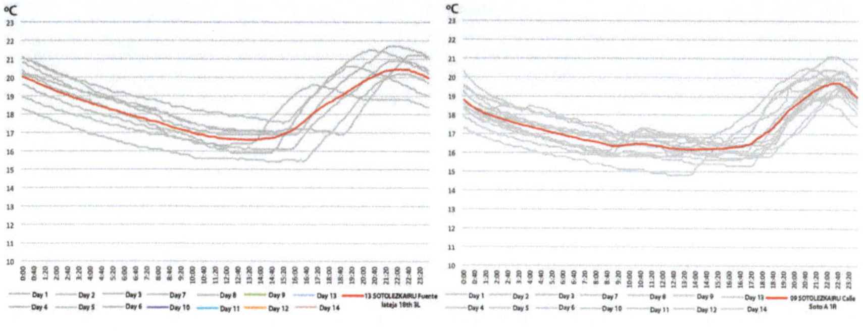

Fig. 5.27 Examples of monitored dwellings with a pattern similar to Use 2. AWARE AND ACTIVE

USE 3. LUNCH AND DINNER

This pattern of use is similar in the number of hours to the previous one, but the heating is switched on twice a day, at lunch and dinner times. In Pamplona and in general in small- and mid-sized cities, people quite frequently go home for lunch. The established heating times are from 14.00 to 16.00 and from 20.00 to 23.00, with a heating setpoint of 20 °C. The remainder of the time the system is switched off. This is also used in dwellings with individual or central heating with individual regulation. Figure 5.28 shows two examples of dwellings similar use to Use 3.

It corresponds to the homes of people with a working social profile with a split timetable, who on weekdays are only at home at lunchtime and in the evening/at night.

USE 4. OCCASIONAL AND ECONOMICAL

This use is to be found in family groups at potential risk of 'energy poverty'. The dwellings have individual heating, with some cases of central heating with individual regulation, or no heating at all. In the latter situation, they use auxiliary movable butane or electric heaters, etc. Despite the fact that the family is in the dwelling, it is only heated at certain moments of the day, generally in the evening, and sometimes only one room is heated. These are, then, very cold and uncomfortable patterns of use and may cause serious health problems especially in older people, children and those who are ill, due to the habitual low temperatures in the dwelling, which depend very much on the outdoor temperatures. Low-quality construction dwellings without renovation built between 1940 and 1980 are the unhealthiest as a result of this pattern of use and because they are inhabited by a population with lower economic resources.

For this study, the times established are from 20.00 to 22.00 (2 h), with a 21 °C heating setpoint. Figure 5.29 gives two examples of dwellings with a similar use to Use 4.

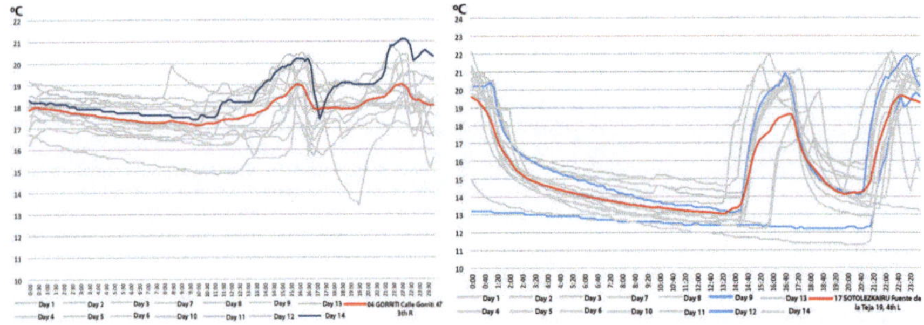

Fig. 5.28 Examples of monitored dwellings with a pattern similar to Use 3. LUNCH AND DINNER

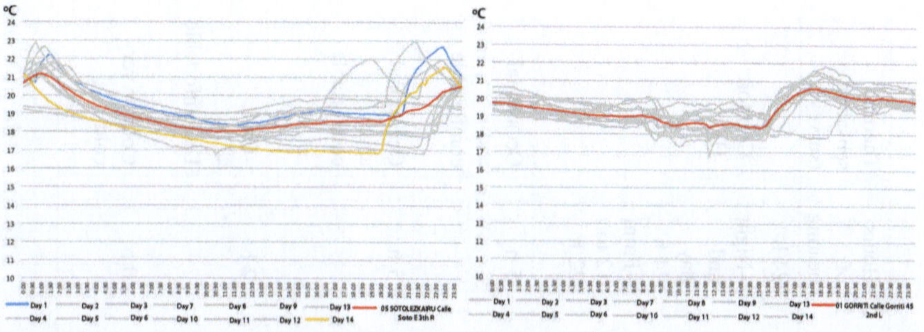

Fig. 5.29 Examples of monitored dwellings with a pattern similar to Use 4. OCCASIONAL AND ECONOMIC

USE 5. WITHOUT HEATING

Finally, we must not forget that we have also found dwellings in which heating is not used, even in this severe winter climate zone. The surveys report a cold or very cold thermal sensation. These are clear situations of 'energy poverty' with important consequences for the physical and mental health of the residents. This situation is particularly serious in dwellings with very poor quality construction or with no renovation, and is habitual in buildings with individual heating or no heating system, as the families in this situation cannot afford the community expense of heating. The temperature is stable but very low, depending on the construction characteristics of the envelope and the outdoor conditions (e.g. in the graphs, about 12–14 °C). Figure 5.30 gives two examples of dwellings which are similar in use to Use 5.

Table 5.9 summarizes the characteristics of each of the patterns of use of the heating systems considered.

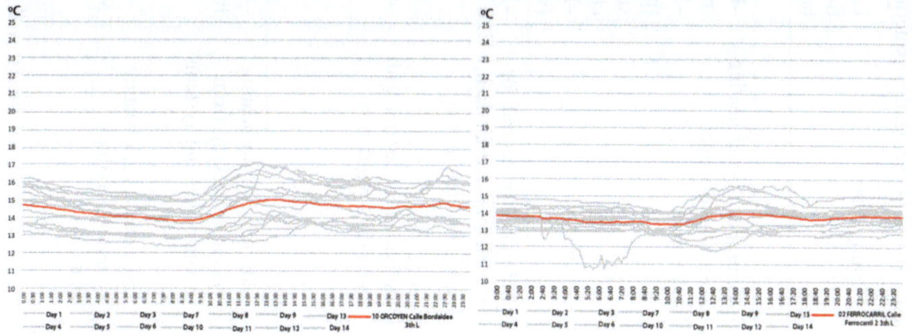

Fig. 5.30 Examples of monitored dwellings with a pattern similar to Use 5. WITHOUT HEATING

Table 5.9 Characteristics of the heating patterns of use detected

Use	Heating type	Length of time in the home	Economic situation	Work situation	Simulated ° T setpoint	Monitored °T	Hours of heating	Thermal sensation (surveys)
U0. CTE					20°–17°		24 h	
U1. Intensive	Central or district heating with no individual regulation	–High (retired) –Medium (presence of small children) –Low (working)	–High	–Retired –Active	23° (5 h) 20° (19 h)	24–20 °C	–Night and day 24 h –From 12 to 22 h	–Comfort –Heat
U2. Aware and active	Individual or central for building or district, with individual regulation	–Low	–Medium	Working one shift	20° (5 h de 18–23 h) –OFF (remainder)	21–16 °C	5 h	Occasional and comfortable
U3. Lunch and dinner	Individual or central with individual regulation	–Low	–Medium	Working with split shift	20° (14–16 h y 20–23 h) –OFF (remainder)	21–14 °C	5 h	Occasional and comfortable
U4. Occasional and economical	Individual or no heating system + backing from auxiliary heating systems	–Medium–high	–Low	–Unemployed –Retired	21° (20–22 h) –OFF (remainder)	14–12 °C	2 h	–Cold –Uncomfortable
U5. No heating	Individual or no heating system + auxiliary systems	–Medium-high	–Very low	Unemployed	–OFF	14–12 °C	0 h	Cold or very cold

5.5.2 *Energy Demands in Accordance with Patterns of Use*

By means of the simulation, we have verified the energy demands for heating in each of the typologies studied in the prestaRener research project, considering the above patterns of use. The study was carried out in the city of Pamplona with buildings facing south (living rooms).

Within each typology, we have simulated the energy demands for heating of the typical building with its original characteristics (referred to as Case Study, CS from now on) and the demands of the building with the envelope retrofitted through an intervention (M1). All simulations have included ventilation in accordance with CTE. We must not forget that the M1 intervention means placing 8 cm insulation on the outside of the façades (rFei08); 12 cm insulation in the roof space (rRic12), generally in the attic, except for flat roofs where it is placed on the outside (rRei12); substitution of windows for others with low-emission double glazing, rGLoE6.16.8B; 8 cm insulation in the ceiling of the ground floor rLei08, except in the single-family type dwelling, where 4 cm is placed on the ground slab; and a general improvement in the airtightness of the dwellings (which varies depending on typologies, height of the building, etc.). The construction characteristics by typologies are given in Table 5.10. The codes of

Table 5.10 Construction characteristics of the typologies: current state (CS) and retrofitting intervention (M1)

Elements characteristics according to CS or M1			Façade	Windows	Roof	Airtightness (h^{-1})	Floor of F1 (or GF in T5)
T1	T1 A	CS	F1	G1	R2	0.70	L2
		M1	F1+rFei08	rGLoE6.16.8B	R2+rRic12	0.30	L2+rLei08
	T1B	CS	F3	G1	R2	0.70	L2
		M1	F3+rFei08	rGLoE6.16.8B	R2+rRic12	0.30	L2+rLei08
T2	T2 A	CS	F2	G1	R2	0.55	L2
		M1	F2+rFei08	rGLoE6.16.8B	R2+rRic12	0.20	L2+rLei08
	T2 B	CS	F3	G1	R3	0,55	L2
		M1	F3+rFei08	rGLoE6.16.8B	R3+rRic12	0.20	L2+rLei08
T3	T3 A	CS	F3	G1	R1	0.80	L2
		M1	F3+rFei08	rGLoE6.16.8B	R1+rRei12	0.35	L2+rLei08
	T3 B	CS	F3	G1	R1	0.80	L2
		M1	F3+rFei08	rGLoE6.16.8B	R1+rRei12	0.35	L2+rLei08
T5	T5 A	CS	F2	G1	R2	0.80	L5
		M1	F2+rFei08	rGLoE6.16.8B	R2+rRic12	0.20	L5+rLei04
	T5 B	CS	F4	G1	R1	0.80	L5
		M1	F4+rFei08	rGLoE6.16.8B	R1+rRei12	0.20	L5+rLei04

the elements are detailed in Table 5.4 for the CS (Case Study) and in Table 5.6 for M1 rehabilitation measures.

The energy simulations have been modelled and simulated with Design Builder (EnergyPlus), with the file for Pamplona of EnergyPlus (SWEC). The table summary of the energy demands is given in Table 5.11 and in Fig. 5.31. As was to be

Table 5.11 Annual heating demands (kWh/m^2 yr), depending on typologies and uses: current state (CS) and retrofitting intervention (M1), for Pamplona, in accordance with Table 5.9

Heating			USE 0 CTE	USE 1 Intensive	USE 2 Aware and active	USE 3 Lunch and dinner	USE 4 Occasional and economical	USE 5 Without heating
T1	T1 A	CS	184.93	235.50	62.55	66.44	34.12	0.00
		M1	49.60	73.09	26.25	26.80	16.17	0.00
		% Energy savings[a]	73.18	68.96	58.03	59.66	52.62	0.00
	T1B	CS	138.26	178.76	48.58	51.18	27.35	0.00
		M1	42.87	61.25	21.24	21.72	13.47	0.00
		% Energy savings[a]	68.99	65.74	56.28	57.57	50.74	0.00
T2	T2 A	CS	164.59	209.07	61.44	64.28	33.62	0.00
		M1	49.98	71.93	26.25	26.85	15.99	0.00
		% Energy savings[a]	69.63	65.60	57.28	58.23	52.42	0.00
	T2 B	CS	125.16	158.55	51.84	53.27	28.77	0.00
		M1	43.73	61.85	24.04	24.38	14.76	0.00
		% Energy savings[a]	65.06	60.99	53.64	54.23	48.69	0.00
T3	T3 A	CS	179.81	229.25	65.55	69.52	35.70	0.00
		M1	55.03	72.06	29.17	29.55	17.50	0.00
		% Energy savings[a]	69.40	68.57	55.51	57.50	50.99	0.00
	T3 B	CS	167.47	203.07	62.86	65.57	34.06	0.00
		M1	55.72	79.34	31.99	32.12	18.53	0.00
		% Energy savings[a]	66.73	60.93	49.11	51.02	45.60	0.00
T5	T5 A	CS	151.12	221.79	56.06	58.92	31.30	0.00
		M1	56.28	80.01	27.43	28.60	16.70	0.00
		% Energy savings[a]	62.76	63.93	51.07	51.47	46.66	0.00
	T5 B	CS	166.61	223.65	61.44	64.58	33.84	0.00
		M1	50.97	74.08	25.38	25.99	15.74	0.00
		% Energy savings[a]	69.41	66.88	58.70	59.75	53.50	0.00

Note[a] One façade to street, the other to small courtyard

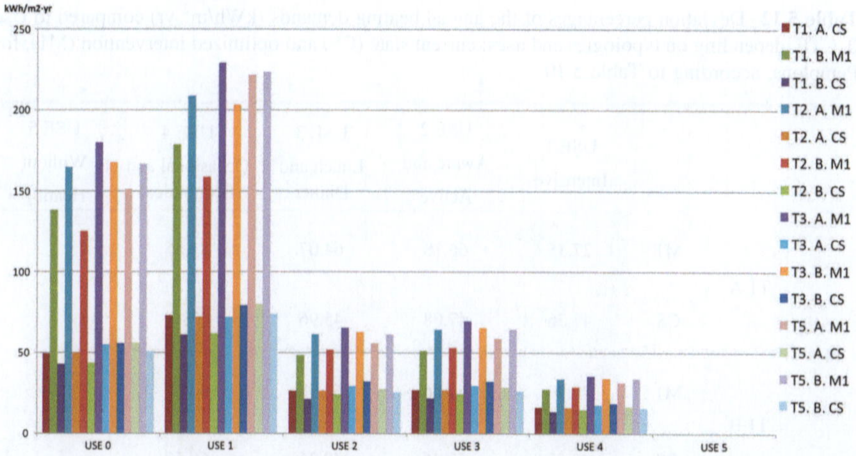

Fig. 5.31 Summary figure, of annual heating demands (kWh/m^2 yr), depending on typologies and uses (rFei08+rRei12+rGLoE6.16.8B+rLei08)

expected, Use 1 (Intensive) has the greatest energy demands for heating, both in its current state (CS) and in the retrofitting intervention (M1), followed by Use 0 (CTE).

Uses 2 and 3 offer similar results, although the demand of Use 3 in CS is slightly higher, because, although the first two hours of heating are during the mildest time of day, the heating must be switched on twice with greater energy consumption. However, in the M1 the difference between uses is a maximum of 0.6 kWh/m^2·yr. Use 4 has very little energy demands, at the expense of the comfort and health of the residents. Use 5 stresses the reality of energy poverty.

On retrofitting with the M1 intervention, the uses which present a greater percentage of reduction in demand are Uses 0 and 1, with a mean of 68 and 65% in all the typologies, followed by Uses 2 and 3, with a mean of 55 and 56%, while Use 4 has a 50% mean reduction.

Finally, the deviations of the energy demands for heating compared to Use 0 of the residential type CTE (Table 5.12). The box is highlighted in grey when the percentage is higher than the CTE and in white when it is lower. We consider that the CTE type use is, in general, appropriate for the typical regulations of comfort (PMV, UNE-EN 7730), except the temperature of 17 °C at night which may be too low for certain population groups (elderly people or infants) even though it is for a nocturnal timetable. According to Table 5.12, Use 1 has a demand approximately 30% higher than Use 0, but however, in Uses 2 and 3, the demand is approximately 60% less. Use 1, in general, is inadequate (it is 30% higher), and may only be justified if there were disabled people at home. Uses 2 and 3 (approximately 60% less) is adequate if the residents are not at home during the hours when the heating is off, although it would be convenient to fix a lower setpoint (17 °C) so that the

Table 5.12 Deviation percentages of the annual heating demands (kWh/m^2 yr) compared to Use 0. CTE, depending on typologies and uses: current state (CS) and optimized intervention (M1), for Pamplona, according to Table 5.10

			USE 1 Intensive	USE 2 Aware and Active	USE 3 Lunch and Dinner	USE 4 Occasional and Economical	USE 5 Without Heating
T1	T1 A	M1	27.35	66.18	64.07	81.55	-
		CS	47.36	47.08	45.96	67.41	-
	T1 B	M1	29.29	64.86	62.98	80.22	-
		CS	42.87	50.46	49.35	68.57	-
T2	T2A	M1	27.03	62.67	60.94	79.58	-
		CS	43.90	47.49	46.28	68.00	-
	T2B	M1	26.68	58.58	57.44	77.02	-
		CS	41.42	45.04	44.25	66.25	-
T3	T3A	M1	27.50	63.54	61.34	80.15	-
		CS	30.95	47.00	46.31	68.20	-
	T3B	M1	21.25	62.47	60.85	79.66	-
		CS	42.38	42.59	42.36	66.75	-
T5	T5A	M1	46.76	62.90	61.01	79.29	-
		CS	42.18	51.26	49.19	70.33	-
	T5B	M1	34.23	63.13	61.24	79.69	-
		CS	45.33	50.22	49.01	69.13	-

Note: [shaded] % higher than Use 0. CTE [unshaded] % lower than Use 0. CTE temperature would not drop much. Use 4 shows the highest deviation compared to the CTE Use (it is approximately 80% less), but is absolutely unsuitable for people's health.

5.6 Towards Efficient Energy Retrofitting in Winter. Case Studies in Southern Europe

The aim of this section is to compare the energy demands for heating in different Mediterranean cities in the south of Europe which have incorporated retrofitting measures (levels M1 and M2) in the present-day situation and in the future 2050 climate change situation when the outdoor temperatures will be higher all over Southern Europe. Six cities have been selected and are represented in Fig. 8.2 in Chap. 8: Mostar (Bosnia), Nîmes (France), Oporto (Portugal), Rome (Italy), Athens (Greece) and Valencia (Spain). The study has been made for typologies T1 Linear Block, T3 Tower and T5 Detached House, since they present diverse building, constructive and energetic characteristics.

The energy simulations have been developed with Design Builder software and EnergyPlus, both well tested and widely recognized. They have been done with climate data IWEC2 of ASHRAE (Huang, 2011). The climate data for the climate change scenario of 2050 have been generated with CCWorldWeatherGen of the University of Southampton (Jentsch, James, Bourikas, & Bahaj, 2013), because, for the locations in Southern Europe, there is no other common climate data source for energy simulation.

The characteristics of the non-retrofitted building Case Study (CS) and all the different levels of retrofitting (M1 and M2) are given in Table 5.13. With these parameters, the demands of the different typologies in the selected cities are simulated. It must be highlighted that in level M1 the ventilation conditions are 0.63 h^{-1} (0.63), and in M2 we have chosen 0.4 h^{-1} with heat recovery systems (0.4 h). Additionally, in typology T5, 4 cm insulation has been placed on the ground slab which separates the ground floor from the land so as not to raise the floor excessively, as it is impossible to do so underneath. In the other typologies, the insulation between the first floor and the ground floor can be placed under the floor slab separation and so 8 cm is used in M1 and 12 cm in M2.

The results of the current energy demands of the non-retrofitted building (CS) and those of the two action levels (M1 and M2), in the present scenario and the 2050 future are shown in Tables 5.14, 5.15, 5.16 and 5.17. These results have been analysed based on the parameters of the greatest influence on energy demands, as is next shown: Climate and typologies (Table 5.14), Orientation (Table 5.15), Position in the building: intermediate floor and under roof (Table 5.16) and ventilation conditions (Table 5.17).

Table 5.13 Simulation parameters and characteristics of the simulated buildings by typology: T1, T3 and T5

Codes of typologies and measures		Façade	Roof	Windows	Floor 1stF (T1) GF (T5)	Airtightness (h^{-1})	Codes for ventilation rate (h^{-1})
T1	**CS**	F3	R3	G1	L2	0.7	0.63/4SN
	M1	F3+ rFei08	R3+ rRic12	rGLoE	L2+ rLei08	0.3	0.63
							0.4
							0.63HR
							0.4HR
	M2	F3+ rFei16	R3+ rRic20	rGLoE	L2+ rLei12	0.2	0.63
							0.4HR
T3	**CS**	F3	R1	G1	L2	0.8	0.63
	M1	F3+ rFei08	R1+ rRei12	rGLoE	L2+ rLei08	0.35	0.63
							0.4
							0.63HR
							0.4HR
	M2	F3+ rFei16	R1 + rRei20	rGLoE	L2+ rLei12	0.2	0.63 0.4HR
T5	**CS**	F3	R1	G1	L5	0.8	0.63
	M1	F3+ rFei08	R1+ rRic12	rGLoE	L5+ rLei04	0.2	0.63
							0.4
							0.63HR
							0.4HR
	M2	F3+ rFei16	R1+ rRic20	rGLoE	L5+ rLei04	0.2	0.63
							0.4HR

Heating Demand, Pattern of use: Setpoints (low) 20 °C (8–24 h), 17 °C (0–8 h), from October to May included (CTE-HE, 2013)
Notes and Legend:
HR heat recovery system
ventilation (V), pattern of use:
0.63 general ventilation rate: 0.63 h^{-1} (CTE-HE, 2013)
0.4: general ventilation rate 0.4 h^{-1}
0.63HR 0.63 h^{-1}, but with HR
0.4HR: 0.4 h^{-1}, but with HR

Regarding the climate and the typologies (Table 5.14):

- In the present scenario, the cities with the greatest demand, in order from the most to the least, are Mostar, Nîmes, Oporto, Rome, Athens and Valencia. In all of them, the energy demand of typology T5 (Detached House) is greater than that of T3 (Tower) and T1 (Linear Block). After renovation at levels M1 and M2, the demands of T1 and T3 were more similar, and T5 continued to have greater demands. These greater demands are due to the fact that the envelope in

Table 5.14 Heating demand (kWh/m^2 yr) per climate and typologies

		2010			2050		
		CS	M1	M2	CS	M1	M2
Country Location Climate	Typology	Heating demand	Heating demand	Heating demand	Heating demand	Heating demand	Heating demand
Mostar BIH Cfb	T1	103.8	45.2	8.2	74.2	29.9	4.1
	T3	115.9	44.5	6.5	85.2	31.2	4.0
	T5	133.4	54.2	17.5	98.7	37.7	10.9
Nimes FRA Csa	T1	89.5	34.5	4.0	65.1	23.6	2.0
	T3	103.7	36.6	4.1	77.1	26.4	2.7
	T5	119.8	43.9	12.6	90.4	31.3	8.4
Oporto PRT Cfb	T1	66.9	20.7	1.0	42.3	11.4	0.4
	T3	81.4	25.9	1.8	54.6	16.4	0.8
	T5	95.7	29.4	6.3	65.4	18.1	3.5
Rome ITA Csa	T1	65.2	21.5	1.1	41.3	11.4	0.3
	T3	78.5	25.4	1.8	52.7	15.8	0.8
	T5	91.9	29.7	6.5	63.5	18.0	3.2
Athens GRC Csa	T1	46.9	14.8	0.60	27.7	7.1	0.1
	T3	57.6	18.2	1.2	36.6	10.6	0.4
	T5	68.7	20.9	4.2	44.2	11.4	1.3
Valencia ESP BSk	T1	34.5	8.3	0.10	18.9	3.0	0.02
	T3	46.9	13.0	0.45	28.2	6.6	0.2
	T5	58.5	15.2	2.6	37.5	7.7	1.0

Notes Results for the floor in an intermediate position (or best position in terms of demand): P2 in T1; P4 in T3, and PB in T5
M1 = rFei08+rR12+rW+rL+0.63 h^{-1}
M2 = rFei16+rR20+rW+rL+0.4HR

T5 has a larger surface in contact with the exterior, and also with the ground. In addition, the thickness of the insulation placed on the floor slab that separates it from the ground (4 cm) is thinner than that placed in T1 and T3 between the ground floor and the first floor (8 cm in M1 and 12 cm in M2). Although the differences are not very great, we can see that both in M1 and in M2 the energy savings increase is in the same order as the cities described. That is to say, Mostar has the greatest demands in the Case Study (between 103.8 kWh/m^2 yr in T1 and 133.4 kWh/m^2 yr in T5), and it is where less savings are made: in M1, there is a saving of between 56% (T1) and 59% (T5), and in M2, it is between 92 and 86% because the resulting energy demands are still higher than in the other cities due to the outdoor temperatures. In Valencia, low demands are achieved: M1 saves between 76 and 74%, and M2 between 99 and 95%, based on the Case Study whose demands were between 34.5 and 58.5 kWh/m^2 yr.

- In the future scenario 2050 of climate change with warmer exterior temperature conditions, we see that there is a drop of 33% in the demand for heating in all

Table 5.15 Heating demand (kWh/m^2 yr) per orientation and climate

			2010			2050		
			CS	M1	M2	CS	M1	M2
Country Location Climate	Typology	Orientation	Heating demand	Heating demand	Heating demand	Heating demand	Heating demand	Heating demand
Mostar BIH Cfb	T1	W (90°)	115.0	52.7	15.0	84.5	36.5	8.6
		SW (45°)	111.4	50.0	9.5	80.8	34.0	4.9
		S	103.8	45.2	8.2	74.2	29.9	4.1
	T3	W (90°)	131.1	52.4	11.2	97.7	37.7	7.5
		SW (45°)	122.5	47.8	8.4	90.7	34.0	5.4
		S	115.9	44.5	6.5	85.2	31.2	4.0
	T5	W (90°)	137.9	57.5	20.8	103.4	41.1	13.6
		SW (45°)	137.6	57.4	20.5	102.5	40.5	13.1
		S	133.4	54.2	17.6	98.7	37.7	10.9
Nimes FRA Csa	T1	W (90°)	101.0	42.2	10.5	75.0	29.7	6.4
		SW (45°)	96.4	39.1	5.0	70.8	27.1	2.7
		S	89.5	34.5	4.0	65.1	23.6	2.0
	T3	W (90°)	117.9	44.2	8.5	88.1	32.2	5.7
		SW (45°)	109.6	39.7	5.9	81.7	28.8	3.9
		S	103.7	36.6	4.1	77.1	26.4	2.7
	T5	W (90°)	124.7	47.5	16.0	95.0	34.7	11.1
		SW (45°)	123.6	46.8	15.2	93.6	33.7	10.3
		S	119.8	43.9	12.7	90.4	31.3	8.4

(continued)

Table 5.15 (continued)

			2010			2050		
			CS	M1	M2	CS	M1	M2
Oporto PRT Cfb	T1	W (90°)	73.5	25.1	3.5	48.8	15.2	1.3
		SW (45°)	69.9	22.7	1.6	45.3	13.1	0.5
		S	66.9	20.7	1.0	42.3	11.4	0.4
	T3	W (90°)	91.3	30.8	3.8	63.0	20.6	2.0
		SW (45°)	84.7	27.3	2.4	57.5	17.8	1.1
		S	81.4	25.9	1.8	54.6	16.4	0.8
	T5	W (90°)	97.9	31.3	8.0	68.3	20.2	4.8
		SW (45°)	97.0	30.6	7.5	67.0	19.4	4.3
		S	95.7	29.4	6.4	65.4	18.1	3.6
Rome ITA Csa	T1	W (90°)	73.4	26.8	4.1	48.7	15.7	1.7
		SW (45°)	69.3	24.1	1.7	44.7	13.5	0.5
		S	65.2	21.5	1.1	41.3	11.4	0.3
	T3	W (90°)	89.8	31.2	4.3	62.2	20.5	2.1
		SW (45°)	82.5	27.2	2.6	56.0	17.3	1.2
		S	78.5	25.4	1.8	52.7	15.8	0.8
	T5	W (90°)	95.0	32.1	8.4	66.8	20.3	4.5
		SW (45°)	94.0	31.4	8.0	65.4	19.4	4.1
		S	91.9	29.7	6.6	63.5	18.0	3.3
Athens GRC Csa	T1	W (90°)	53.2	18.4	1.8	32.9	9.7	0.32
		SW (45°)	50.5	16.8	0.9	30.5	8.5	0.14
		S	46.9	14.8	0.6	27.7	7.1	0.1
	T3	W (90°)	66.4	22.4	2.6	43.2	13.7	0.9
		SW (45°)	61.0	19.7	1.6	39.1	11.8	0.5
		S	57.6	18.2	1.2	36.5	10.6	0.4

(continued)

Table 5.15 (continued)

			2010			2050		
			CS	M1	M2	CS	M1	M2
	T5	W (90°)	71.4	22.9	5.2	47.1	13.1	1.88
		SW (45°)	70.8	22.5	5.1	46.1	12.6	1.81
		S	68.7	20.9	4.2	44.2	11.4	1.3
Valencia ESP BSk	T1	W (90°)	43.2	13.3	1.0	26.1	6.4	0.18
		SW (45°)	39.2	10.8	0.21	22.7	4.7	0.02
		S	34.5	8.3	0.10	18.9	3.0	0.02
	T3	W (90°)	59.9	19.3	1.9	38.4	11.3	0.8
		SW (45°)	51.6	15.1	0.8	31.9	8.3	0.3
		S	46.9	13.0	0.4	28.2	6.6	0.2
	T5	W (90°)	62.4	17.9	3.5	41.3	9.9	1.49
		SW (45°)	61.3	17.3	3.5	40.1	9.3	1.41
		S	58.5	15.2	2.6	37.5	7.7	1.0

Note Results for the floor in an intermediate position (or best position in terms of demand): P2 in T1; P4 in T3, and PB in T5
M1 = rFei08+rR12+rW+rL+0.63 h^{-1}
M2 = rFei16+rR20+rW+rL+0.4HR

Table 5.16 Heating demand (kWh/m^2 yr) per climate and position in the building: intermediate floor and under roof

			2010			2050		
			CS	M1	M2	CS	M1	M2
Country Location Climate	Typology	Floor	Heating demand	Heating demand	Heating demand	Heating demand	Heating demand	Heating demand
Mostar	T1	P4	148.5	51.5	10.8	106.4	34.3	5.3
		P2	103.8	45.2	8.2	74.2	29.9	4.1
BIH	T3	P9	172.8	53.1	9.5	126.0	37.3	5.9
		P4	115.9	44.5	6.5	85.2	31.2	4.0
Cfb	T5	P1	170.8	57.2	18.3	123.0	38.9	11.1
		PB	133.4	54.2	17.5	98.7	37.7	10.9
Nimes	T1	P4	129.3	40.0	5.9	94.6	27.4	3.0
		P2	89.5	34.5	4.0	65.1	23.6	2.0
FRA	T3	P9	157.3	44.0	6.4	116.9	31.7	4.3
		P4	103.7	36.6	4.1	77.1	26.4	2.7
Csa	T5	P1	152.8	45.8	12.9	112.6	32.1	8.1
		PB	119.8	43.9	12.6	90.4	31.3	8.4
Oporto	T1	P4	97.0	24.1	1.5	62.6	13.4	0.5
		P2	66.9	20.7	1.0	42.3	11.4	0.4
PRT	T3	P9	122.4	31.5	3.1	82.3	20.2	1.5
		P4	81.4	25.9	1.8	54.6	16.4	0.8
Cfb	T5	P1	115.7	29.8	5.5	76.0	17.7	2.7
		PB	95.7	29.4	6.3	65.4	18.1	3.5
Rome	T1	P4	95.7	25.1	1.6	62.4	13.4	0.4
		P2	65.2	21.5	1.1	41.3	11.4	0.3

(continued)

Table 5.16 (continued)

			2010			2050		
			CS	M1	M2	CS	M1	M2
ITA	T3	P9	119.3	30.8	3.0	80.9	19.4	1.4
		P4	78.5	25.4	1.8	52.7	15.8	0.8
Csa	T5	P1	114.0	30.7	6.0	75.9	18.0	2.5
		PB	91.9	29.7	6.5	63.5	18.0	3.2
Athens	T1	P4	68.7	17.0	0.8	41.0	8.0	0.1
		P2	46.9	14.8	0.6	27.7	7.1	0.1
GRC	T3	P9	87.8	22.3	2.0	55.4	13.0	0.6
		P4	57.6	18.2	1.2	36.6	10.6	0.4
Csa	T5	P1	83.4	21.3	3.5	50.7	11.1	0.7
		PB	68.7	20.9	4.2	44.2	11.4	1.3
Valencia	T1	P4	55.4	10.1	0.1	31.9	3.8	0.02
		P2	34.5	8.3	0.1	18.9	3.0	0.02
ESP	T3	P9	75.4	16.5	0.8	47.8	8.8	0.3
		P4	46.9	13.0	0.4	28.2	6.6	0.2
BSk	T5	P1	70.6	15.2	1.5	44.2	7.0	0.4
		PB	58.5	15.2	2.6	37.5	7.7	1.0

Notes
M1 = rFei08+rR12+rW+rL+0.63 h^{-1}
M2 = rFei16+rR20+rW+rL+0.4HR

Table 5.17 Heating demand (kWh/m^2 yr) per climate and ventilation conditions

			2010			2050		
			CS	M1	M2	CS	M1	M2
Country Location Climate	Typology	Ventilation	Heating demand	Heating demand	Heating demand	Heating demand	Heating demand	Heating demand
Mostar BIH Cfb	T1	0.63	103.8	45.2	34.1	74.2	29.9	21.6
		0.4		31.2			19.5	
		0.63HR		19.6			12.1	
		0.4HR		16.5	8.2		9.7	4.1
	T3	0.63	115.9	44.5	30.3	85.2	31.2	20.5
		0.4		32.61			22.13	
		0.63HR		18.1			12.4	
		0.4HR		16.6	6.5		11.1	4.0
	T5	0.63	133.4	54.2	48.2	98.7	37.7	33.2
		0.4		39.3			26.3	
		0.63HR		27.1			17.8	
		0.4HR		23.0	17.5		14.8	10.9
Nimes FRA Csa	T1	0.63	89.5	34.5	24.7	65.1	23.6	16.3
		0.4		22.4			14.7	
		0.63HR		13.7			8.9	
		0.4HR		10.9	4.0		6.7	2.0
	T3	0.63	103,7	36.6	23.9	77.1	26.4	16.9
		0..4		26.01			18.34	
		0.63HR		14.2			10.3	
		0.4HR		12.7	4.1		8.9	2.7

(continued)

Table 5.17 (continued)

			2010			2050		
			CS	M1	M2	CS	M1	M2
	T5	0.63	119.8	43.9	38.3	90.4	31.3	26.9
		0.4		30.5			21.2	
		0.63HR		20.8			14.6	
		0.4HR		17.2	12.6		11.8	8.4
Oporto PRT Cfb	T1	0.63	66.9	20.7	12.9	42.3	11.4	6.4
		0.4		11.4			5.4	
		0..63HR		7.0			3.5	
		0.4HR		4.7	1.0		1.9	0.4
	T3	0.63	81.4	25.9	15.7	54.6	16.4	9.4
		0.4		17.25			10.27	
		0.63HR		9.3			5.4	
		0.4HR		7.7	1.8		4.2	0.8
	T5	0.63	95.7	29.4	24.6	65.4	18.1	14.9
		0.4		18.6			10.5	
		0.63HR		12.6			7.8	
		0.4HR		9.6	6.3		5.5	3.5
Rome ITA Csa	T1	0.63	65.2	21.5	14.0	41.3	11.4	6.7
		0.4		12.1			5.6	
		0.63HR		7.1			3.3	
		0.4HR		4.8	1.1		1.9	0.3
	T3	0.63	78.5	25.4	15.6	52.7	15.8	9.1
		0.4		17.09			9.92	
		0.63HR		9.0			5.1	
		0.4HR		7.6	1.8		4.0	0.8

(continued)

Table 5.17 (continued)

			2010			2050		
			CS	M1	M2	CS	M1	M2
	T5	0.63	91.9	29.7	25.1	63.5	18.0	14.6
		0.4		19.0			10.6	
		0.63HR		12.7			7.1	
		0.4HR		9.8	6.5		5.1	3.2
Athens GRC Csa	T1	0.63	46.9	14.8	9.3	27.7	7.1	3.9
		0.4		8.1			3.1	
		0.63HR		4.7			1.6	
		0.4HR		3.1	0.6		0.8	0.1
	T3	0.63	57.6	18.2	11.1	36.6	10.6	5.9
		0.4		12.13			6.51	
		0.63HR		6.6			3.3	
		0.4HR		5.4	1.2		2.5	0.4
	T5	0.63	68.7	20.9	17.5	44.2	11.4	9..2
		0.4		13.1			6.2	
		0.63HR		8.8			4.0	
		0.4HR		6.6	4.2		2.6	1.3
Valencia ESP BSk	T1	0.63	34.5	8.3	4.2	18.9	3.0	1.2
		0.4		3.2			0.8	
		0.63HR		2.0			0.7	
		0.4HR		1.0	0.1		0.2	0.0
	T3	0.63	46.9	13.0	6.9	28.2	6.6	3.1
		0.4		7.40			3.24	
		0.63HR		3.7			1.9	
		0.4HR		2.6	0.4		1.1	0.2

(continued)

Table 5.17 (continued)

			2010			2050		
			CS	M1	M2	CS	M1	M2
	T5	0.63	58.5	15.2	12.3	37.5	7.7	6.0
		0.4		8.1			3.6	
		0.63HR		5.9			2.8	
		0.4HR		3.9	2.6		1.6	1.0

Notes
Ventilation conditions
−0.63: 0.63 h^{-1} all day
−0.4: 0.4 h^{-1} all day
−0.63HR: 0.63 h^{-1} all day heat recovery
−0.4HR: 0.4 h^{-1} all day heat recovery
M1 = rFei08+rR12+rW+rL+0.63 h^{-1}
M2 = rFei16+rR20+rW+rL+0.4HR

the cities. The demands of the Case Study in 2050 are, depending on the places, between 25 and 45% lower than in the present-day Case Study. We also see that the order described for the cities coincides with greater to lesser demands. In this situation, Mostar in the Case Study has demands of between 74.2 and 98.7 kWh/m^2 yr and Valencia has between 18.9 and 37.5 kWh/m^2 yr. The demands of T5 are always higher than T3 and the latter is higher than T1. With the M1 level, savings are produced of between 59 and 79%, with the demands for Valencia being between 3 and 7 kWh/m^2 yr and the remaining cities having demand values of under 30 kWh/m^2 yr, except for Mostar and Nîmes which are slightly above this value. With M2, savings of between 85 and 100% are achieved, and all the cities have demands below 5 kWh/m^2.yr except Mostar and Nîmes, which, in T5, have values of 10 and 8 kWh/m^2 yr, respectively. This shows not only the effectiveness of the insulation but also of the incorporation of heat recovery systems in ventilation.

Regarding orientation (Table 5.15):

- Southern orientation always produces the lowest demands, followed by South-West and West, although the percentages change on applying the M1 and M2 measures, both for the present scenario and for that of 2050.
- In the present-day scenario: we can see the percentages of demand for the southwestern (SW) and Western (W) orientations in comparison with the southern one (S). In the Case Study, typology T1, SW has between 4 and 14% greater demands and W has between 10 and 25%. In typology T3, SW demand is between 4 and 10% greater, and W between 12 and 28%. In T5, the differences are lower, SW only demands between 1 and 5%, and W between 2 and 9% more than the south-facing. Here, the first value corresponds to Oporto and the second to Valencia. If we analyse the M1 intervention in the present-day scenario: in typology T1, the SW orientation demands between 10 and 31% more than south-facing; W between 17 and 61% more. In typology T3, SW requires between 6 and 17% more than southerly; W requires between 18 and 49% more. In T5, the differences are smaller, SW requires between 4 and 14% more than south-facing and W between 6 and 18% more than southerly. Here, the first value corresponds to Mostar and the second to Valencia. The remaining cities have intermediate percentages.
- In the 2050 scenario: there is a drop in demand in all cases compared to the present-day situation. On comparing the differences between orientations, the percentages remain similar to those described for the present-day situation, although they are a little higher.

Regarding the position of the apartment in the building (Table 5.16):

- The Tables show the results for the sub-roof floor and the intermediate floor of the building. The energy demands for heating are always higher for the sub-roof floor than the intermediate floors, as the former has the most disadvantages because it has a greater surface of the thermal envelope exposed to the elements, and so the conditions of comfort are worse.

- In the present-day situation and in the Case Study, the sub-roof floor has approximately 45% greater demands than the intermediate floors in T1, 50% in T3, and 25% in T5, approximately. After retrofitting with M1 and M2: in T1 and T3, the sub-roof floor continues to have worse conditions, but the demands and differences between the two floors are much reduced. However, in T5, in the coldest cities, Mostar and Nîmes, the first floor has worse conditions than the ground floor, while in milder climates such as Oporto, Rome, Athens and Valencia, in general the ground floor has greater demands than the first. This may be due to the fact that in M1, 12 cm insulation is placed on the roof and 4 on the ground slab in contact with the land; and in M2, it is 20 cm on the roof and 4 cm on the ground slab in contact with the land. Whatever the case, the differences between floors in T5 are far smaller than in the other typologies.
- Comparing the future 2050 scenario (with much higher outdoor temperatures) with the present day, in the Case Study, we find that the heating demands of the sub-roof floor drop by between 28 and 78% in typology T1, between 15 and 67% in T3 and between 17 and 70% in T5. The first value corresponds to Mostar, the coldest city, and the second to balmy Valencia.
- In the 2050 scenario, the M1 and M2 renovation levels offer improvements compared to the Case Study (non-renovated building). In the coldest city, Mostar, with M1, the drop in demand on this floor is 67% in T1, 70% in T3 and 68% in T5. In addition, the differences are reduced between the sub-roof floor and the intermediate floor. With M2 measures, the demands of the highest floor are reduced by 95% in T1, and 91% in T3 and T5. In Valencia, with M1, the drop in demand is 88% in T1, 81% in T3 and 84% in T5. With the M2 measures, the drop in demand for this floor reaches 99% in the three typologies. This shows the effectiveness of the thicker insulation used on the roof and of the use of heat recovery systems.

Regarding the ventilation conditions (Table 5.17):

- In these Tables, both in the present-day scenario and in 2050, four measures have been assessed in M1, two with natural ventilation (0.63 h^{-1} and 0.4 h^{-1}), and another two that incorporate heat recovery systems to the two previous rates. In M2, a high ventilation rate has been considered (0.63 h^{-1}) and another with less ventilation rate and a heat recovery system (0.4 h^{-1} h).
- We can see how the demands lessen when the natural ventilation flow is reduced from 0.63 h^{-1} to 0.4 h^{-1} and when ventilation is incorporated through heat recovery systems.
- In the present-day scenario, with M1 in all the cities and T1 and T5 typologies, the demand from higher to lower depends on the conditions of ventilation: 0.63 > 0.4 > 0.63 h > 0.4 h.
- In typology T1, in Mostar, the reduction in demands compared to the CS varies between 56.5% with 0,63 natural ventilation and 81.2% when the heat recovery

system is applied to the same flow (0.63 h), and it is 84.1% less when it is 0.4 h. When M2 was applied in the same typology, the drop in demand compared to the CS varies between 67% with 0.63 natural ventilation and 92% when it is 0.4 h. In Valencia, beginning with a demand of 34.5 kWh/m^2 yr, with M1 and 0.63 natural ventilation, the demand drops to 8.3 kWh/m^2.yr (76% less) and with a heat recovery system (0.63 h), it is 2 kWh/m^2 yr (90% less); if we apply 0.4 h, it reaches 1 kWh/m^2 yr (97% reduction). When M2 is applied, with 0.63 ventilation the demand drops to 4.2 kWh/m^2 yr. (87%) and with use of a heat recovery system in 0.4 h, it reaches 0.1 kWh/m^2 yr (99%).

- In the typology with the greatest demands, T5, the application of heat recovery systems is also very effective, particularly in the coldest cities. In Mostar, on the basis of the Case Study with 133.4 kWh/m^2 yr, using ventilation with a heat recovery system 0.4 h, it reaches 23 kWh/m^2 yr (82% less) with M1, and 17.5 kWh/m^2 yr (87% reduction) with M2. In Valencia, on the basis of the Case Study with 58.5 kWh/m^2 yr, using ventilation with a heat recovery system 0.4 h, it reaches 3.9 kWh/m^2 yr (93%) with M1 and 2.6 kWh/m^2 yr (96%) with M2.
- In the future scenario 2050, as has already been stated, all the demands of the Case Study are reduced by approximately 33%. With the application of heat recovery systems, in M1 the demand values are reduced in T1 to 9.7 kWh/m^2 yr in Mostar and 0.2 kWh/m^2 yr in Valencia. In M2 they drop to 4.1 in Mostar and 0 kWh/m^2 yr in Valencia.

In short, the cities with greater to lesser demands are as follows: Mostar, Nîmes, Oporto, Rome, Athens and Valencia. In all of them, the demand for typology T5 (Detached House) is greater than that of T3 (Tower) and of T1 (Linear Block). A southern orientation always has lower demands followed by south-west and west. The sub-roof floor has greater energy requirements, but when different thicknesses of insulation are placed on the roof, the demands drop as do the differences between this floor and the intermediate ones. It is shown that ventilation with heat recovery systems in winter is very effective in all the cities.

Finally, it is clear that in the future scenario, the demands for heating will be much less, and for this reason, one might think that there is no need to use so much insulation in buildings. However, present-day conditions do require such insulation in order to guarantee the necessary indoor comfort levels.

5.7 Discussion

The results obtained from the case study monitoring detect key aspects for the retrofitting of residential buildings in order to reduce the energy demands for heating in winter and to increase user comfort, particularly among more vulnerable groups.

Regarding the measures to be adopted we consider it is relevant to:

- Give priority to the retrofitting of the envelope rather than that of the heating system, as the latter depends on the user, who can choose not to use the heating in cases of 'energy poverty'. The retrofitting of the envelope in itself does not guarantee comfort inside the dwellings, but it does reduce energy consumption and improves the cold wall effect and comfort. It results in temperatures between 2 and 4 °C higher than in the buildings with no envelope retrofitting (in the monitored case studies) and may raise temperatures above levels which are dangerous for health.

Regarding the validation of the measures adopted, it is advisable to:

- Promote post-occupational studies which should analyse not only the reduction in energy consumptions but also the consequences for the residents, particularly in the collective residential sector in social housing areas with old, inefficient dwellings. Measurement of the indoor temperatures detects, on the one hand, the cases of energy poverty and its consequences on people's health due to unsuitable indoor thermal environments. On the other hand, it detects situations of excess energy consumption, both in buildings which have been retrofitted and which have not, in the former case CE (with a retrofitted thermal envelope) due to habit or preferred thermal sensation (especially in elderly people who are accustomed to high temperatures with unregulated centralized systems), and, in the latter case, with non-retrofitted thermal envelopes SE, in an unfulfilled attempt to reach comfortable temperatures, increasing consumption due to the inefficiency of the dwelling.
- Insert the design and cost of monitoring the indoor conditions of the dwellings into retrofitting projects (preferably before and after the intervention), which would detect whether the measures taken have been effective and whether the conditions of comfort, that is, temperature, better airtightness, etc. have been reached. Although this may be expensive and the budgets for social housing rehabilitation are very limited, it is one of the items may be subsidized by the Public Administration.

Regarding the blower-door tests carried out, it is important to stress that:

- Blower-door tests are costly, at both a technical and human level, but they give important information on the quality of the dwelling as regards the airtightness and the source of infiltrations (based on the thermography). The improvement in airtightness is fundamental to improve user comfort and reduce the energy consumed in heating, particularly in the south of Europe where controlling infiltrations has traditionally never been a priority due to the mild climate conditions.
- The values obtained in Pamplona, Spain in the blower-door tests are in general between 2 and 5 h^{-1} (average level of airtightness according to UNE 12831:2003), both in renovated and non-renovated buildings; therefore, we can see that improving airtightness is an objective that is not being achieved. These

are not construction solutions which can be improvised, but rather must be specified in the project and require exhaustive building control during their implementation. They should be tested by blower door, preferably before and after retrofitting, in order to assess the improvement.

- The 'key factor' in the improvement of airtightness of the buildings is proper decision-making regarding the space which includes the window, the box cassette of the blinds and the joint with the façade. This should be decided in the project and the construction should be controlled to achieve the energy-saving objectives.
- The greatest improvements in airtightness are produced when the original windows and box cassettes are replaced with new ones with built-in monoblock blinds. With these systems, values close to 2 h^{-1} can be reached with average-quality, affordable windows. Although more demanding standards such as those of the Passivhaus reduce these values to 0.6 h^{-1} in new builds and 1 h^{-1} in retrofitting, the objective is to get as close as possible to these values by checking the benefits of the window and its placement, with affordable solutions for the residents. In addition, the potential effects that the increased airtightness may produce, such as damp caused by condensation, should be studied if the thermal insulation of the envelope and suitable ventilation are not carried out simultaneously.
- The solution of placing double windows on the outside while keeping the original windows and box cassettes on the inside are economical and easy to carry out in order to avoid rehousing the resident, but they do not bring about major improvement in the airtightness of the dwelling, at least if this is done as at present, with low-performance sliding windows. This does not mean that there are no added values such as acoustic enhancement, improvement in the sensation of thermal comfort and the aesthetic upgrading of the façade when it is carried out for the building as a whole. But this solution would be more effective if the frame was of low air permeability and were installed after the manufacturers had resolved the airtightness of the joint.

Regarding reaching suitable indoor temperatures it must be stressed that:

- The type of heating system, its regulation and control, has a great effect on the range of indoor temperatures which may be reached. In the homes monitored in Pamplona, we have found varied situations: dwellings with district heating (with or without individual controls), individual heating in each dwelling, dwellings with no heating system and/or auxiliary devices such as butane or electric heaters.
- The system and use of the heating depend very much on the socio-economic characteristics of the residents. Among the groups with higher incomes, we have found a higher proportion of homes with central, district or one-building heating, and also individual gas heating. The number of these systems drops gradually in the lower income groups where, in critical conditions, they are mainly replaced or complemented by auxiliary heating devices such as electric

radiators, gas heaters, firewood, etc. Consequently, this worsening of the systems used has an impact on the level, quality and distribution of heat in the home, together with a potential health risk depending on the temperatures reached and accidents caused by fire.

- In the unregulated central heating systems, higher temperatures are found, much higher than the standards recommended. With reference to individual heating systems, the possibility of controlling these favours greater variation in the patterns of use, from high levels to very low ones with which it is difficult to maintain comfortable temperatures in the home.

On the analysis of possible cases of energy poverty,

- Monitoring temperatures is the key to detection of these cases. We have found cases of inability of the homes to maintain suitable and consistent levels of temperature throughout the dwelling. As has been seen, there are dwellings whose permanent temperature is about 12 or 14 °C which undoubtedly affects health.
- In other cases, some people are accustomed to switching off radiators in some rooms, reducing the hours of heating, and consequently, the average temperature is below the level considered comfortable. They attempt to reduce this effect with more clothing or use of auxiliary heating systems which, economically, respond to specific needs but are also close to the limits for health. This corresponds to families with more limited incomes, who are unemployed or retired, spend more time at home.
- Also noteworthy is that in buildings with central or district heating, we find situations in which the dwelling may have comfortable temperatures, even excessively high temperatures in winter, at a fixed price which does not take into account whether the dwelling is in use or not, but is shared out over the year. In this way, families with economic difficulties can afford it and do not suffer the consequences for health of an inadequately heated home. However, when the building is renovated and individual heating or heating with individual regulation is installed, some families '*decide*' not to switch on the heating or to have it at temperatures below the level of comfort, permanently or for a few hours; thus, some residents may be in a situation of energy poverty. For this reason, it is necessary to promote passive energy rehabilitation measures of the thermal envelope so that the buildings demand and consume as little heating as possible, particularly in social housing, as we cannot know if the heating will be used appropriately, or if the temperatures will be below safe health standards.

On the subject of the patterns of use,

- Proper use of the dwellings is considered fundamental. It is not a question of reducing energy consumption at the expense of not reaching comfortable temperatures indoors. Nevertheless, this use must match the family structure, their habits, the age of the residents (elderly people and babies are more vulnerable), their socio-economic and employment situation, health conditions (people who are ill or disabled require higher temperatures), the time spent in the dwelling, etc.

- The retrofitting interventions must include 'Building's User Guide' or instructions that explain the effect the user's actions have on the energy demands, comfort and health. The more knowledge the occupier has of these points, the more the consumption and economic cost of the heating drops.

Regarding the results obtained from the simulation

- In the climate change scenario, through simulation, it has been shown that there will be a drop in energy demands for heating in all the Mediterranean cities in the south of Europe. If some criteria in designing the retrofitting of the envelope and the use and ventilation conditions are taken into account, very low values of demand may be reached, in some cases almost zero level demands for heating. However, buildings must also respond to the current demands for heating so, from now until the future situation, they must offer healthy and comfortable conditions for their occupiers.
- These design criteria have, first, to do with the typology. Single-family T5 homes require more energy than T3 and T1 towers. For this reason, more efficient retrofitting measures must be applied to the T5 envelopes. Second, south-facing dwellings have less demand for heating, so, in the retrofitting, we must attempt to increase window surface facing south although the orientation may be different, even though these solutions imply difficulties. Third, the dwelling under the roof is the one that requires most energy, so in the retrofitting, a demand similar to that of the remainder of the floors must be attempted, by insulating the roof very well and reducing air infiltrations. Finally, the placement of heat recovery systems in the ventilation significantly reduces demand. In these ways, in some locations, almost zero heating demands could be achieved.

5.8 Conclusions

The struggle against climate change demands investment in the retrofitting of existing buildings, particularly those considered most vulnerable (with uninsulated thermal envelopes, without heating systems), whose occupiers are more susceptible to energy poverty. In these, we must actively focus on the reduction of energy demands to minimum levels by intervening on the thermal envelope of the buildings. In the cases monitored in Pamplona, the positive impact of refurbishment of the envelope on the indoor temperatures and comfort has been shown. With this action (retrofitting of the façades, roof, windows, separation from unheated spaces and a reduction in infiltrations) with very minor use of heating, an increase in the indoor temperature of between 2 and 4 °C is achieved, thus distancing it from the temperature thresholds which negatively affect health. It is also recommended that appropriate use of the dwellings be stressed. This includes heating timetables and setpoints, together with ventilation guidelines, in order to reach temperatures matching the standards of comfort.

In the present situation and in that of climate change in the year 2050, the proposed retrofitting measures of the thermal envelope would, in some European locations, allow for buildings with almost zero heating demands. The key factors that contribute to this objective are design criteria for the envelope taking into account: the climate and the differences between building typologies which will require a greater or lesser levels of intervention (thickness of insulation, etc.), orientation towards the south for greater solar gains, the position of the dwelling in the building so that all the apartments have the same energy demands, and ventilation incorporating heat recovery systems.

Acknowledgements We would like to thank the Spanish Ministry of Economy and Competitiveness for financing the project 'Performance Buildings Envelope Rehabilitation Protocol (prestaRener) through BIA 2012-38666, which is the basis of this chapter. We wish to thank the participants from Grupo SAVIArquitectura in this project: Purificación González, Silvia Domingo, Germán Ramos, Juan Echeverría, and specially Ana Castillejo and Jorge San Miguel, likewise to Brian Meacham, from Worcester Polytechnic Institute (USA). Also thanks to Isabel Izcue and Ana Bretaña from NASUVINSA, as well as Alberto Calvo from Department of Building Retrofitting of the City Hall of Pamplona. Finally, special thanks to Cristina Güell for her assistance in this chapter.

References

Comisión Europea. (2010). Directiva 2010/31/UE relativa a la eficiencia energética de los edificios. *Boletin Oficial, L135*, pp. 13–35.

CTE-HE. (2013). Documento Básico CTE-HE Ahorro de energía. *Septiembre*.

Dascalaki, E. G., Droutsa, K. G., Balaras, C. A., & Kontoyiannidis, S. (2011). Building typologies as a tool for assessing the energy performance of residential buildings—A case study for the Hellenic building stock. *Energy and Buildings, 43*(12), 3400–3409. https://doi.org/10.1016/j.enbuild.2011.09.002.

Deponds, D. (2010). The Urban and Social challenges concerning an operation of urban renewal in the parisian periphery. *The Example of la Croix-Petit in Cergy, 13,* 83–102.

Domingo-Irigoyen, S., Sánchez-Ostiz, A., & San Miguel-Bellod, J. (2015). Cost-effective renovation of a multi-residential building in Spain through the application of the IEA Annex 56 Methodology. *Energy Procedia, 78,* 2385–2390. https://doi.org/10.1016/j.egypro.2015.11.194.

Gutiérrez Palomero, A. (2010). La iniciativa comunitaria urban y la construcción inconclusa de una política urbana para la unión europea. *Papeles de Geografía, 1*(25003), 159–167.

Huang, J. (2011). *ASHRAE Research Project 1477-RP Development of 3,012 typical year weather files for international locations. Final Report.*

IEA Annex 56. (2016). Annex 56. Retrieved October 10, 2017, from http://www.iea-annex56.org/.

International Organization for Standardization. (2000). *ISO 6781:1983 Thermal performance of buildings. Qualitative detection of thermal irregularities in building envelopes. Infrared method.* (ISO 6781:1983, modified) (2000).

International Organization for Standardization. (2002). *ISO 9972:2015 Thermal performance of buildings—Determination of air permeability of buildings—Fan pressurization method.* (ISO 9972:1996, modified) (2002).

International Organization for Standardization. (2014). *ISO 9869-1:2014 Thermal insulation—Building elements. In-situ measurement of thermal resistance and thermal transmittance—Part 1: Heat flow meter method.*

Jentsch, M. F., James, P. A. B., Bourikas, L., & Bahaj, A. S. (2013). Transforming existing weather data for worldwide locations to enable energy and building performance simulation under future climates. *Renewable Energy, 55,* 514–524. https://doi.org/10.1016/j.renene.2012.12.049.

López de Lucio, R. (2008). *Ordenar el territorio, proyectar la ciudad, rehabilitar los tejidos existentes. La relevancia del planeamiento a través de los premios nacionales de urbanismo.*

Ma, Z., Cooper, P., Daly, D., & Ledo, L. (2012). Existing building retrofits: Methodology and state-of-the-art. *Energy and Buildings, 55,* 889–902. https://doi.org/10.1016/j.enbuild.2012.08.018.

Meacham, B., Poole, B., Echeverria, J., & Cheng, R. (2012). *Fire Safety Challenges of Green Buildings*. Berlin: Springer.

Ministero de Fomento. (2014). Estrategia a largo plazo para la rehabilitación energética en el sector de la edificación en España.

Morandi, C., Pessina, G., & Scavuzzo, L. (2010). Strumenti innovativi per la riqualificazione dei quartieri residenziali in Italia: tre casi esemplari. *Ciudades: Revista Del Instituto Universitario de Urbanística de La Universidad de Valladolid, 13*, 103–122. Retrieved from http://dialnet.unirioja.es/descarga/articulo/3309118.pdf%5Cn, http://dialnet.unirioja.es/servlet/extart?codigo=3309118.

Naciones Unidas. (1998). Protocolo de kyoto de la convención marco de las naciones unidas sobre el cambio climático. *Protocolo de Kyoto, 61702*, 20. Retrieved from http://unfccc.int/resource/docs/convkp/kpspan.pdf.

Prat Navarro, F. X., & Wadel, G. (2010). La experiencia de rehabilitación de viviendas en España bajo parámetros ambientales. Un estado del arte. *SB10Mad. Edificación Sostenible, Revitalización Y Rehabilitación de Barrios*, pp. 1–12.

Rubio del Val, J. (2011). Rehabilitación Urbana en España (1989–2010). Barreras actuales y sugerencias para su eliminación. *Informes de La Construcción, 63*(Extra), 5–20. http://doi.org/10.3989/ic.11.060.

Rubio del Val, J., & Molina Costa, P. (2010). Estrategias, Retos y Oportunidades en la Rehabilitación de los Polígonos de Vivienda Construidos en España entre 1940 Y 1980. *Ciudades, 13,* 15–37.

Sánchez-Ostiz, A., Meacham, B. J., Domingo-irigoyen, S., Echeverria, J. B., & González, P. (2014). Implications of Thermal Envelope Retrofitting in Other Building Requirements' Performance. In *40th IAHS World Congress on Housing*, pp. 1–9.

TABULA Episcope. (n.d.). IEE Project TABULA. Retrieved October 11, 2017, from http://episcope.eu/iee-project/tabula/.

UNE-EN 12831:2003. (2003). *Heating systems in buildings. Method for calculation of the design heat loaf.*

UNE-EN 13829. (2002). *Thermal performance of buildings. Determination of air permeability of buildings. Fan pressurization method* (ISO 9972:1996, modified).

UNFCCC. (2015). Paris Climate Change Conference-November 2015, COP 21. In *Adoption of the Paris Agreement. Proposal by the President., 21932* (December), 32. http://doi.org/FCCC/CP/2015/L.9/Rev.1.

Vandentorren, S., Bretin, P., Zeghnoun, A., Mandereau-Bruno, L., Croisier, A., Cochet, C., et al. (2006). August 2003 heat wave in France: Risk factors for death of elderly people living at home. *European Journal of Public Health, 16*(6), 583–591. https://doi.org/10.1093/eurpub/ckl063.

Chapter 6
The Scope of Retrofitting on an Urban Scale. Use of Geographic Information Systems, GIS, for Diagnosis of Energy Efficient Interventions at an Urban Level

6.1 The Importance of Working on an Urban Scale to Achieve Energy-Saving and Emission-Reducing Objectives

The European Union (EU) has set targets to progressively reduce greenhouse gas emissions up to 2050, with the aim of maintaining the global rise in temperature under 2 °C. Scientists consider that a 2 °C increase over the preindustrial era temperature marks the limit beyond which there is a much greater risk of dangerous and catastrophic changes for the global environment. At present, the mean temperature has risen by 0.85 °C, with unprecedented temperature increases in the last three decades.

These targets aim to place the EU on the pathway to a transformation towards a low-carbon economy by 2050 (Commission of the European Communities, 2008). To do so, it has set a climate and energy package of measures up to 2020, and a climate and energy framework for 2030.

There are three specific targets for 2020: 20% cut in greenhouse gas emissions from 1990 levels, 20% of EU energy from renewables and 20% improvement in energy efficiency. These targets affect the sectors of housing, agriculture, waste and transport (excluding aviation).

For 2030, there are three targets: 40% cuts in greenhouse gas emissions (from 1990 levels), 27% share for renewable energy and 27% improvement in energy efficiency.

Adapting cities to global warming demands a new research approach to the built environment, from the study of new developments to the analysis of transformations in the existing buildings (Mavrogianni et al., 2009).

Action on an urban scale rather than on a buildings scale is based, on the one hand, on the need, in pre-existing cities, for intervention areas by types of building with the same characteristics and adaptation requirements which can be updated together so as to improve broader areas and reach more ambitious objectives

© Springer International Publishing AG 2018

A. Monge-Barrio and A. Sánchez-Ostiz Gutiérrez, *Passive Energy Strategies for Mediterranean Residential Buildings*, Green Energy and Technology,
https://doi.org/10.1007/978-3-319-69883-0_6

directed towards these European strategies of the H2050 prospect. On the other hand, the synergies of intervening on a higher scale offer benefits: lower investment costs, the potential for installing joint heating (district heating), greater ease of administration procedures, urban and social regeneration, etc. Additionally, the updation of the built environment and its surroundings is always more sustainable than the expansion of the city towards new developments, due to better use of the available resources (land and materials) and a reduction of waste and of daily commuting, with the corresponding cuts in energy consumption and emissions.

6.1.1 City Produced Impacts on Energy and Emissions

Cities consume 75% of all energy and produce 80% of greenhouse gas emissions (Soriano, n.d.). Urbanization rates are increasing and will continue to increase. Over half of the global population live in cities, and it is expected that by 2030 over 5 billion people will live in cities, that is, approximately 60% of the population (Alhamwi, Medjroubi, Vogt, & Agert, 2017). This percentage will continue to rise until 2050 when 70% of the world population will live in urban areas (Inter-American Development Bank, 2010).

Many aspects are involved in urban sustainability, such as mobility and transport, building and housing, waste, pollution, and socio-economic and institutional factors, but one of the most crucial is energy consumption in the building sector (Braulio-Gonzalo, Bovea, & Ruá, 2015).

At a national and international level, different actions and commitments have been taken in order to reduce the impact of cities on the environment and to harmonize human social and economic development with the earth and its resources. Examples of these are Agenda 21, passed at the United Nations Conference on Environment and Development (CNUMAD), in Río de Janeiro in June 1992; the European Network of Sustainable Cities and Towns supported by the International Council for Local Environmental (ICLEI); the Aalborg Charter, signed by the participating towns at the European Conference on Sustainable Cities and Towns in May 1994; the Kyoto Protocol, passed in 1997, which introduced to legally binding objectives for emission reduction in the developed countries; the EU Revised Strategy for Sustainable Development created in June 2006; the Leipzig Charter on Sustainable European Cities signed in 2007 by the Member States of the European Union; the Europe 2020 Strategy, signed in June 2010; the Declaration of Toledo, signed by the Ministers responsible for Urban Development in June 2010; the Spanish Strategy for Urban and Local Sustainability (EESUL), produced by the Ministries of Housing and the Environment, the Rural and Marine Environment in 2011; the Reference Framework for Sustainable Cities RFSC created in 2012; and the 'Green Paper' (European Commission, 2013) in 2013, an EU-wide public consultation on climate and energy aims for the year 2030, the results of which were published in 2014 in the Communication 'A policy framework for climate and energy in the period from 2020 to 2030' (European Commission, 2014). Since

2012, different United Nations conferences on climate change have been held (European Commission, 2016): Doha 2012, Warsaw 2013, Bonn 2014. The year 2015 culminated with the Paris Agreement which recognizes the global role of cities and urban authorities in the struggle against climate change and the global reduction of CO_2 emissions (UNFCCC, 2015). This agreement establishes a worldwide action plan to limit the global temperature rise to 2 °C or less.

As a result of these programmes and agreements, specific actions have been taken, outstanding amongst which, in the energy area, are those designed to improve the energy efficiency of buildings, to reduce pollutant emissions and to increase the use of renewable energies.

The creation of more sustainable cities demands a radical change both in the shape and structure of city design and in the consumer habits of those who live in the urban areas. Compared to the extensive city which makes use of large areas of productive land and causes greater daily commutes, the intensive city stands out from the sustainable perspective due to its lesser use of land and lower energy consumption.

6.1.2 Renovation of the Existing City Compared to New Urban Development

On this point, it is necessary to slow down urban expansion by means of the retrofitting of the existing city and adapting its buildings to the new energy efficiency demands by improving their thermal envelopes and their deteriorated energy systems. Moreover, this retrofitting must be supported by economic aid or specific incentives.

There is consensus on the need to assess the energy consumption of the residential building stock. Cities must diagnose the current energy characteristics of their building stock in order to establish a starting point from which stakeholders and policymakers can promote sustainable development and retrofitting cities.

The 40% of the total EU energy consumption is used in buildings (EPDB, 2010), and 29% of this percentage corresponds to the residential sector. These data show the residential sector to be one of the fundamental pieces for intervention in order to achieve the above-mentioned European objectives for the reduction of external energy dependency on fossil fuels and to reduce greenhouse gas emissions.

Given that the majority of the building stock will still be standing within the above-mentioned timeline, the restoration of buildings and neighbourhoods is one of the most important challenges and obligations we now face in the European area, in line with the promotion of the energy retrofitting of homes and reduction of their energy consumption and CO_2 emissions. Particular attention is demanded by the over 11 million (56%) pre-1980 dwellings in Spain of a total of 25 million (INE, 2011) which, in most cases, do not have the necessary conditions for energy efficiency and make a critical environmental impact due to their high consumption rates. Additionally, in Europe, houses built before 1979 are 143 millions, of 211 millions surveyed until 2008, resulting in the 68% of total (Entranze, n.d.).

It has been calculated that in Spain by 2050, for a perspective of 10 million retrofitted dwellings, with a total version of 260.000 million euros (Cuchí & Sweatman, 2014), it would be possible to save 68.000 GWh of energy per year, and we would avoid the emission of 8.600.000 metric tonnes of CO_2. This would mean savings of 390.000 million euros, more than the total invested (Cuchí, 2013), and the impact would be beneficial not only from the **environmental** perspective but also from an economic and social one.

6.1.3 Methodology for the Analysis of Energy Efficiency on an Urban Scale: Different Approaches

Different approaches may be used to assess energy efficiency at a neighbourhood and/or city scale. Swan and Ugursal, and Kavgic reviewed the modelling techniques used to predict residential energy consumption (Swan & Ugursal, 2009). Two distinct approaches were identified: top-down and bottom-up. The top-down approach is based on macroeconomic indicators such as price, income and climate data. It treats the residential sector as an energy sink and is not concerned with individual end uses (Braulio-Gonzalo et al., 2015). It utilizes historic aggregate energy values and regresses the energy consumption of the housing stock as a function of top-level variables (macroeconomic indicators such as gross domestic product, unemployment, inflation, energy prices and general climate). The bottom-up approach, on the other hand, extrapolates the estimated energy consumption of a representative set of individual houses to regional and national levels, and consists of two distinct methodologies: the statistical method and the engineering method (Johansson, Vesterlund, Olofsson, & Dahl, 2016).

The key differences between the approaches are the use of different parameters, the levels of input information and the calculation techniques (Swan & Ugursal, 2009). Bottom-up engineering approaches provide detailed profiles of individual dwellings in a housing stock with relatively uniform archetypes (Min, Hausfather, & Lin, 2010).

In their work, Swan and Ugursal concluded that the bottom-up engineering approach 'is the only method that can fully develop the energy consumption of the sector without any historical energy consumption information' and that 'these techniques have the capability of determining the impact of new technologies' (Swan & Ugursal, 2009).

In addition, the majority of the previous studies that incorporate GIS tools and focus on climate change have been conducted in Europe, at city or neighbourhood scale and in the residential sector, and use the bottom-up approach. The fact that the majority of studies have been conducted at neighbourhood or city scale is due to the availability of energy data and increased awareness, which are key to create effective municipal energy plans and energy-reduction measures in the building stock. Moreover, the majority of the population live in suburban homes (Gupta & Gregg, 2013). However, the articles highlight organizational issues and problems associated with interoperability and data collection (Johansson et al., 2016). There is currently no common model for collecting data on building stock characteristics, and no specific efforts are being put into laws that demand the creation of energy plans. There is also a particular focus on the residential sector, since it is a substantial consumer of energy.

6.2 Urban Vulnerability and Social Vulnerability in Cities

Chapter 5 has addressed the issue of the residential vulnerability of the apartment buildings constructed between 1940 and 1980 in Spain, from after the Civil War (1936–1939) to the advent of the first regulations on thermal conditions of buildings. In Europe, a similar situation occurred after the Second World War (1945). These buildings are noteworthy for the low-quality construction of their thermal envelopes, for their very insecure energy systems and because many of them do not have lifts. In general, they are found in suburban areas of cities and create very distinctive neighbourhoods in which we find two issues: first, the ongoing deterioration of the buildings together with that of the urban setting and, second, the changes in social conditions due to the age and activity of the original inhabitants.

In addition, during this time period, residential buildings were constructed on the limit of the city centres. Although their building and constructive characteristics are similar to the buildings in the social neighbourhoods, they later became very central areas because of the progressive construction processes of the city. Today, these districts are being restored to a great extent because the area attracts a more affluent population group. Thus, the restoration process is more effective in these areas because of the availability of economic resources among their residents. However, in these more central or urban zones, we find the problem of gentrification, that is, the increase in value of the neighbourhood after public investment, which is echoed in the increase in price of the dwellings and rents, often displacing the original inhabitants of the area who, frequently, cannot afford the increased rents or the retrofitting itself. This results in the arrival of new, more affluent inhabitants who displace the original ones and the breakdown of the existing social networks. The effect of this process could be reduced by better regulation of public aid and control of the rising rents.

6.2.1 Urban Vulnerability

We here define vulnerable neighbourhoods as those which have a high proportion of social housing constructed between 1940 and 1980 which are located in sub-urban areas of the city. In these areas, there are problems not only with the features of construction but also because of a lack of public buildings and equipment, deficiencies in the installation network and in the development of vehicle and pedestrian thoroughfares. The public spaces and gardens are usually wasteland. On the subject of accessibility and mobility, the problems go from the existence of obstacles that are impassable for people with reduced mobility to the lack of public transport and/or parking areas. Some of these residential complexes stand out due to their distance from, dependence on and difficult relationship with the town centre, which leads to isolation and, to a certain extent, 'invisibility' in the eyes of the policymakers. Given the need for homes in the time period we are studying, the construction in general was carried out, not building by building, but rather in groups of buildings in a single project which might include from 50 to 500 dwellings. Many of these suburban complexes spring from public or private initiatives, with a social purpose of providing homes for the immigrants who came from rural to urban areas in search of jobs in industry.

These dwellings do not answer to modern demands as they were built with criteria which gave priority to quantity rather than quality. Over time they have fallen gradually into physical decay, simultaneous with the ageing of their inhabitants and their occupation by a low-profile socio-economic group. Today, the retrofitting of the thermal envelope of these social dwelling complexes is indispensable in order to reduce energy consumption and ensure the well-being of their users in consonance with present-day standards and to improve the conditions of their installations and accessibility.

As Juan Rubio del Val states (Rubio del Val, 2011), if the twentieth century was the time for restoration of historic centres, the twenty-first century should be dedicated for renovating the 1940–1980 urban fabric in vulnerable areas. The objective is to put in place regeneration strategies which will stop the deterioration of the urban and social fabric, to preserve its heritage values, to strengthen the social cohesion and to favour economic activity.

The aim of arriving at sustainable urban development and achieving zero-emission buildings and neighbourhoods demands the retrofitting of this plentiful building stock and also that of its urban setting, which affects both the conditions of the buildings and those of their occupiers. It is also the moment to consider how this retrofitting should contemplate climate change scenarios in order to obtain climate-ready buildings.

6.2.2 *Social Vulnerability in Vulnerable Neighbourhoods*

Cities are facing major challenges when it comes to social aspects such as poverty, energy vulnerability or the segregation or concentration of groups of old people, immigrants and homes with low resources, as these influence the harmonic development and social cohesion needed in a sustainable city. These challenges must be kept in mind as important variables regarding the relationship of the city with future climate change scenarios and a lower carbon economy (H2050), together with the retrofitting of existing buildings and the social transformation (demographic and socio-economic) which will have to be undertaken with the passage of time.

Housing and its urban setting are decisive factors in the satisfactory development of citizens' lives. The adaptation of the habitat to the needs of the people is a basic aspiration, and the lack of this adaptation produces frustration, malaise and even conflict among neighbours, which may impede the community agreements necessary to tackle the improvements.

The consequences are not merely reduced well-being among the residents but also the process of decadence and marginalization which may occur in these areas, provoking the abandonment of the dwellings or their use by a more vulnerable population with little economic resources and low cultural and professional preparation, or by groups which are not socially integrated, maladjusted and in many cases excluded. All this complicates taking on the cost of restoring buildings and improving the urban setting.

Social reality, so frequently believed to be a secondary or limiting factor in the driving force for urban development, is becoming more important as a key factor in future restoration scenarios with the intention of achieving zero-emission and climate-ready buildings. An example of this is the slow evolution of retrofitting in the most rundown and oldest areas of our cities, despite the efforts of national and European institutions to motivate improvements, both by means of new regulation and economic assistance. Thus, solutions must be sought which clearly identify the causes that impede retrofitting: first, the great economic effort involved, particularly for the most vulnerable groups; second, the high number of dwellings that need updating; and third, the absence of sufficient public resources to tackle both residential and social vulnerabilities.

The social structure of these neighbourhoods is characterized by a predominantly low or lower middle-class population, with high levels of unemployment and many elderly people, and a growing immigrant population. To this social reality, we must add the economic impact of the household expenses that the families must assume (dwelling and energy). Among the most vulnerable groups, most at risk are those who hold mortgages or are paying rent; an impact that must be added to energy costs which is one of the most common negative factors in the suburban neighbourhoods, characterized by the low energy efficiency of buildings from the studied period (1940–80). As consequence, this situation places these families at risk of social and economic vulnerability, in addition to their residential vulnerability.

There is a very clear relationship between the lack of resources and residential insecurity. Financial shortage obliges the families to find lower rents which are associated with old dwellings in worse conditions, without heating or a lift, located in the suburban neighbourhoods of our cities, which, little by little, bring together those with the lowest incomes together with the most vulnerable groups. It is here that we find the cases of people at risk of suffering energy poverty, which has four causes basically: shortage of resources; the high cost of energy (electricity, heating, hot water, etc.); the deficient characteristics of the thermal envelope of the buildings where they live; the absence of heating or having individual boilers and/or isolated auxiliary heating elements.

Therefore, the analysis at social neighbourhood and/or city level, with the identification of building types and their study in accordance with energy risk factors (year of construction, structural features of the thermal envelope, orientation, type of heating: individual or with no heating, fuel used, etc.), together with a socio-economic study of the householders (family structure, age, income, occupation, etc.), is a useful tool to detect cases of energy–social vulnerability.

Likewise, the relationship between residential energy and the social characteristics of dwellings is therefore a key element in policymaking in the residential sector (Santamouris et al., 2007). Thus, for policymakers, social and constructive mapping on an urban scale allows for the making of realistic decisions and finding solutions with room for private initiatives and funding, for example, by raising building heights (rooftop extension, upward extension) for the development of new dwellings whose sale will permit the retrofitting of the whole building.

6.3 Use of GIS in Urban Renovation

According to National Geographic Society (Caryl-Sue, 2017), a geographic information system (GIS) is a computer system for capturing, storing, checking and displaying data related to positions on Earth's surface.

Geographical information system (GIS) tools are particularly well suited for managing energy efficiency in urban environments. A robust methodology to map and model our cities would enhance the effectiveness of the local planning of energy efficiency and resilience improvement measures in the future. Using GIS tools with an energy map would provide a straightforward way of revealing whether or not our neighbourhoods and cities are ready to be zero emissions.

Energy-reduction measures aimed at making cities more sustainable begin with city districts and neighbourhoods, which represent the optimal scale for analysing urban climate-ready retrofit measures. Systematic approaches that integrate these city district and neighbourhood energy models and link them to real-world data are required.

Research related to integrated energy planning in spatial frameworks has drawn significant interest from scientists, engineers and policymakers to meet the increasing targets for the share of renewable energy sources (RES) and the decreasing carbon emissions targets (Alhamwi et al., 2017). Accurately diagnosing and modelling current energy consumption at urban scale is the crucial starting point for any urban-scale low-carbon energy policy (Nouvel et al., 2015). Urban spatial databases contain geometric data and a range of information related to the characteristics of buildings and cities; this makes GIS an indispensable tool for handling such datasets (Theodoridou, Karteris, Mallinis, Papadopoulos, & Hegger, 2012).

GIS-based models assist stakeholder and policymaker decision-making in this field and contribute to the sustainable development of cities. GIS-based models can be used to build strategies, research future scenarios, develop retrofit measures and analyse ways in which the transition towards resilient cities can be made.

New approaches that integrate large-scale urban energy models with existing building stock data at an urban scale and the most appropriate retrofit measures for achieving adaptation and mitigation objectives are greatly needed. Geographic information systems (GIS) present many advantages in this regard. In addition, GIS tools help policymakers justify their intervention policies in this field, since they allow the current energy consumption of cities to be represented in multilayer maps, thus identifying the results of their policies.

To date, the integration of energy planning into the GIS mapping setting is under development (Alhamwi et al., 2017; Mentis et al., 2016). Some examples do exist of GIS use for the diagnosis of energy efficiency in neighbourhood and city buildings. For example, (Fabbri, Zuppiroli, & Ambrogio, 2012) have presented to the map of the centre of the city of Ferrara (Italy) showing the energy rating of its buildings; this map indicates the need for renovation, given the high number of E-G energy performance certificates together with areas with high density of buildings with the same energy demands.

Likewise, a large-scale energy model of the suburban neighbourhoods of the cities which demand energy retrofitting such as those we have described could be produced. This model is based on building types and on their energy systems, and would allow us to obtain their energy rating. In addition, it would allow its extrapolation to other neighbourhoods with the aim of analysing the implementation of specific energy improvement both for winter and summer.

6.4 Case Study: Pamplona

We present here the study carried out in the city of Pamplona, specifically in the social housing suburban neighbourhoods constructed between 1940 and 1980, before the first regulations on the thermal fitness of buildings. This study diagnosed the present state of the building stock as regards its energy efficiency and the socio-economic situation of its inhabitants, in order to establish effective retrofitting actions and to detect the areas with most potential for energy retrofitting and those which might have more difficulty in funding this retrofitting. All data obtained was processed using a GIS (Geographic Information System) tool.

Pamplona is a city located in the north of Spain and is part of the Mediterranean area. It has approximately 200.000 inhabitants and an extension of 23.55 km^2. It is a compact city with a clear urban structure and open spaces which cover 20% of its surface; each of its neighbourhoods has different features in its urban fabric, construction types, open spaces and mobility. The city of Pamplona, due to its size and variety of building types, is a suitable case for study to validate the methodology used in the application of GIS to the energy retrofitting of buildings and neighbourhoods. In addition, this methodology can be extrapolated to other medium-sized cities, as approximately 80% of European cities have characteristics similar to those of Pamplona, regarding climate, mean size, uses and residential typologies.

Pamplona has 90,175 dwellings, of which 48,050 were constructed between 1940 and 1980, that is, approximately 53% (INE, 2011). Some of these dwellings (21,331 homes) are located in social housing suburban neighbourhoods around the city, and their characteristics are similar to those described in point Sect. 6.2.1. Basically, these are the neighbourhoods of Chantrea, Rochapea, San Jorge, Santa Engracia, Echavacoiz and Milagrosa, which are represented by a red circle in Fig. 6.1.

There are more dwellings with the same characteristics located in the urban area of Pamplona, which, due to their situation, bordering the centre of the city, are in great demand nowadays among people with economic resources and the possibility of tackling retrofitting. These neighbourhoods, marked with a pink circle in Fig. 6.1, are Ensanche (Fig. 6.2), San Juan, Iturrama and Ermitagaña. These neighbourhoods share social-type dwellings with other non-social ones from the same period. The other neighbourhoods indicated in pink are the Historic Centre (*Casco Antiguo*), with particular characteristics, and the new neighbourhoods developed after 1980. In contrast, Fig. 6.3 shows some examples of social housing in suburban neighbourhoods.

6.4.1 Description of the Applied Methodology

In order to analyse the energy demands of the social housing constructed between 1940 and 1980, a neighbourhood-scale diagnosis was made from the energy perspective, based on the building typologies defined in Chap. 5. This permits assessment of the potential for renovation and energy saving of the existing construction and the study of the different retrofitting scenarios to assess their effectiveness.

The methodology adopted in this study follows the bottom-up approach and was composed of the following phases:

1. Identification of the different building typologies and construction sub-typologies of residential buildings from between the years 1940 and 1980.
2. Selection of social housing suburbs, which are representative of the greatest percentage of residential buildings of these types, to apply and validate the methodology.
3. Inspection of a significant sample of buildings for each of the existing typologies. Preparation of index cards to arrange the data obtained.

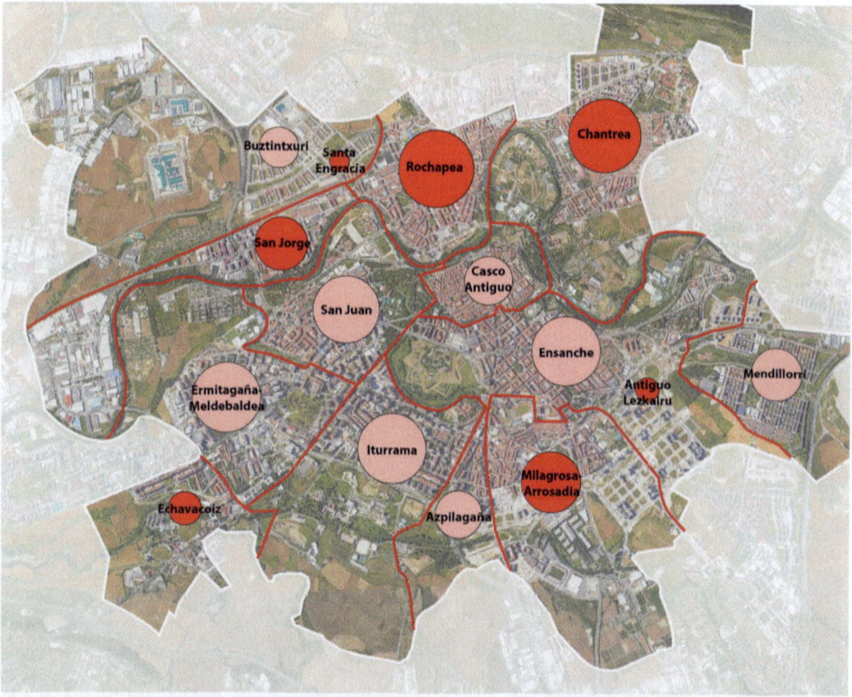

Fig. 6.1 Map of Pamplona showing the different neighbourhoods

Fig. 6.2 Photograh of Ensanche neighbourhood

Fig. 6.3 Photograph of different suburban neighbourhoods as Milagrosa (**a**) Echavacoiz neighbourhood with renovated buildings (**b**) and Soto Lezkairu (**c**)

4. Diagnosis of the current state. Definition of residential vulnerability.
5. Definition of energy vulnerability and the potential for retrofitting. Procurement of the current energy rating of the buildings, to obtain a clear view of the quality of construction from the perspective of energy demand. In this way, the greatest potential for intervention is detected. Besides, two scenarios for retrofitting of the thermal envelope were proposed: one fulfilling the demands of the Spanish regulations (CTE-HE1) for restoration and another, more demanding proposal, obeying the regulations for new constructions.
6. Gathering of socio-economic data of the residents.
7. Assessment of energy and social vulnerability. Detection of the priority areas for intervention.
8. Extrapolation to the buildings constructed in the city of Pamplona between 1940 and 1980 as a whole.

The programme used was ArcGIS which permitted the gathering, storing, analysis and representation of the data obtained. The collection of the building data was carried out through study of the original project documentation, of the building inspection and of the surveys carried out, together with the data from SITNA (Gobierno de Navarra, n.d.).

6.4.2 *Identification of the Typologies of Dwellings Built Between the Years 1940 and 1980*

Table 6.1 shows building typologies identified in the prestaRener project for 1940–1980 residential buildings: linear block (T1), H-shaped linear block (T2), H-shaped tower (T3), other blocks (T4) and single-family home (T5). The definition of the typologies was made with reference to the year of construction, the morphology of the block, the number of floors in the building and the number of dwellings per floor. Additionally, the sub-typologies were based on the constructive features of the main components of the thermal envelope.

The application of these typologies and sub-typologies to the city of Pamplona was carried out by visiting the difference suburbs and analysing the buildings with the SITNAMAP application of the Government of Navarra, together with confirmation from Google Maps and Google Earth. Besides, a representative sample of buildings were selected and inspected in order to verify their constructive features and their fitting in with the corresponding sub-typologies.

Due to the repeatability of the typologies found, a diagnosis may be established of the features of the existing buildings and the most suitable measures for their retrofitting may be evaluated.

Figure 6.4 shows the map of Pamplona with identification of the typologies and sub-typologies of the buildings constructed between 1940 and 1980 in the suburbs.

Table 6.1 Characteristics of typologies and sub-typologies defined in prestaRener project

	Year of construction	Height	Dwellings per floor	M^2 dwelling	Orientation	Building envelope characteristics					
						F	R	W	L	C	P
T1 linear block											
T1.1 Grouped in quarter	1940–1980	≤ GF + 4	2	60–118 80	Double 180°	F3	R2 R3	G1	L3 L2	C3 C2	P2
T1.2 Linear not grouped	1940–1980	≤ GF + 4	2	54–125 80	Double 180°	F1, F2, F3	R2 R3	G1	L3 L4	C3 C2	P2
T1.3 Linear with 1 orientation	1950–1971	≤ GF + 4 GF +10	4	59–94 85	One	F2	R3	G1			
T1.4 Linear with minimum courtyard	1962–1979	≤ GF + 6	2	70–130 95	Double 180°	F2 F3	R3	G1			
T2 H-shaped block with linear group											
T2.1 h ≤ PB + 4	1940–1960	GF +4	4	49-96 70	Double[a] 180°	F1 F2	R2 R3	G1	L2 L3	C1 C3	P1 P2
T2.2 h > PB + 4	1968–1979	>GF + 4 <GF + 11	4	80–100 85	Double[a] 180°	F3	R2	G1	L2	C2	
T2.3 Block in H with minimum courtyard	1954–1979	>GF + 4 <GF + 8	4	70–123 75	Double[a] 180°	F3	R2	G1		C2	
T3 tower											
T3.1 H-tower with lateral courtyard	1960–1975	≥ GF + 8	4	55–107 80	Double 180°	F2 F3	R1	G1	L3 L2	C2	
T3.2 Without courtyard	1970–1980	>GF + 8	4	<90	Double 90°	F4	R1	G1 G2	L1 L3	C2	
T3.3 Cross towers	1966	GF + 10	4	67–95 80	Triple			G1		C2	
T4 other towers											

(continued)

Table 6.1 (continued)

	Year of construction	Height	Dwellings per floor	M^2 dwelling	Orientation	Building envelope characteristics					
						F	R	W	L	C	P
T4.1 Block in 'T'	1964–1978	≤GF + 7	3	67–95 80	Double or triple			G1			
T4.2 Block in 'U'	1963–1973	≤GF + 8	2	80–159 100	Triple	F2	R3	G1	L2	C3 C2	P4
T5 single-family houses											
T5.1 Terraced house	1940–1960	GF	1	80	Double 90°			G1			
T5.2 Terraced house	1949–1980	GF + 1	1	80	Double 90°	F2	R2 R3	G1	L4	C3 C2	P2

Note [a]means one facade facing to a street and the other to a courtyard

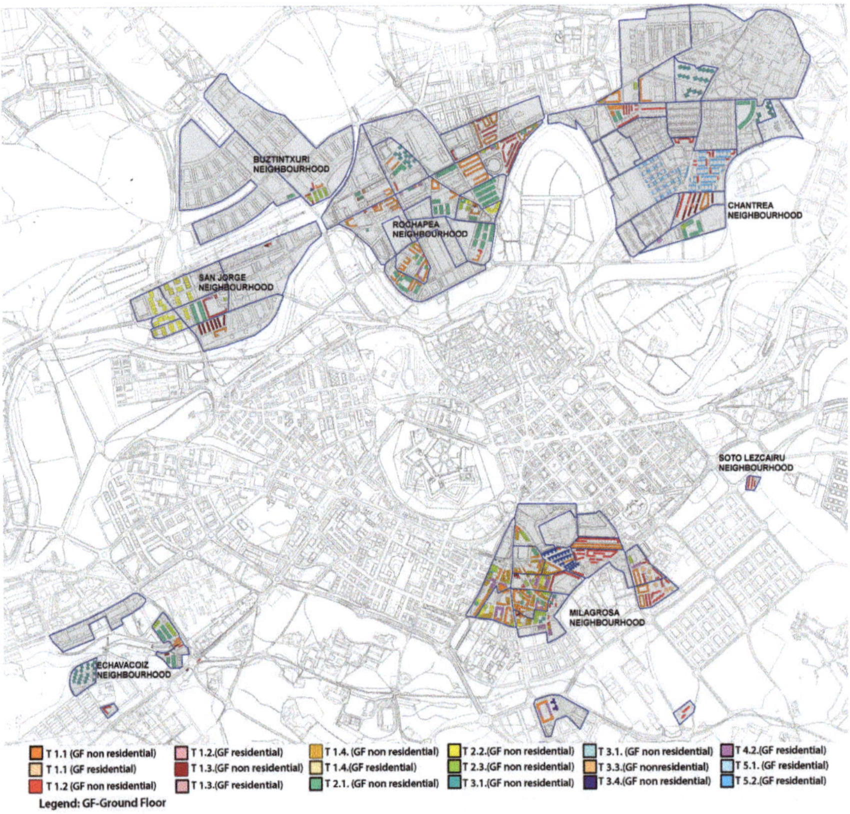

Fig. 6.4 Map of Pamplona showing the studied neighbourhoods and surburb and different buildings typologies built between 1940 and 1980

6.4.3 Selection of Representative Social Neighbourhoods for Analysis

Pamplona has six main suburban neighbourhoods with social housing: Chantrea, with 6,214 homes from 1940–80, Rochapea with 4,721, San Jorge (2,720 homes), Santa Engracia (325 homes), Echavacoiz (1,118 homes) and Milagrosa (6,135 homes). In addition, Soto Lezkairu, which is a small group of 98 homes from the same time period and is now part of a new suburban development with great demand.

Table 6.2 shows the amount of social housing between 1940 and 1980 by construction typologies which is located in suburban areas of Pamplona, and the percentage of each typology. Table 6.3 details construction of sub-typologies and neighbourhoods of Rochapea and Chantrea.

Table 6.2 Number of social housing dwellings located in different suburbs of Pamplona. Breakdown by 1940–80 construction and typologies

Typologies		Number of dwellings 1940–1980	%
T1	Linear block	9,942	46.60
T2	H-shaped block with linear agrupation	6,292	29.49
T3	Tower	3,076	14.42
T4	Other towers	1,075	5.03
T5	Terraced house	946	5.46
Total		21,331	100

To carry out an in-depth analysis of the thermal envelope, the energy efficiency of the systems and the socio-economic characteristics of the residents, the suburbs of Chantrea (Fig. 6.5) and Rochapea (Fig. 6.6) were selected as being representative for the application of the methodology. These suburbs were chosen because of the high percentage of social housing constructed between 1940 and 1980 and because they have very repetitive building typologies. The dwellings from this time period occupy most of the area of the Chantrea suburb, as also occurs in Echavacoiz, Santa Engracia and Milagrosa. In Rochapea, areas with 1940–1980 dwellings coexist with more modern ones, as in other suburbs such as Lezkairu and San Jorge. In this way, two types of suburbs with results which could be extrapolated to the remainder were studied.

Figures 6.7 and 6.8, respectively, show the GIS maps of Chantrea and Rochapea, with the typologies of the buildings that were built between 1940 and 1980.

Here, the case of Rochapea is shown, as it is a more deteriorated suburb, which has problems with the integration of the new buildings being constructed; the restoration of the older buildings could enhance this integration (Figs. 6.6 and 6.8).

The Rochapea suburb has 10,743 dwellings. There are 4,721 homes between 1940 and 1980, that is, 43.94% of the total. It is the oldest neighbourhood in Pamplona (with the exception of the Historic Quarter) and was founded by market gardeners who worked and lived outside the city walls. In the early twentieth century, some working-class dwellings were constructed with no urban planning; foundries and new industries were set up together with the existing market gardens. As of 1940, due to the industrial boom, new dwellings were constructed in different parts of the neighbourhood, somewhat anarchically. New post-1980 buildings have filled in the empty spaces.

In Rochapea, we find different construction typologies from the 1940–1980 period. Only a few blocks have undergone modification since their construction (installation of lifts, some refurbishment of façades, etc.). It is an old, deteriorated and very disadvantaged patrimony which is also home to complex situations of social need which complicate urban regeneration.

Table 6.3 Number of social housing dwellings located in Rochapea and Chantrea neighbourhoods built in 1940–80, breakdown by sub-typologies

Sub-typologies		Rochapea		Chantrea		Total houses of studied suburban social neighbourhoods in Pamplona
		Studied houses	% of total studied houses	Studied houses	% of total studied houses	
T1.1	Linear block. Grouped in quarter	1,845	44.4	336	8.1	4,158
T1.2	Linear block not grouped	470	11.1	2,176	51.5	4,223
T1.3	Linear block with 1 orientation	8	1.1	192	25.4	756
T1.4	Linear block with minimum courtyard	48	6.0	–	–	805
T2.1	H-shaped block. H $\leq$ PB + 4	1,768	62.4	556	19.6	2,832
T2.2	H-shaped block. H > PB + 4	190	8.0	–	–	2,385
T2.3	H-shaped block with minimum courtyard	104	9.7	80	7.4	1,075
T3.1	H-tower	184	6.7	1,904	68.9	2,764
T3.2	Tower without courtyard	–	–	–	–	0
T3.3	Cross-shaped. Tower	–	–	–	–	240
T3.4	Tower with stairs	–	–	–	–	72
T4.1	Block in 'T'	–	–	–	–	369
T4.2	Block in 'U'	104	14.7	24	3.4	706
T5.1	Terraced house (1 floor)	–	–	78	100.0	78
T5.2	Terraced house (2 floor)	–	–	868	100.0	868
	Total	4,721		6,214		21,331

Fig. 6.5 Photograph of the Chantrea neighbourhood

Fig. 6.6 Photograph of the Rochapea neighbourhood

6.4.4 Selection and Inspection of Buildings

Given the identification of the typologies of each of the suburbs (Figs. 6.7 and 6.8), a representative building sample of the different typologies and sub-typologies found was selected. In this way, we verified the building and constructive characteristics, particularly of the thermal envelope, the energy systems and the refurbishment carried out in the buildings.

In Rochapea, we identified and inspected a sample of 172 blocks of a total of 458, which is the 37.55% of the 1940–1980 buildings. First, we studied the original project documentation held in the Registry of Pamplona City Hall; this was followed by the in situ inspection of the buildings and the dwellings.

An inspection card was designed with a list of the 219 information points needed for each of the buildings. These cards were used by the fieldwork teams to gather in situ information on:

Fig. 6.7 Typologies and sub-typologies of the Chantrea neighbourhood

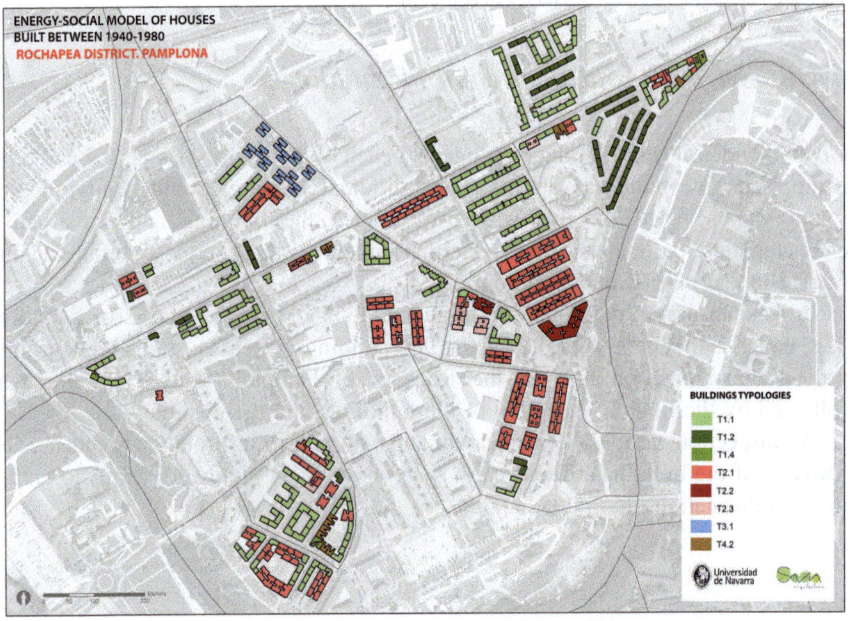

Fig. 6.8 Typologies and sub-typologies of the Rochapea neighbourhood

- Building, urban and location characteristics: orientation, geometry, exposure to sunlight, crossed ventilation, etc.
- Constructive characteristics of the thermal envelope (façades, windows, roofs, separations with staircases or other blocks and with the ground slab). Modifications made to the façades such as replacement of window frames, etc.
- Thermal installations: individual, collective, district heating and energy type.
- Other characteristics of the building which might affect its renovation: for example, accessibility of the building, with or without a lift.

In addition, surveys were carried out among the residents on aspects of the use of the building in winter and summer, together with the socio-economic profile of the occupier.

Once the data for the selected buildings had been obtained, a comparison of their typology and construction was carried out, assigning the same characteristics to a greater number of buildings.

Using all the information gathered, a systematic database was designed which was used for input for the Geographic Information System (GIS).

6.4.5 Diagnosis of Current Condition: Residential Vulnerability

The data gathered provided us with a diagnosis of the current situation of the buildings in each suburb. The GIS permits visualization at suburb and city level, and the building vulnerability situation.

The key factors of this vulnerability are the year of construction (Fig. 6.9), the building typology and the construction sub-typology (Fig. 6.8), the type of thermal envelope (with no thermal insulation and poor airtightness), the heating system (Fig. 6.10) and the non-existence of a lift.

After the typology analysis shown in Fig. 6.8, there is a majority of T1 linear blocks (T1.1, T1.2 and T1.4) and T2 H-shaped linear blocks (T2.2 and T2.3). The percentage of T1 typologies compared to the total buildings in the area is 50.22%, and that of typology T2 is 43.68%.

As can be seen in the same figure, the 1940–1980 buildings in the area of Rochapea are separated in the suburb as a whole. Nevertheless, the level of separation varies substantially depending on the census section.[1] In the most easterly

[1]The census section is defined, with exceptions, as a continuous territory inhabited by minimum of 500 inhabitants and a maximum of 2.500, clearly limited by territorial, geographical and/or urban features. For this reason, and from a social perspective, due to its small extension, it ensures evident uniformity for its inhabitants and their families—particularly in terms of its relative position compared to the remaining sections. Therefore, the mean characteristics of all the other individuals in the section can be taken as valid for each individual unit of which it is composed (Veres, 1999).

On the maps, the continuous lines which divide the suburb mark the different census sections.

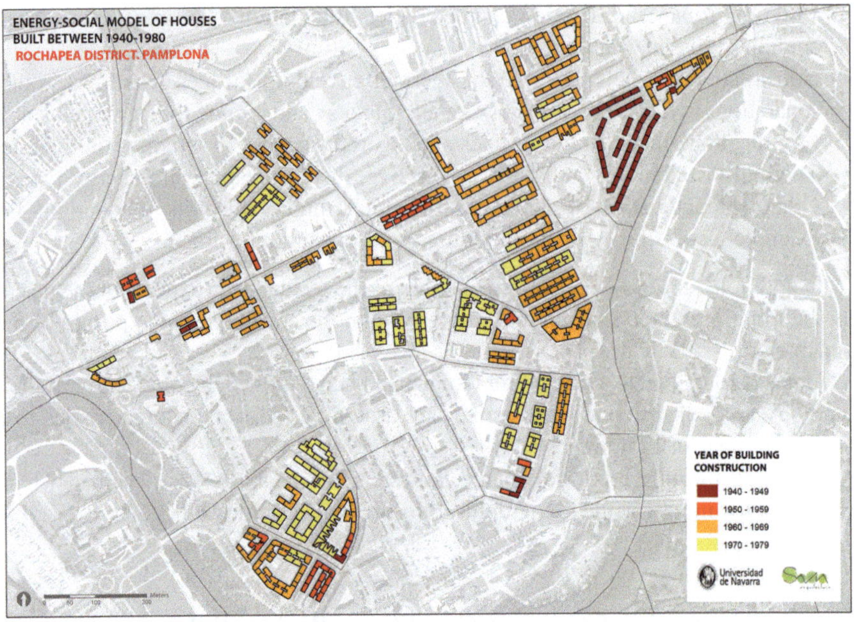

Fig. 6.9 Year of building construction. Rochapea

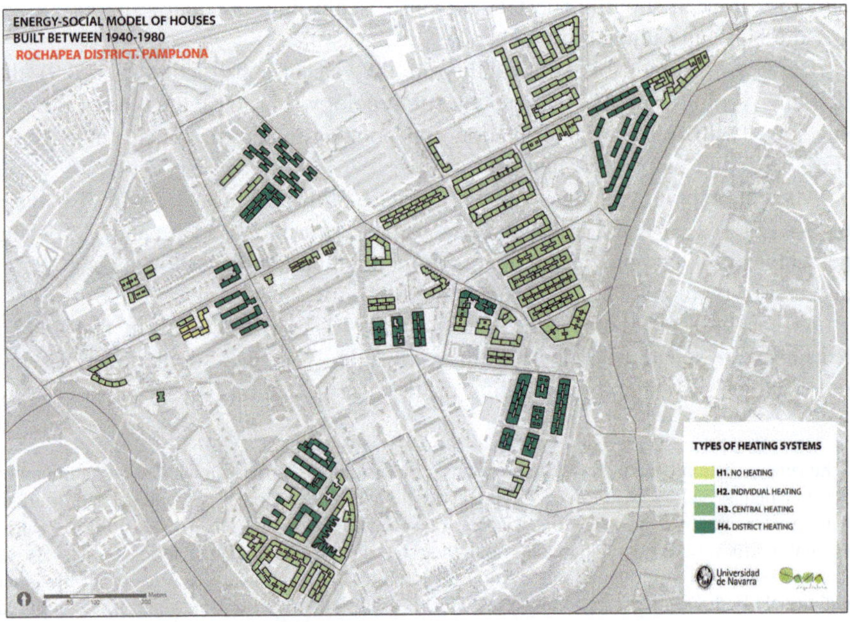

Fig. 6.10 Heating system in buildings in the area. Rochapea

areas, the 1940–1980 buildings have a majority presence, while in the more westerly sections we find greater separation of the area, forming isolated groups close to a majority of new developments.

Up to the present moment, the percentage of renovated buildings with improvements in the thermal envelope is very low.

Regarding the heating system, we must stress that we have verified that the most vulnerable population group lives in buildings with no heating or with individual heating. Thus, their options are not to use heating due to their lack of resources or to use occasional cheaper energy production systems (heaters, braziers, etc.). In fact, in the tracking carried out in various dwellings (Chap. 5), it was found that the families who live in buildings with no heating or with individual heating are at risk of energy poverty. In comparison, belonging to district heating groups reduces the probability of suffering energy poverty because the purchase, production and collective distribution of the energy, together with the community management of the collective installation, has a more beneficial cost/comfort ratio. It permits a high standard of thermal comfort for an annual mean cost. In this sense, the annual distribution of the energy costs into smaller monthly quotas (flat rates) is of benefit to the regular economic management of the families and reduces the risk of non-payment or delays in at-risk economies. Despite this, there are frequent non-payments which force families to leave these homes in order to live in those which have no heating.

In Fig. 6.10, we can see that in this suburb there are small district heating groups, standing out buildings with individual heating which originally had none, and indeed that there are still some buildings which have no heating.

6.4.6 *Energy Vulnerability and Potential for Energy Retrofitting*

The energy vulnerability may be assessed in view of the energy rating of the current building. In turn, this depends on the characteristics of the thermal envelope and of the energy system. The improvement of that mentioned rating may be assessed when different measures are applied and analysis is carried out to find the optimal ones.

Energy Performance Certification (*Certificación Energética de los Edificios*) is a demand of Directive 2002/91/CE. This directive and the later 2010/31/UE, of May 19, regarding the energy efficiency of buildings, is transposed to each of the countries (in the Spanish legal system by the Royal Decree 235/2013 of April 5), to establish the basic procedure for the certification of the energy efficiency of buildings, both newly constructed and older ones. In this certificate, by means of an energy efficiency assessment, each building is assigned a Score, which varies from Score A for the most efficient, to Score G for the least efficient.

The energy rating of the buildings inspected has been carried out in order to offer a clear view of the quality of the construction from the perspective of energy demands and CO_2 emissions.

In addition, to assess the potential for energy refurbishment of these buildings, two levels for the retrofitting of the thermal envelope were established and a new energy rating was calculated. The first level (called M1) contemplates the minimum demands of the Spanish Regulations for Restoration (Código Técnico de la Edificación) (CTE-HE, 2013) and consists of placing 12 cm insulation in the under-roof space, 8 cm insulation in the separation from the ground floor and replacement of the windows with others with a low transmittance frame and low-E double glazing in addition to the original 8 cm of insulation on the façades. The second level (called M2) would fulfil the demands of the CTE for new building works and is more in consonance with the new demands designed to achieve almost zero-energy buildings; in this case, the following is added to the original envelope: 16 cm insulation on façades, 20 cm insulation in under-roof space and 8 cm insulation in the separation from the ground floor and windows with low transmission profiles and low-E glazing.

Thus, for each building inspected, we shall have three different ratings:

- The energy efficiency rating of the building in its current state.
- The energy efficiency rating of the building with M1-type retrofitting.
- The energy efficiency rating of the building with M2-type retrofitting.

Figures 6.11 and 6.12 show the energy-rating maps for demand and CO_2 emissions, respectively, of the buildings in their current state. Figures 6.13 and 6.14 show the maps for heating-demand ratings after applying the retrofitting measures to the thermal envelope, in the two intervention levels, M1 and M2. Comparing the three ratings, a diagnosis of the potential energy savings for each suburb is obtained. Additionally, these maps permit the comparison between suburbs, to establish priorities at city level. For example, Fig. 6.15 shows the energy efficiency ratings for the Chantrea neighbourhood where we can see a higher density of 1940–1980 buildings with high energy-demand ratings (F-G).

We can see on Rochapea's maps (Figs. 6.11, 6.13 and 6.14) that most of the analysed buildings currently have a heating-demand energy rating of letters E (9,9%), F (0,9%) and G (78%). The energy rating coincides with that of the emissions, as none of the buildings has systems for the production of renewable energy. When M1-type retrofitting of the envelope is applied, the ratings obtained are C (10.1%), D (23.2%) and E (39.2%). On applying M2-type retrofitting, most of the buildings remain at ratings C, D and E. There is a rise in the ratio of the buildings graded as D (35.9%) and C (15.8%), and decrease of those graded as E (29.8%), in relation to M1. For this reason, taking into account only the retrofitting of the thermal envelope, without changing the energy systems, M1-type retrofitting would be the best for the current climate scenario. Besides, the cost would be more economical and would allow for intervention in a greater number of dwellings.

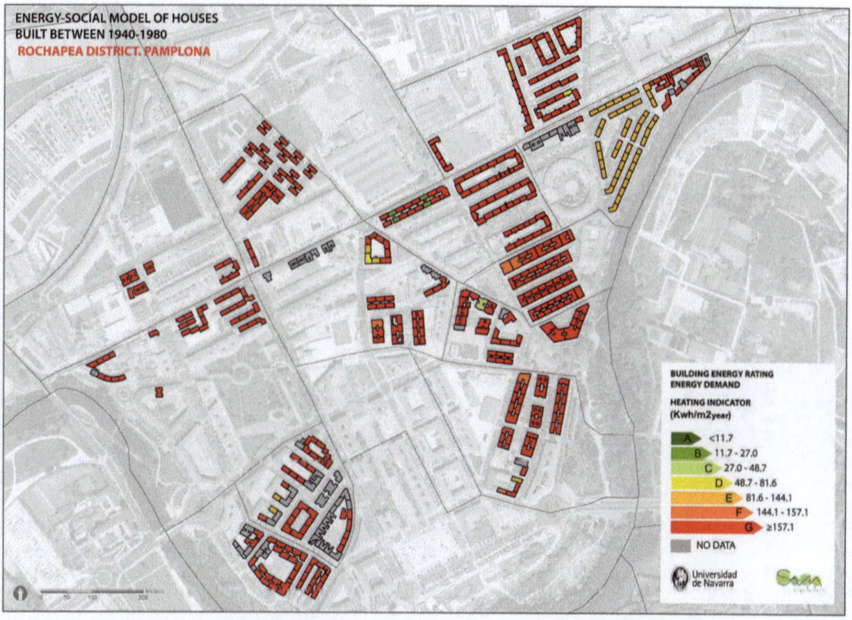

Fig. 6.11 Energy rating of buildings in original state. Demand. Rochapea

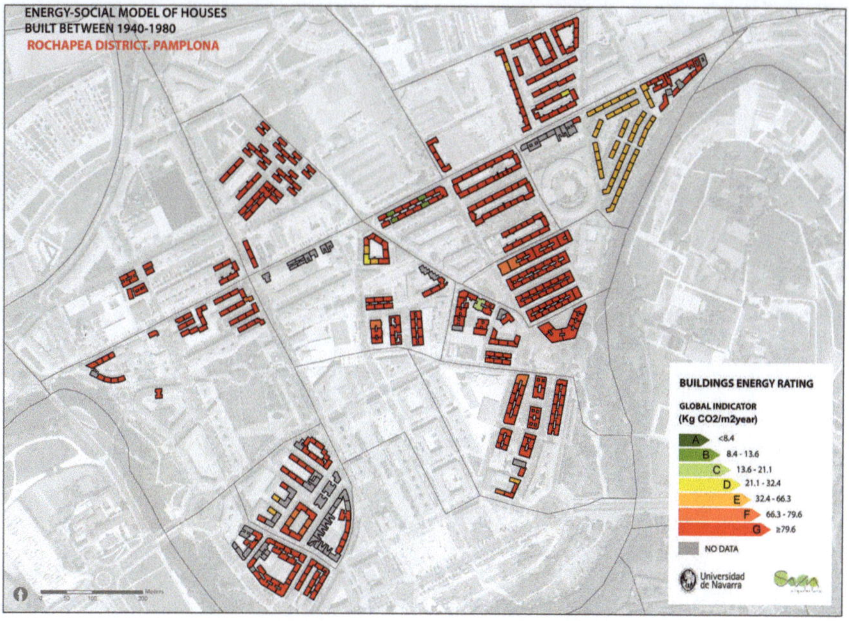

Fig. 6.12 Energy rating of buildings in original state. CO_2 emissions. Rochapea

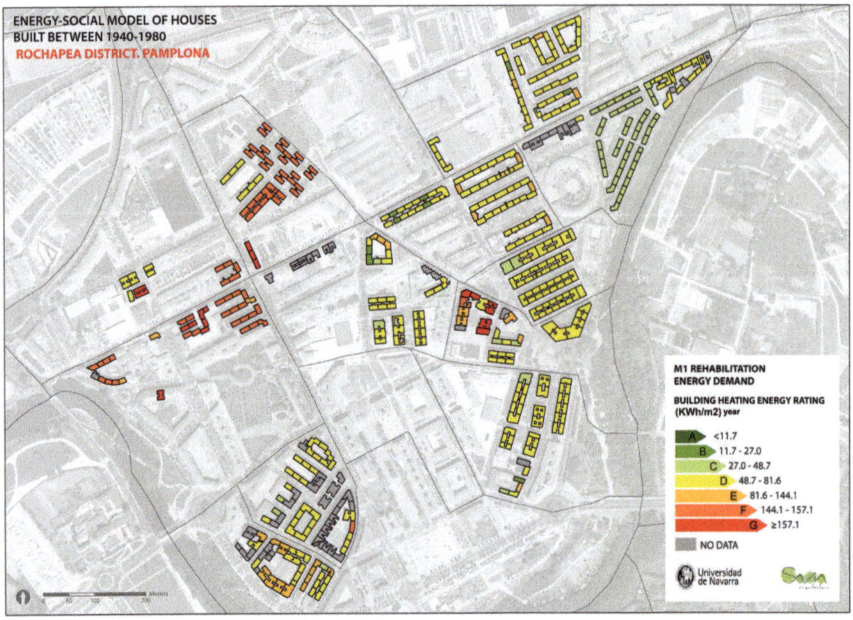

Fig. 6.13 Energy rating of buildings with M1 retrofitting. Demand. Rochapea

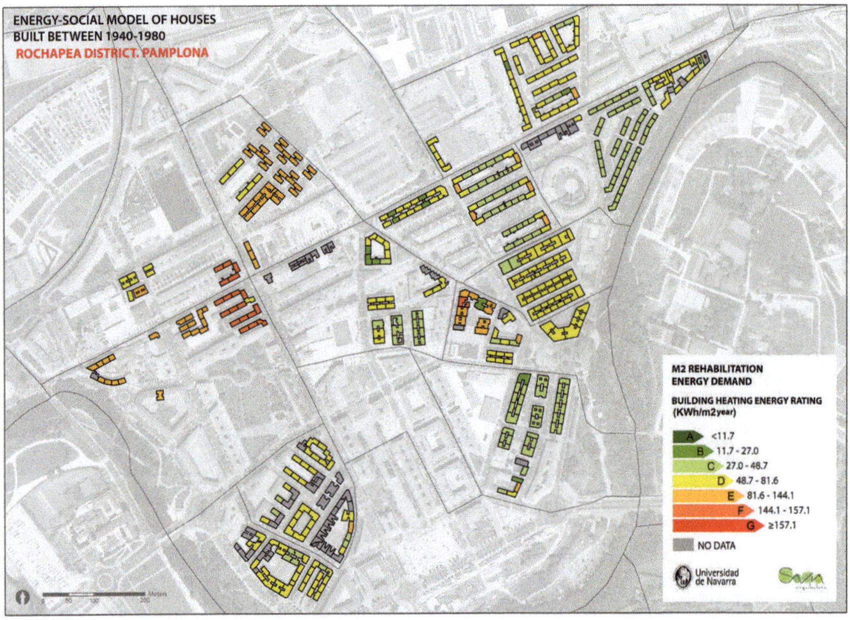

Fig. 6.14 Energy rating of buildings with M2 retrofitting. Demand. Rochapea

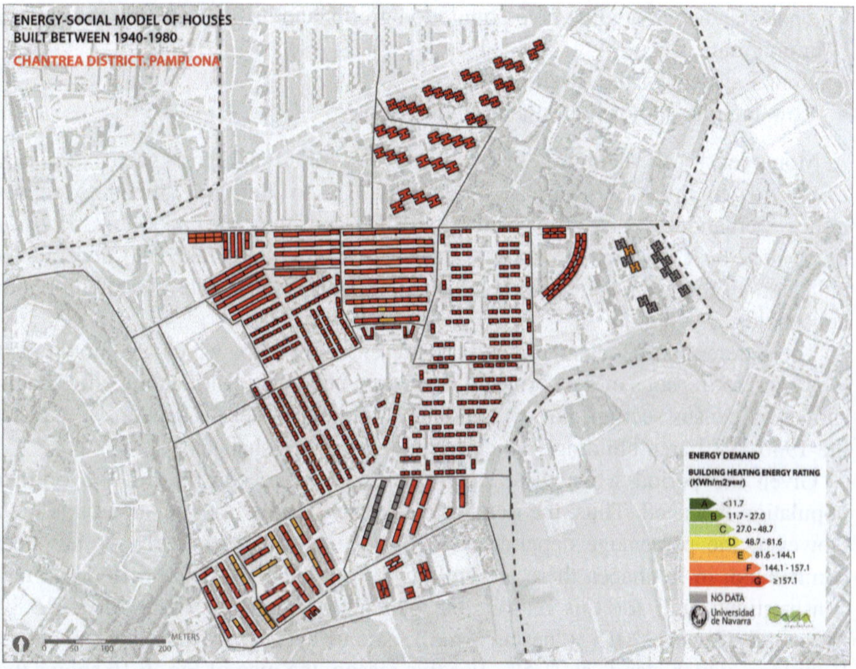

Fig. 6.15 Energy rating of buildings in original state. Demand. Chantrea

Thus, with retrofitting the potential for improvement is high, not only at the level of retrofitting of the thermal envelopes but also of the energy systems and the use of renewable energies, which would permit the attainment of higher rating levels.

The San Pedro group of houses, in the eastern section, stands out with a current rating which is better than the remainder (E) as it was renovated in 1985. However, although the first regulations on thermal conditions of buildings already existed, it was only applicable to new constructions, so only 2 cm of thermal insulation was placed on the façades. The group still has ample potential for renovation.

6.4.7 Consideration of Socio-economic Aspects

Surveys were carried out to gather the socio-economic data on the residents of the selected dwellings, and analysis was also carried out for the data from the Municipal Electoral Register, the *National Statistics Institute (Instituto Nacional de Estadística)* Housing and Population Census, together with those on employment and unemployment.

Some of the social economic variables detected as having more influence on the impossibility of tackling retrofitting are

- The existence of a vulnerable population group, and its high density: the elderly, immigrants;
- Rental system of the dwelling; and
- Low income and/or high unemployment rate.

Moreover, for the definition of energy poverty, different reports such as *Pobreza Energética* in Spain, *Análisis de tendencias* (Tirado Herrero, Jiménez Meneses, López Fernandez, & Martín García, 2014), and the *Informe sobre Pobreza Energética en Guipúzkoa* (Vasco & de Álava, 2013), also take into account the indicators of the mean education level, single-member households and single-parent households.

The different variables mentioned are presented on maps which are shown next. However, for reasons of data protection to avoid personal identification, the data are shown by census section, not by building. Within each census section we can see the 1940-1980 area buildings which are more brightly shaded.

Given the period when these neighbourhoods were constructed, the original population has aged. Thus, the **density of elderly people** in these suburbs is high. However, the percentage depends on the structure of each suburb. As has been commented, in Rochapea there are buildings from the 1940–1980 period and new constructions and, for this reason, the percentage of people over 65 is 15.8% following Census 2015 ("nastat," n.d.), somewhat lower than the mean for Pamplona, which stands at 21.5%. For this reason, the map of Fig. 6.16 shows that when the density of these buildings is higher, there is a greater number of elderly people in the area. The sections with a low percentage of elderly people coincide with a low density of buildings from the period.

In contrast, Chantrea, a suburb that was established mainly in the 1940–1980 period, has a higher number of older buildings and in addition the percentage of elderly people is 24.4%. Figures 6.16 and 6.17 show the percentage of people over the age of 65 in the different census sections of the suburbs of Rochapea and Chantrea, respectively.

Additionally, for these people, the absence of a lift is also a critical factor. On this issue, we have found examples of elderly people with reduced mobility who have had to move house due to the lack of accessibility, which makes them feel uprooted on leaving the area in which they have always lived. Others choose to stay in their home, due to a lack of resources, and live as 'prisoners in their own home', unable to access their urban environment or their neighbourhood. For these people, a priority would be to invest in a lift rather than retrofitting the thermal envelope.

Moreover, the homes they left behind are occupied by the new **immigrant population**, which results in a process of social renovation in these neighbourhoods. This means that, little by little, the resident group is changing (Fig. 6.18). These immigrant families occupy the buildings that are in the worst of conditions with the additional problem of overcrowding, as several families live together in a single dwelling, each of them in a single room. Depending on the areas, the level of integration and neighbourhood coexistence is good, although friction or tension

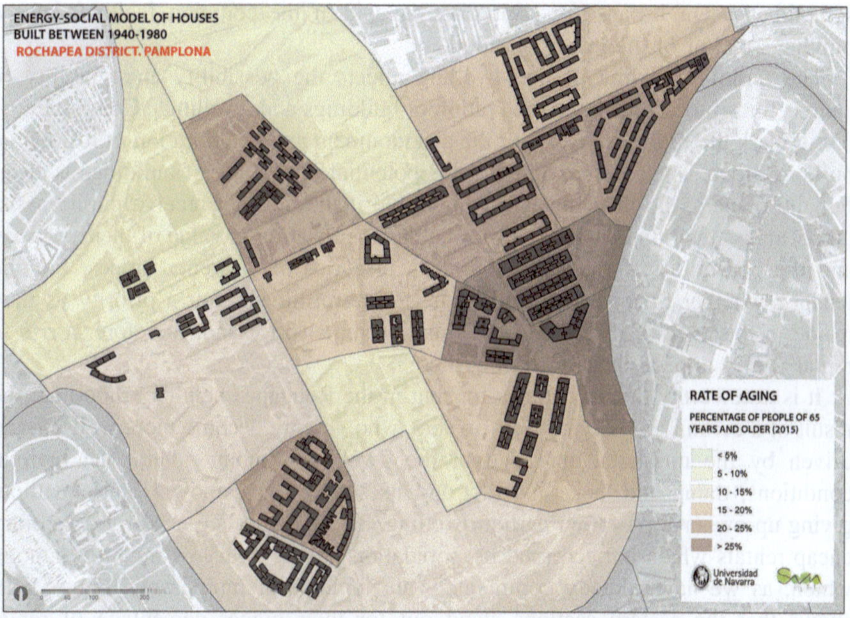

Fig. 6.16 Map showing percentage old people over the age of 65. Rochapea

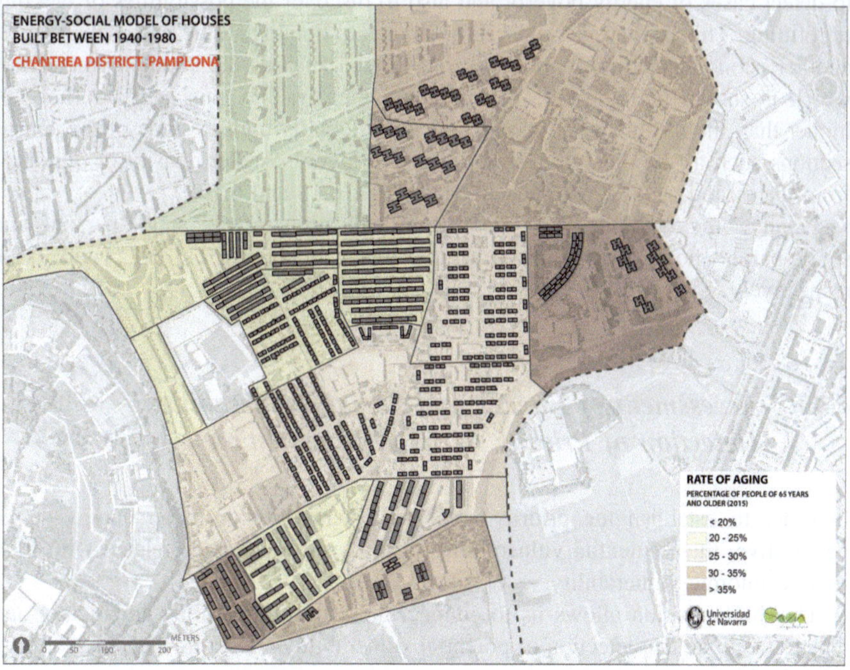

Fig. 6.17 Map showing percentage old people over the age of 65. Chantrea

sometimes occur because of their different use of the common facilities of the buildings and/or of the neighbourhood.

The **tenancy regime** of the dwelling affects the possibility and viability of actions for refurbishment and renovation of buildings and dwellings. Consequently, it will mark the future recovery, or the abandonment and deterioration of residential building stock. The repercussions for the potential retrofitting of buildings is clear. For the owners, the investment is impossible if they do not receive public assistance, unless they find a compensation in the appreciation in value of their property and the possibility of reflecting the costs in a rise in the rental prices. For the tenants, whether Spanish or immigrants, the retrofitting is not a priority as they reside in this type of dwelling due to its low rental price and, therefore, a rise in rental would impede their remaining there.

It is noteworthy that the homes for rent in the Rochapea neighbourhood are the result of a certain mobility of their owners, who, having become more well-off and driven by the unsatisfactory state of the dwelling (among others, the thermal conditions, damp and lack of a lift) and its setting, have moved house without giving up ownership of their earlier dwelling. These homes are used for temporary cheap rentals which are accessed by population groups with low incomes, among which, as we have already commented, are families of immigrants. Figure 6.19 shows that the eastern sections stand out for their higher percentage of rental properties.

Being aware of the real **income** of the residents of these buildings is fundamental to detect cases of energy poverty and also to discover the possibilities of tackling retrofitting. This is, then, the most difficult data to obtain and the most sensitive, so another economic indicator may be used instead such as the unemployment rate, which is shown in Fig. 6.20. The unemployment rate in this neighbourhood is higher than the mean for Pamplona, and is particularly high in some of the census sections such as the most easterly ones where there is higher density of buildings from the period, as also occurred in Map (Figs. 6.18 and 6.19).

Finally, the variables of the mean education level, single-person and single-parent homes are shown in the maps in Figs. 6.21, 6.22 and 6.23, respectively.

6.4.8 Assessment of Energy and Social Vulnerability: Detection of Priority Areas of Intervention

In order to establish the priority areas for intervention from an energy–social perspective, the residential vulnerability of the residents was correlated with their socio-economic vulnerability.

This juxtaposition allows us to connect these two dimensions graphically and thus to facilitate the detection of potential vulnerability situations where retrofitting is of the highest priority.

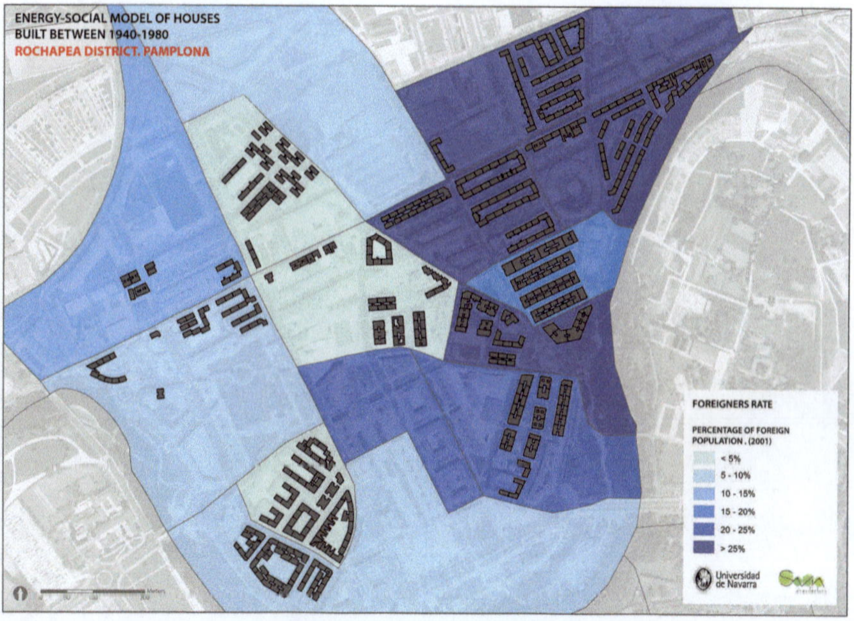

Fig. 6.18 Map showing percentage of foreigners. Rochapea

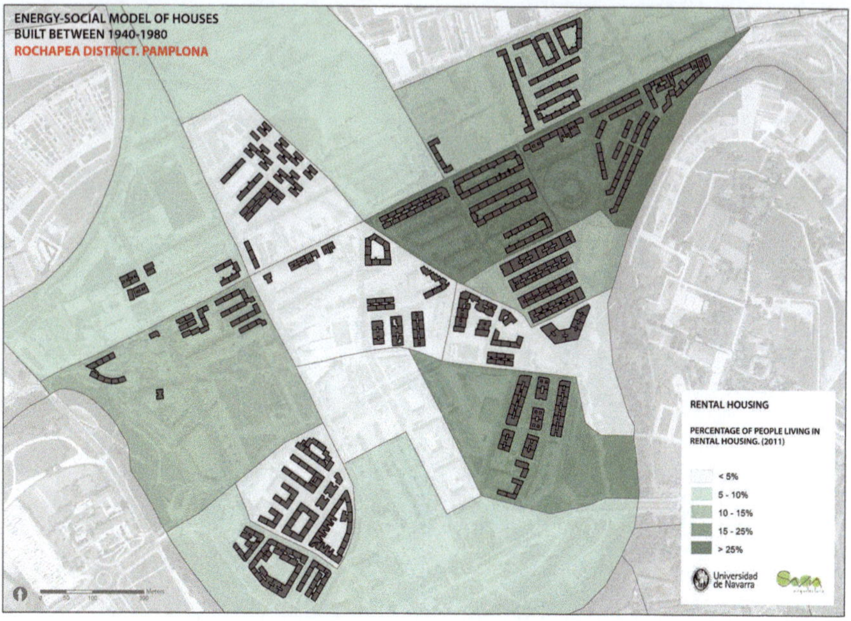

Fig. 6.19 Map percentage rental homes section. Rochapea

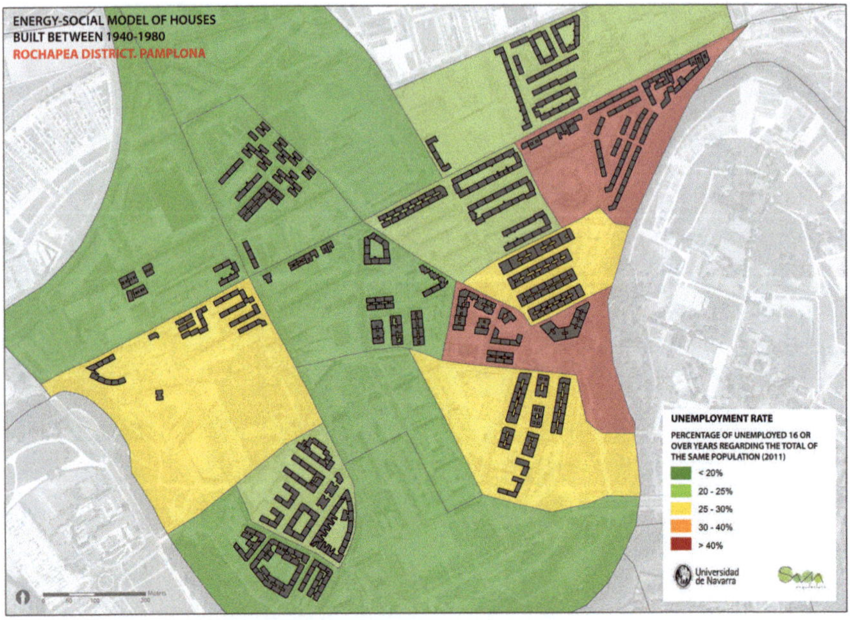

Fig. 6.20 Map of unemployment rate. Rochapea

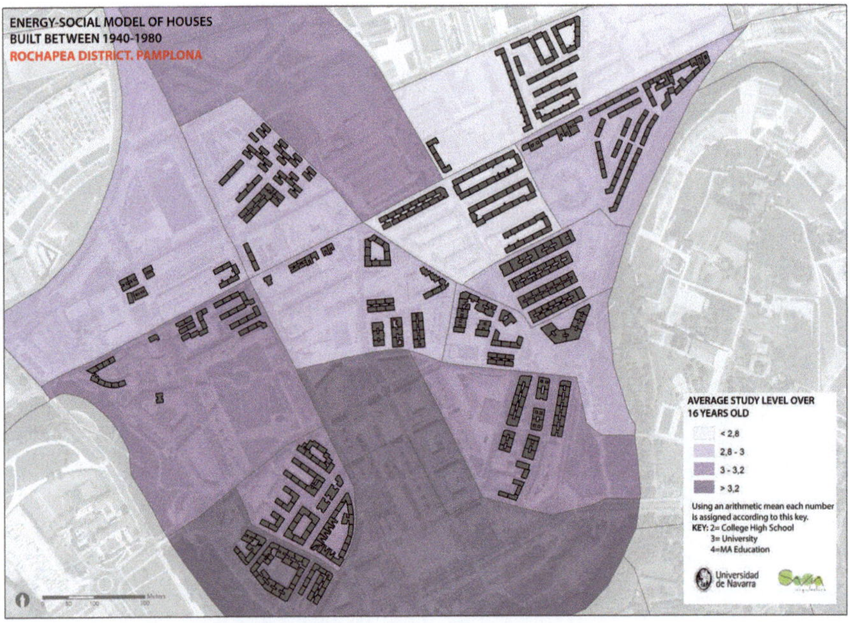

Fig. 6.21 Map of the mean education level. Rochapea

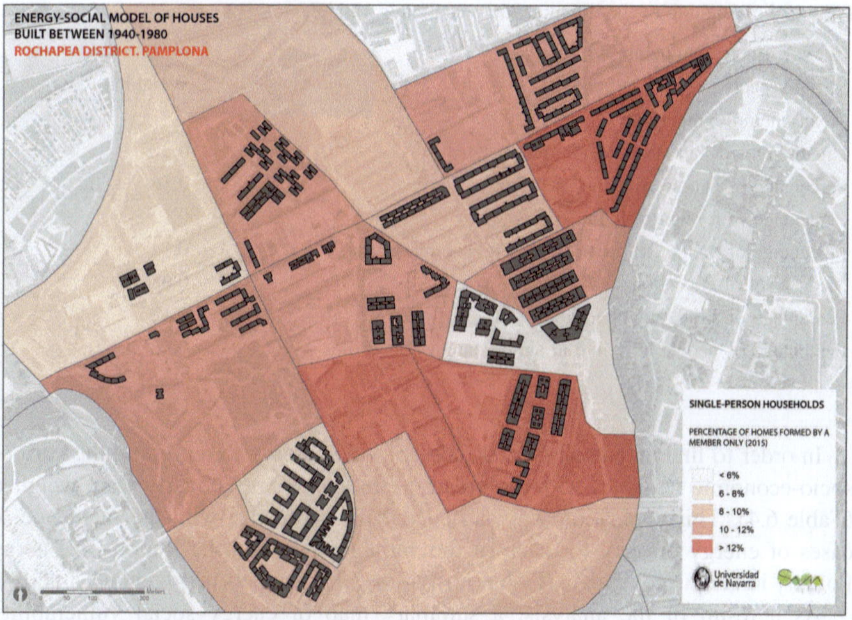

Fig. 6.22 Map single-person homes. Rochapea

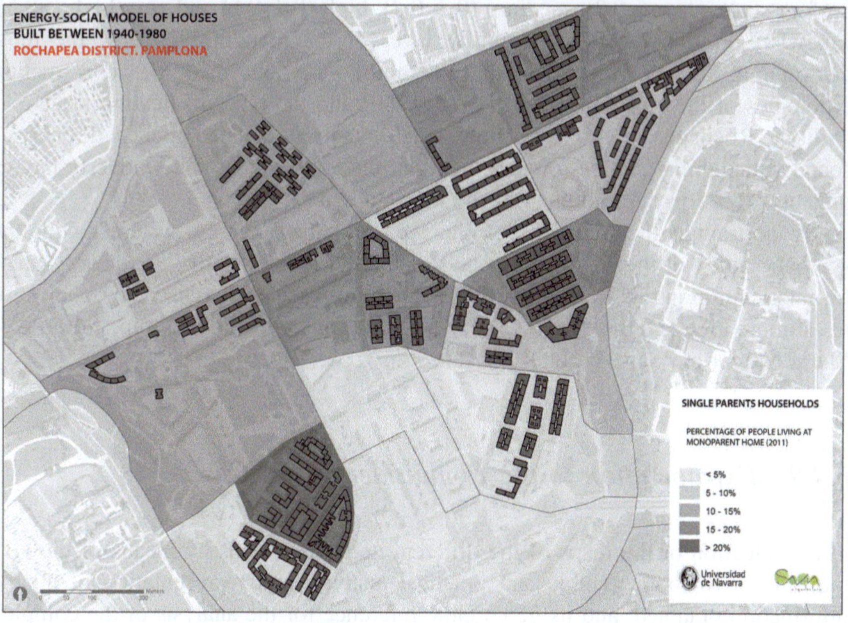

Fig. 6.23 Map single-parent households. Rochapea

Table 6.4 Weighing of the different socio-economic variables

Chart 1. Energy vulnerability	
Social aspects	Weighing
Mean unemployment rate	
Unemployment rate 1940–1980 area	0.22
Percentage of people over the age of 65	0.16
Percentage of foreign residents	0.12
Percentage of people who live in rented accommodation	0.14
Mean level of education	0.08
Percentage of single-member households	0.12
Percentage of people who live in single-parent households	0.16

In order to find an energy–social index of vulnerability to represent, the various socio-economic variables with the greatest effect on energy poverty were weighed (Table 6.4). This evaluation was carried out in accordance with the incidence of cases of energy poverty for each of the variables given in the Report on Energy Poverty in Guipuzcoa, Spain (Gobierno Vasco, 2013), which are given in Table 6.4.

As a result of the analysis, a summary map of energy–social vulnerability (Fig. 6.24) was obtained, and corresponds to an index which carefully gathers the weight of the variables shown in the seven maps (Figs. 6.16, 6.18, 6.19, 6.20, 6.21, 6.22 and 6.23), and, following Table 6.4, indicates the level of vulnerability accumulated by a census section. This degree of vulnerability has been categorized on four scales, low (<0.3), medium (0.3–0.4), high (0.4–0.5) and very high (>0.5). Likewise, on this social vulnerability layer of the map in Fig. 6.24, the most determining factors of residential vulnerability are presented: the construction typologies of the area considered (1940–1980) and the types of heating system.

In the Rochapea suburb as a whole, small localized, isolated pockets of energy–social vulnerability can be found, given that the indexes that affect poverty have a greater impact on the buildings of the area studied.

By portraying the buildings of the area and their energy systems, with the synthetic index of the socio-economic variables, we can identify the priority areas for intervention in this Rochapea suburb that requires special assistance policies for retrofitting; these are the Grupo Oscoz, the Salvador Cooperative group, Grupo Ave María and Grupo San Pedro.

6.4.9 *Extrapolation to the Remainder of the City*

The production of the maps of the different neighbourhoods by means of the GIS tool allows for both graphic and practical analysis and representation of all the information obtained, and its geographic reference for the analysis of the complex problems which the retrofitting faces. In this way, we can carry out a technical

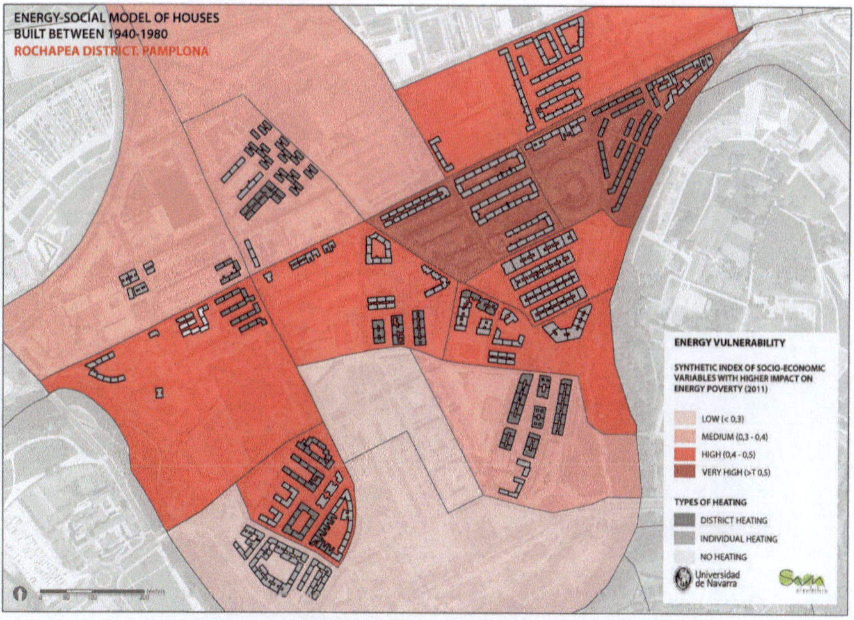

Fig. 6.24 Map of energy and social vulnerability. Rochapea

diagnosis of the situation of the buildings and their shortcomings, propose efficient energy retrofitting solutions and detect cases of energy poverty, all this with the intention of reducing energy and CO_2 emission consumption, besides gaining awareness of where investment of public and social assistance would be more effective.

Thus, Fig. 6.25 shows the different typologies of the buildings constructed between 1940 and 1980 in Pamplona, both in the suburban and urban areas. This identification by typology gives an insight into the economic cost involved in the retrofitting of all the buildings constructed during this period (Table 6.5). The economic assessment has been carried out with the M1 retrofitting scenario in mind, as it was considered the best (see Sect. 6.4.6).

Furthermore, in Pamplona as a whole, areas of energy–social vulnerability have been detected not only in the cases mentioned in Sect. 6.4.8, but also other suburbs such as Santa Engracia, other groups of buildings such as Urdánoz in Echavacoiz and the old Soto Lezkairu group, which demand special attention.

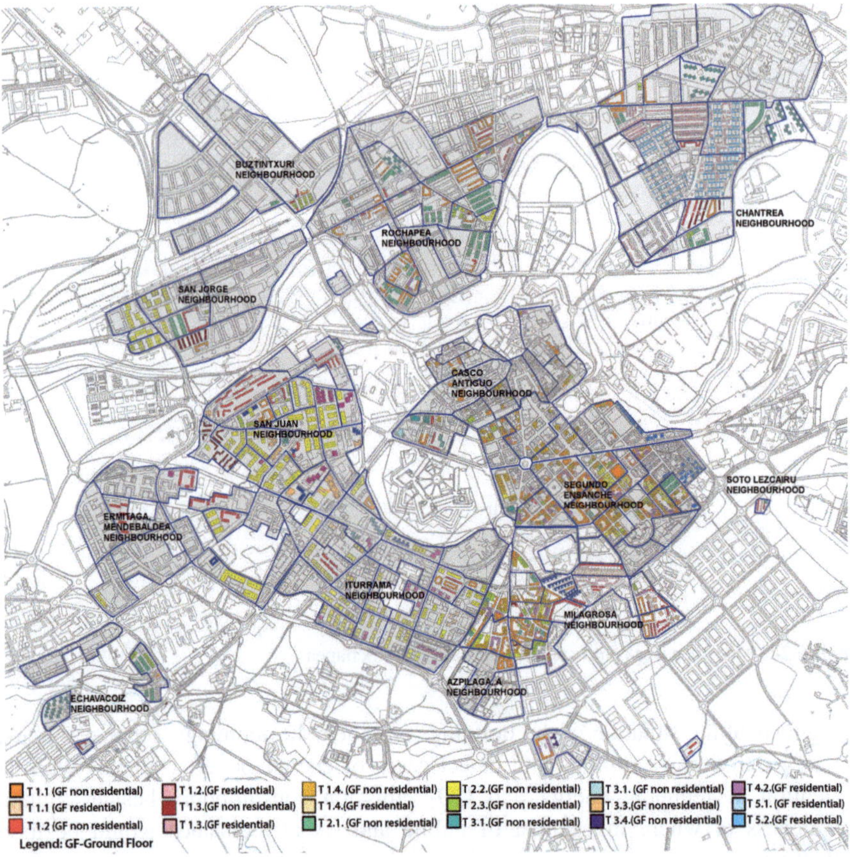

Fig. 6.25 Map of Pamplona with the building typologies constructed between 1940 and 1980

6.4.10 Case Study Conclusions

In the case of the Rochapea neighbourhood, the renovation of the 1940–1980 buildings could bring about, for the whole of the area, greater parity of conditions between the old and the new buildings, thereby achieving the wished-for integration of both. In this way, we could avoid situations of isolation of the older buildings and the exclusion and marginalization of their vulnerable residents.

The density of vulnerable buildings and of a population with a low socio-economic level in certain census sections, as can be seen on the map in Fig. 6.25, stresses the need for priority action with public assistance and the search for solutions in which private initiative may be involved.

Table 6.5 Cost of the retrofitting of buildings constructed in Pamplona between 1940 and 1980

Typology	Dwellings in social neighbourhood 1940–1980		Total dwellings 1940–1980		Average area dwellings (M2)	Cost/ Dwelling (€/Viv.)	Total cost dwellings in social neighbourhood 1940–1980 (€)	Total cost dwellings 1940–1980 (€)
T1	9,855	46%	17,143	38%	73	16.323	160,863,165	279,825,189
T2	6,292	29%	17,284	39%	69	18.329	115,326,068	316,798,436
T3	3,076	14%	6,902	15%	71	17.728	54,531,328	122,358,656
T4	1,075	5%	2,233	5%	85	21.223	22,814,725	47,390,959
T5	1,033	5%	1,072	2%	131	45.583	47,087,239	48,864,976
Total	21,331	100%	44,634	100%			400,622,525	815,238,216

6.5 Discussion and Conclusions

A diagnosis of the current energy characteristics of the building stock of cities is required in order to establish a starting point from which stakeholders and policymakers can promote sustainable development and energy efficiency in the cities. By improving GIS tools, retrofit strategies and measures can be developed, in order to reach the objectives of energy efficiency, reduction of energy consumption and emissions. GIS tools represent a robust and particularly well-suited option for enhancing the effectiveness of their decisions and policies.

By means of graphics, the GIS can show what the building and construction typologies are on the neighbourhood or city scale and what their energy ratings are; it can define those buildings and areas with greater potential for retrofitting in order to reduce energy consumption and CO_2 emissions, and can detect cases of socio-economic vulnerability in the population. In this way, areas for intervention may be established and prioritized.

Public assistance for works on accessibility, the thermal envelopes and the energy efficiency of the buildings can be seen as the foundation for the renovation of social housing suburban neighbourhoods, on a higher scale. It can also be seen as a preventative measure to reduce the effect of the urban and social deterioration of same. There are synergies between the retrofitting of buildings and the improvement of the poor level of services and open spaces; these would increase the socio-economic level of the neighbourhoods and slow down their physical and functional decline.

Simultaneously, the synergies must be used to attract private investment to intervene in these buildings. For example, through actions such as increases in buildability (rooftop or upward extensions) for the creation of new dwellings in the buildings in return for their retrofitting, or even the demolition of buildings in very bad conditions and new construction with greater densification. In this way, relying on public and private initiatives, there will be advances in the adaption to climate change of existing buildings and in the urban and social regeneration of the different city neighbourhoods.

Acknowledgements We would like to thank the Fundación Caja Navarra for founding the Project 'Energy-Social Model of Houses Built between 1940 and 1980 in Rochapea and Chantrea Neighbourhoods, Pamplona'. We wish to give special thanks to Juan José Pons-Izquierdo, Ana Castillejo and Jorge San Miguel for their collaboration in the GIS programme and collecting data. Finally, we appreciate the assistance of Cristina Güell in this chapter.

References

Alhamwi, A., Medjroubi, W., Vogt, T., & Agert, C. (2017). GIS-based urban energy systems models and tools: Introducing a model for the optimisation of flexibilisation technologies in urban areas. *Applied Energy, 191,* 1–9. https://doi.org/10.1016/j.apenergy.2017.01.048.

Braulio-Gonzalo, M., Bovea, M. D., & Ruá, M. J. (2015). Sustainability on the urban scale: Proposal of a structure of indicators for the Spanish context. *Environmental Impact Assessment Review, 53,* 16–30. https://doi.org/10.1016/j.eiar.2015.03.002.

Caryl-Sue. (2017). GIS (geographic information system)—National Geographic Society. Retrieved November 28, 2017, from https://www.nationalgeographic.org/encyclopedia/geographic-information-system-gis/.

Commission of the European Communities. (2008). Package of implementation measures for the EU's objective on climate change and renewable energy for 2020, 27. http://doi.org/10.1007/s13398-014-0173-7.2.

CTE-HE. (2013). Documento Básico CTE-HE Ahorro de energía. *Septiembre.*

Cuchí, A. (2013). Mesa Redonda sobre Rehabilitación y Empleo, organizada por PU Europe e IPUR - YouTube. Retrieved October 10, 2017, from https://www.youtube.com/watch?list=UUVJI0TzX1BlJsL7hduc_7RQ&time_continue=3&v=uWIf1Fk9-j8.

Cuchí, A., & Sweatman, P. (2014). Informe GTR 2014. Estrategia para la rehabilitación. Claves para transformar el sector de la edificación en España, 84.

Entranze. (n.d.). ENTRANZE: *Welcome to ENTRANZE project page*. Retrieved October 10, 2017, from http://www.entranze.eu/.

EPDB. (2010). Directive 2010/31/EU of the European Parliament and of the Council of 19 May 2010 on the energy performance of buildings (recast). *Official Journal of the European Union,* 13–35. https://doi.org/doi:10.3000/17252555.L_2010.153.eng.

European Commission. (2013). Green paper—A 2030 framework for climate and energy policies. *COM(2013) 169 Final,* 1–16. Retrieved from http://ec.europa.eu/clima/policies/strategies/2030/documentation_en.htm.

European Commission. (2014). Comunicación de la comisión al Parlamento Europeo, al consejo, al Comité Económico y social europeo y al Comité de las regiones, pp. 1–21.

European Commission. (2016). *Acuerdos internacionales sobre actuación en materia climática.* Retrieved October 10, 2017, from http://www.consilium.europa.eu/es/policies/climate-change/international-agreements-climate-action/.

Fabbri, K., Zuppiroli, M., & Ambrogio, K. (2012). Heritage buildings and energy performance: Mapping with GIS tools. *Energy and Buildings, 48,* 137–145. https://doi.org/10.1016/j.enbuild.2012.01.018.

Gobierno de Navarra. (n.d.). SITNA: Sistema de Información Territorial de Navarra. Retrieved October 10, 2017, from http://sitna.navarra.es/geoportal/?lang=.

Gobierno Vasco, D.F. de G. (2013). Pobreza energética.

Gupta, R., & Gregg, M. (2013). Preventing the overheating of English suburban homes in a warming climate. *Building Research & Information, 41*(3), 281–300. https://doi.org/10.1080/09613218.2013.772043.

INE. (2011). Demografía y población. Cifras de población y censos demográficos. Censos de Población y Viviendas 2011. Retrieved October 10, 2017, from http://www.ine.es/censos2011_datos/cen11_datos_inicio.htm.

Inter-American Development Bank. (2010). Inter-American Development Bank, (March). Retrieved from http://www.iadb.org/en/inter-american-development-bank,2837.html.

Johansson, T., Vesterlund, M., Olofsson, T., & Dahl, J. (2016). Energy performance certificates and 3-dimensional city models as a means to reach national targets—A case study of the city of Kiruna. *Energy Conversion and Management, 116,* 42–57. https://doi.org/10.1016/j.enconman.2016.02.057.

Mavrogianni, A., Davies, M., Chalabi, Z., Wilkinson, P., Kolokotroni, M., & Milner, J. (2009). Space heating demand and heatwave vulnerability: London domestic stock. *Building Research & Information, 37*(5–6), 583–597. https://doi.org/10.1080/09613210903162597.

Mentis, D., Andersson, M., Howells, M., Rogner, H., Siyal, S., Broad, O., … Bazilian, M. (2016). The benefits of geospatial planning in energy access—A case study on Ethiopia. *Applied Geography, 72*, 1–13. http://doi.org/10.1016/j.apgeog.2016.04.009.

Min, J., Hausfather, Z., & Lin, Q. F. (2010). A high-resolution statistical model of residential energy end use characteristics for the United States. *Journal of Industrial Ecology, 14*(5), 791–807. https://doi.org/10.1111/j.1530-9290.2010.00279.x.

Nouvel, R., Mastrucci, A., Leopold, U., Baume, O., Coors, V., & Eicker, U. (2015). Combining GIS-based statistical and engineering urban heat consumption models: Towards a new framework for multi-scale policy support. *Energy and Buildings, 107*, 204–212. https://doi.org/10.1016/j.enbuild.2015.08.021.

nastat. (n.d.). Retrieved from http://www.navarra.es/home_es/Gobierno+de+Navarra/Organigrama/Los+departamentos/Economia+y+Hacienda/Organigrama/Estructura+Organica/Instituto+Estadistica.

Rubio del Val, J. (2011). Rehabilitación Urbana en España (1989–2010). Barreras actuales y sugerencias para su eliminación. *Informes de La Construcción, 63*(Extra), 5–20. http://doi.org/10.3989/ic.11.060.

Santamouris, M., Kapsis, K., Korres, D., Livada, I., Pavlou, C., & Assimakopoulos, M. N. (2007). On the relation between the energy and social characteristics of the residential sector. *Energy and Buildings, 39*(8), 893–905. https://doi.org/10.1016/j.enbuild.2006.11.001.

Soriano, M. (n.d.). Observatorio de la Sostenibilidad - X UN FUTURO + SOSTENIBLE. Retrieved October 10, 2017, from http://www.observatoriosostenibilidad.com/.

Swan, L. G., & Ugursal, V. I. (2009). Modeling of end-use energy consumption in the residential sector: A review of modeling techniques. *Renewable and Sustainable Energy Reviews, 13*(8), 1819–1835. https://doi.org/10.1016/j.rser.2008.09.033.

Theodoridou, I., Karteris, M., Mallinis, G., Papadopoulos, A. M., & Hegger, M. (2012). Assessment of retrofitting measures and solar systems' potential in urban areas using geographical information systems: application to a Mediterranean city. *Renewable and Sustainable Energy Reviews, 16*(8), 6239–6261. https://doi.org/10.1016/j.rser.2012.03.075.

Tirado Herrero, S., Jiménez Meneses, L., López Fernandez, J. L., & Martín García, J. (2014). Pobreza Energética en España. Informe de tendencias. *Asociacion de Ciencias Ambiantales, 1*, 171.

UNFCCC. (2015). Paris Climate Change Conference-November 2015, COP 21. *Adoption of the Paris Agreement. Proposal by the President., 21932*(December), 32. http://doi.org/FCCC/CP/2015/L.9/Rev.1.

Veres, E. J. (1999). Ordenación de secciones censales según un indicador de pobreza. *Estadística Española, 41*(144), 169–201. https://doi.org/10.1016/j.proenv.2010.10.187.

Chapter 7
Facing Heatwaves and Warming Conditions in the Mediterranean Region

7.1 Heatwaves Are Already Affecting Our Buildings and Households

As it has been widely exposed, heatwaves are already affecting the dwellings of the Mediterranean locations (Chap. 2), having a special effect on human health, increasing morbidity and mortality (Chap. 3), although it differs according to the severity of climate conditions and the adaptive ability of population. Residential buildings are the main place where people face these extreme weather events.

7.1.1 Building-Related Risk Factors that Contribute to Overheating During a Heatwave

It is well documented that the effects of heatwaves on health take place mainly at home (Cadot, Rodwin, & Spira, 2007), and diverse studies on previous heatwaves, as, e.g. Vandentorren et al. research (Vandentorren et al., 2006) have found the following risk factors related to the building:

- Construction before the first Building Codes on energy efficiency, which mainly supposes a lack of insulation in façades and roofs.
- Dwellings in top floor under the roof, mainly if a bedroom is located there. Meanwhile, the gradient of risk according to the floor (in the rest of floors different from upper floor) was no statistically significant.
- Few numbers of rooms.
- Lack of *Cool Retreats*.
- High percentage of windows related to the floor area, and their orientation.
- Natural ventilation during the afternoon, instead of during other cooler hours.

© Springer International Publishing AG 2018
A. Monge-Barrio and A. Sánchez-Ostiz Gutiérrez, *Passive Energy Strategies for Mediterranean Residential Buildings*, Green Energy and Technology,
https://doi.org/10.1007/978-3-319-69883-0_7

- Difficulties for natural ventilation in city centres due to high levels of noise, concerns about security, high index of pollution (Díaz et al., 2002) and high minimum night temperatures due to *urban heat island effect* (UHI) (Laaidi et al., 2012; Santamouris, Sfakianaki, & Pavlou, 2010).

7.1.2 Monitored Experiences During Heatwaves in Spain

This section shows different case studies in residential buildings, with monitorings that have been carried out in summer conditions during a heatwave or in very hot periods for the location, in Pamplona, Madrid and Alicante (Spain).

The criterion chosen by AEMET (State Meteorological Agency of Spain) to define heatwaves in Spain is an episode of at least three consecutive days where a minimum of 10% of the given stations register maximum temperatures over 95% percentile of its series of temperatures during July and August in the period 1971–2000. Cesar Rodriguez Ballesteros from AEMET has carried out an interesting report on heatwaves in Spain during 1975–2016, from that 1971–2000 climate series (AEMET, 2017), and the number of episodes and days is summarized in Fig. 7.1. More detailed information on heatwaves in Spain from 2011 to 2016, based on that report, is shown in Table 7.1, period during which the building monitoring illustrated in this chapter has been accomplished.

It is interesting to compare the different thresholds considered for a heatwave definition. As some examples, the thresholds vary from 34 °C in Alicante, 36 °C in Pamplona and 36.4 °C in Madrid (Retiro) to 41.2 °C in Seville.

From these data, the summers of 2016 and 1991 can be highlighted, as the ones that have registered the highest number of heatwaves (four in both cases), the 2015 heatwave for being the longest (26 days), and the 2003 one for being the second longest in the series and for claiming the highest associated mortality index throughout Europe (Fig. 1.2 in Chap. 1).

According to AEMET, in the summer of 2017, in the Spanish Peninsula and Baleares Islands, there have been at least three heatwaves: 14–18 June, 12–14 July and 3–5 August (AEMET, n.d.-a). This paragraph also shows monitoring of residential buildings done during the first heatwave.

On the other hand, Fig. 7.2 shows the maximum temperatures registered in Pamplona during the 2011, 2012 and 2016 heatwaves, during which building monitoring was carried out, shown in paragraphs 7.1.2.a and 7.1.2.b. In 2016, heatwaves did not affect Pamplona but there were days with very high temperatures for the location, as can be seen in Fig. 7.2a. According to the mentioned climatic series 1971–2000, the average maximum temperature in July in Pamplona is 27.6 and 27.8 °C in August.

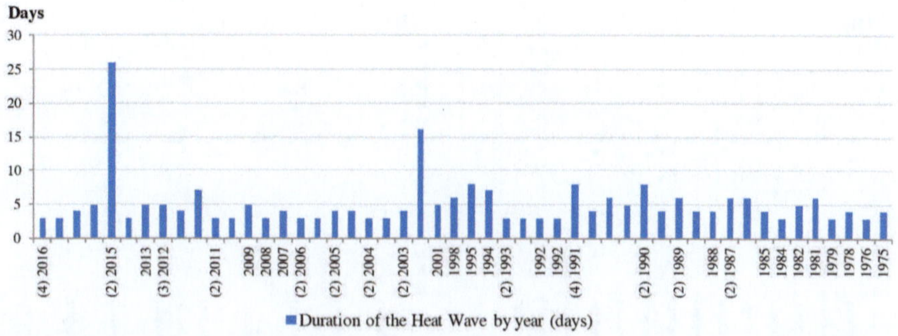

Fig. 7.1 Heatwaves in Spain from 1975 to 2016, including duration (days) and number of heatwaves per year, extracted from AEMET (2017)

Table 7.1 Heatwaves in Spain for the 2010–2016 period, extracted from AEMET (2017)

Year	Start	End	Length (days)	Heatwave anomaly (°C)	Heatwave max. temp. (°C)	Affected provinces
2016	17/07/2016	19/07/2016	3	3.5	37	20
	26/07/2016	28/07/2016	3	1.3	37	13
	22/08/2016	25/08/2016	4	1.8	35.9	11
	03/09/2016	07/09/2016	5	3.3	38.6	23
2015	27/06/2015	22/07/2015	26	3.4	37.6	30
	27/07/2015	19/07/2015	3	2.3	38.7	10
2013	05/07/2013	09/07/2013	5	2.4	37.7	13
2012	24/06/2012	28/06/2012	5	2.1	38.3	25
	08/08/2012	11/08/2012	4	3.7	39.5	40
	17/08/2012	23/08/2012	7	2.8	36.2	30
2011	25/06/2011	27/06/2011	3	1.6	37.8	15
	19/08/2011	21/08/2011	3	2.3	37.1	19

These monitored data aim to illustrate the problematic of heatwaves affecting dwellings and occupants. Case studies cover both single-family houses and multifamily buildings, counting with different social profiles, because although heatwaves are especially dangerous for vulnerable population, they affect the population as a whole. The monitored buildings are located in Pamplona (Spain) where residential buildings do not normally have air conditioning due to traditionally mild summers. In addition, monitored buildings located in Madrid and Alicante (Spain) are shown, where cooling systems have been progressively incorporated to the residential sector during these last years in order to deal with high summer temperatures (see Chap. 4)

Since the objective was to study as many dwellings as possible, both with a thermal retrofit envelope or without any improved measures, with different socio-economic

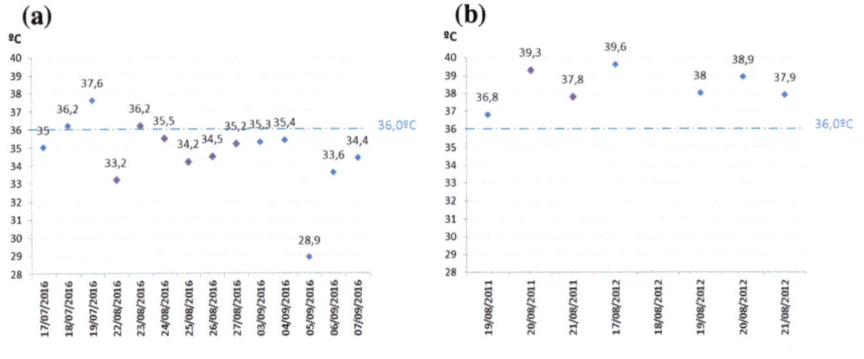

Fig. 7.2 Maximum registered temperatures during building monitoring in Pamplona shown in paragraphs 7.1.2.a y 7.1.2.b, that is, in 2016 (**a**), and in 2011 and 2012 (**b**). Daily data from Meteo Navarra weather stations (METEONAVARRA, n.d.)

and use profiles, the decision was to use *low-cost* equipment that would also disturb as minimum as possible the actions of the user. Since dwellings in different buildings were monitored at the same time, results were studied together with available online data from the nearest weather station, although generally, measurements of exterior conditions were taken too. Data registered in these monitoring campaigns were mainly air temperature and relative humidity, but also globe temperature and CO_2 ppm concentration.

Graphs with the monitoring data registered are shown with the upper limits of adaptive comfort, for the three categories of indoor environment according to the European standard (UNE-EN 15251, 2008), due to the intergenerational profile of occupants in residential buildings (as it is widely exposed in paragraph 3.4 in Chap. 3). Dwellings are codified by data protection and housing, and users' characteristics have been analysed in detail in order to draw conclusions from the buildings studied in real conditions of use.

The general codes used in the monitoring graphs are as follows: CE, with a refurbished envelope; SE, without a refurbished envelope; LR, living room; BR, bedroom; orientations between parenthesis (S, south; N, north; W, west; and E, east). The upper limit for temperature in naturally conditioned buildings in summer for the three categories of indoor thermal environment I, II and III described in the standard (UNE-EN 15251, 2008) is codified in graphs as UpT.C.I, UpT.C.II and UpT.C.III.

In some building's monitoring exposed in this section, a graph is attached with indoor temperature frequency for the given period, and another graph with discomfort hours according to the mentioned adaptive comfort approach UNE-EN 15251, for Categories I and II.

7.1.2.1 Monitoring Social Dwellings in Pamplona (Spain)

Pamplona is a city in the north of Spain that has a Cfb climate, warm oceanic climate, with an average yearly temperature of 12.9 °C and a mean monthly temperature of 21.4 °C (28.3 °C maximum and 14.5 °C minimum) in August, the warmest month, according to AEMET airport meteorological station values from climatic series 1981–2010 (AEMET, n.d.-b). Annual relative humidity is 67%, being in August 58%.

In prestaRener Project (described in Chap. 5), 103 dwellings in 21 case studies involving 33 buildings in social neighbourhoods built between 1940 and 1980 in Pamplona (Spain) were monitored. Summer monitoring of three case studies was carried out under conditions much warmer than usual for the location (see Fig. 7.2a): FERRO, GORRI and SOLE case studies (Fig. 7.3).

Each case study consisted in general in two similar buildings, with and without a refurbished thermal envelope (called CE and SE, respectively). From the monitored cases shown, FERRO and SOLE buildings just rehabilitated the envelope in 2016, and GORRI building was rehabilitated in 2013. SOLE and GORRI buildings belong to linear block typology (T1 Typology), with ground floor and fourth floors each block, GF + 4 (8 flats per block), and FERRO building belongs to H-shaped typology (T2 Typology)[1], with ground floor and four floors, GF + 4 (16 flats per block). In both SOLE and GORRI case studies, CE and SE buildings were monitored at the same time in order to be able to compare results easily.

Monitoring looked for residential buildings with their main façades looking south and with a double orientation, allowing crossed ventilation. It tried to cover representative floors (at least a first floor, an intermediate and a top floor), and inside the dwelling, two rooms with opposed orientations (mainly south and north, which generally correspond to the living room and a bedroom). The main characteristics of thermal envelopes in these case studies are summarized in Chap. 5.

FERRO Case Study is in *Santa Engracia* neighbourhood, an area considered vulnerable by Pamplona's Town Hall, both for being in the outskirts and for its occupiers' socio-economic profile. The envelope of FERRO (CE) has been recently rehabilitated thanks to specific funding from the town hall, which covered up to 80% of the investment, depending on the family's economic situation. The buildings are originally from 1956, and FERRO (CE) building, whose monitoring is shown here, was rehabilitated in 2016. The general aim was the integration of these buildings and their occupants with the new housing development surrounding them.

From surveying and monitoring of nine dwellings (from rehabilitated T2 Typology, with 16 dwellings per block), we extracted data on the family unit, income and unemployment rate, tenure status, and the cooling systems and their use. Family units are generally composed of two or three adults, in two of the dwellings

[1]T2 Typology (H-shaped block, with four dwellings per floor and with interior courtyard) is described in Chap. 5.

Fig. 7.3 Images of FERRO, GORRI and SOLE case studies, with and without thermal rehabilitated envelope

people had health problems (CE_6_4D[2] and CE_4_3B), and in one there was an aged occupant (CE_4_2C). Three of the dwellings are under unemployment situation or on basic social pay, and in four of the dwellings, income is between 500 and 1000€/month. All of them are under ownership and just two are paying mortgage.

Not a single dwelling has air conditioning and only two count with fans (both in the fourth floor). They generally use shading devices during the day and ventilate when it gets cooler, except CE_6_4D and CE_4_3B, who ventilate all day long, according to the survey.

In Fig. 7.4 where FERRO (CE) Case Study is shown, there were very high temperatures on 17–19 July 2016, exceeding Pamplona threshold of 36 °C on 18 and 19 July (threshold ticked in the figures for reference), reaching a maximum temperature of 35 °C on 17 July (see Fig. 7.2).

During the studied week, only top floor CE_6_4D exceeds, on 18 and 19 July, the temperature thresholds for all categories of indoor thermal environment, in the living room on north as in the bedroom on south (courtyard). The other dwellings on the top floor (CE_4_4A and CE_4_4D) are also the ones to overheat the most in relation with the rest, although they are inside acceptable temperature limits, as can be deduced from the graph.

Dwelling CE_6_4D reaches those temperatures because of its height and its incorrect all-day ventilation, being also one of the most vulnerable ones, according to the survey. It is interesting to note how dwelling CE_4_4D ventilates at 21 h on the 17th and 18th when temperatures have not gone down yet, therefore overheating the dwelling, but waiting longer on the 19th until the temperature outside has gone down.

In the monitoring of other case studies, it has also been observed that when summer conditions are more severe than normal, and there is no air conditioning in the dwelling, there are users that do not know what to do to maintain adequate temperatures and carry out actions that worsen indoor conditions, probably following their daily routines for typical summer conditions. In this case, the user keeps ventilation as every day, without being conscious that external temperatures are higher than indoor ones, and therefore, this measure is not only ineffective but also counterproductive, considerably increasing overheating in indoor spaces.

[2]The last number of the identification code of each dwelling is the number of the floor. So in this case, this dwelling is situated in the fourth and upper floor.

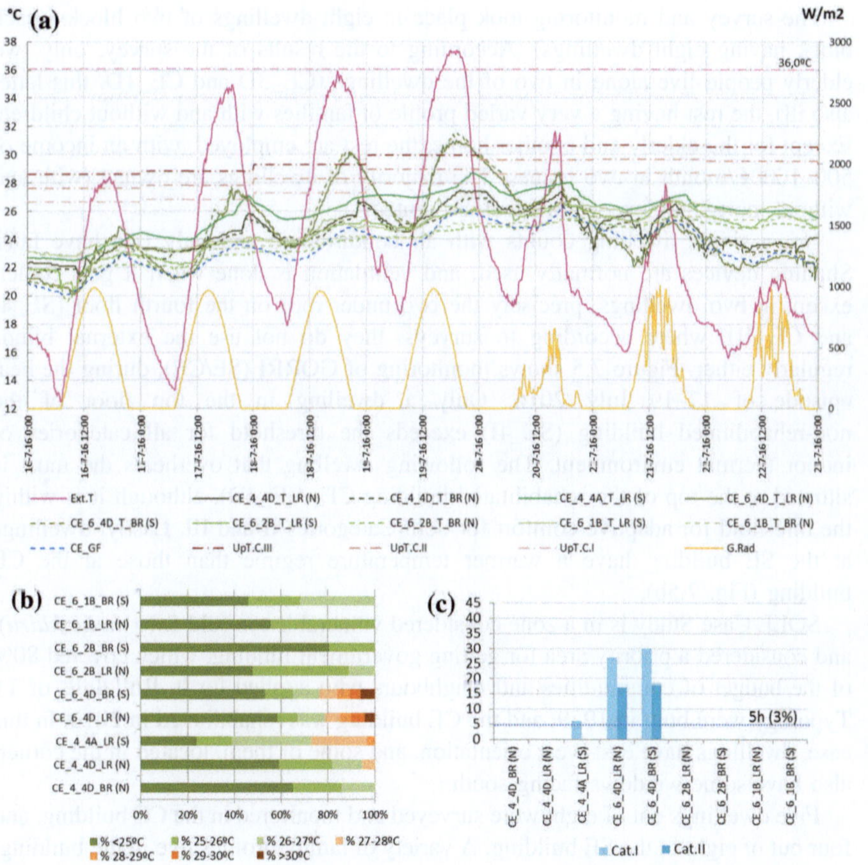

Fig. 7.4 Monitored building FERRO (CE) in Pamplona (Spain), from 16 to 23 July 2016. Graphs of registered temperatures (**a**); temperature frequency (**b**) and discomfort hours according to UNE-EN 15251 for buildings without mechanic cooling systems (**c**)

Lastly, it must be highlighted that dwellings could further take advantage of the strategy of night ventilation to cool their rooms, since although at night indoor temperatures decrease, they could decrease further, if we look at minimum exterior temperatures (between 4 and 6 °C, Fig. 7.4). During a heatwave, it is important to cool the dwelling during the night, so not to start the next morning with temperatures higher than the day before. This is especially relevant if heatwaves have a lasting duration, as is previewed for this climatic zone.

GORRI Case Study is inside Pamplona's *II Ensanche* neighbourhood, one of the most valued and consolidated zones in the city. Two very similar blocks of T1 Typology buildings (linear block) have been monitored, one rehabilitated (CE) and the other one not rehabilitated (SE). The buildings were originally from 1943, and the rehabilitation was undertaken in 2013.

The survey and monitoring took place in eight dwellings of two blocks (each block having eight dwellings). According to the results of the survey, only two elderly people live alone in two of the dwellings (CE_3D and CE_1D, this latter also ill), the rest having a very varied profile of families with and without children. Except for the elderly and another home, the rest are employed, with an income of 500–1000€/month in two homes. The majority of dwellings are owned (with and without mortgage), except two that are rented.

Not a single dwelling counts with air conditioning and only two have fans. Shading devices are normally used, and ventilation is done when it gets cooler, except in two dwellings, precisely the two under roof on the fourth floor (SE_4I and CE_4I), where according to surveys, they do not use the external blinds regularly either. Figure 7.5 shows monitoring of GORRI (SE/CE), during the heat episode of 17–19 July 2016. Only a dwelling in the top floor of the non-rehabilitated building (SE_4I) exceeds the threshold for all categories of indoor thermal environment. The following dwelling that overheats the most is situated at the top of the rehabilitated building CE (CE_4D), although it is within the threshold for adaptive comfort for both Categories II and III. Lastly, dwellings at the SE building have a warmer temperature regime than those at the CE building (Fig. 7.5b).

SOLE Case Study is in a zone considered vulnerable too (old *Soto de Lezkairu*), and considered a priority area for getting government funding, which covered 80% of the budget of communities and neighbours who applied for it. Buildings of T1 Typology were built in 1959, and the CE building was rehabilitated in 2016. In this case, dwellings have east-west orientation, and some of them, located in the corner, also have some windows facing south.

Five dwellings out of eight were surveyed and monitored in the CE building, and four out of eight in the SE building. A variety of family profiles live in the building, being four unipersonal, two of which are elderly people (over 65 years, one of them ill). It must be highlighted that the last two live in the SE building and are receiving widower's pension. There is also an unemployed person and in two dwellings, income is between 500 and 1000€/month. Dwellings are generally owned (with and without mortgage), and only two are rented out.

Two of the dwellings count with air conditioning (CE_G_3D and SE_A_1D), and two of them have fans. Generally, shading devices are used, and ventilation is done in cooler conditions during the night, except for three dwellings that ventilate all day (CE_G_2I, SE_A_1D and SE_A_4D).

SOLE buildings were affected by the high temperatures registered in Pamplona at the same time the heatwave hit Spain on 22–25 August 2016, although Pamplona cannot be considered under it since conditions that define a heatwave were not given.

Monitoring results are shown in Fig. 7.6. None of the dwellings exceed Categories II and III thresholds, although CE_G_4I and SE_A_1D punctually do for Category I. The last one, counting with air conditioning, has the greatest thermal oscillations, probably due to the users inadequate actions (specially ventilation),

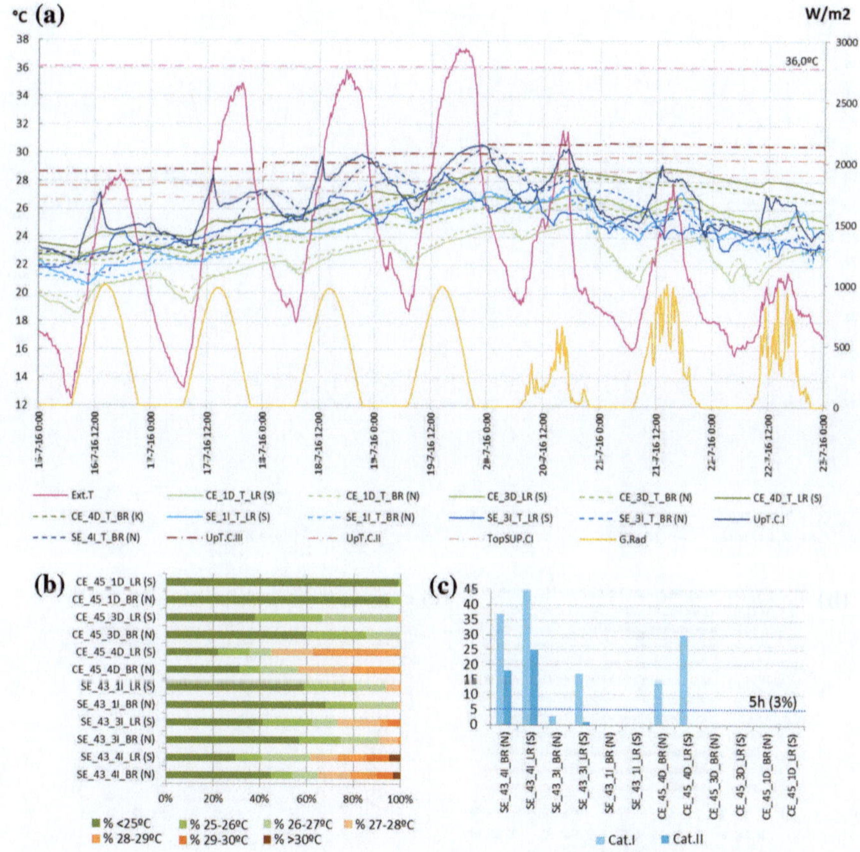

Fig. 7.5 Monitoring of building block GORRI CE/SE in Pamplona, from 16 to 23 July 2016. Graphic of registered temperatures (**a**); temperature frequency (**b**) and hours of discomfort according to UNE-EN 15251 for buildings without mechanic cooling systems (**c**)

while the first one (CE_G_4I) suffers the greater overheating, even though it counts with a rehabilitated thermal envelope.

As a summary, according to monitoring of multifamily social housing in warmer summer conditions exposed in this paragraph, there is overheating risk especially on the upper floor, for buildings with and without rehabilitated thermal envelope. So, although there is a reduction in the registered temperature regime, thermal envelope rehabilitation (mainly roof) may not be guaranteeing a proper behaviour of dwellings under warmer conditions or heatwaves in Pamplona, avoiding overheating, as has already been seen in previous research in other locations (Saman et al., 2013). The need for additional measures in the design, building details and execution of roofs seems evident.

At the same time, user actions are key to avoid overheating both by using solar devices (all dwellings including the most humble count with blinds), or by

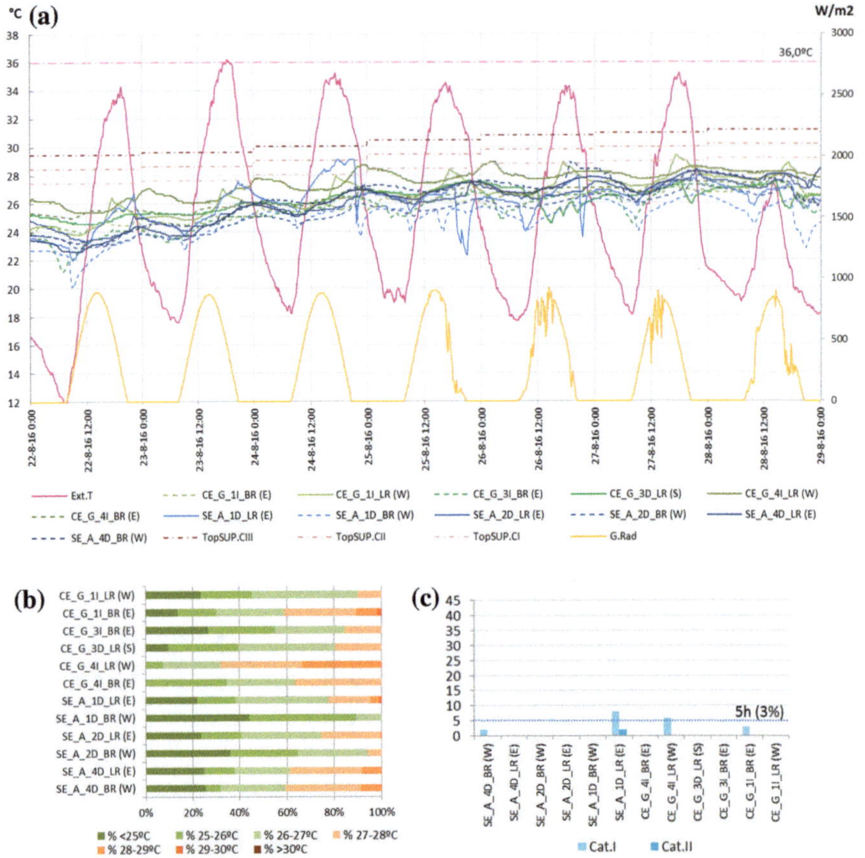

Fig. 7.6 Monitoring of building block SOLE CE/SE in Pamplona (Spain), from 22 to 29 August 2016. Graph of registered temperatures (**a**); temperature frequency (**b**) and hours of discomfort according to UNE-EN 15251 for buildings without mechanic cooling systems (**c**)

ventilating when external conditions are favourable. Precisely, the dwellings under roof with stronger overheating are the ones where users have not acted properly contributing to the worsening of indoor temperatures.

Lastly, double orientation in dwellings allows for an effective ventilation and has the advantage that even in small dwellings, they always have a room looking north (or in opposite orientation), and therefore cooler, as a '*Cool Retreat*'. As a guidance, differences in temperature between rooms with different orientation in the same dwelling, from FERRO and GORRI monitoring data, are attached in Fig. 7.7.

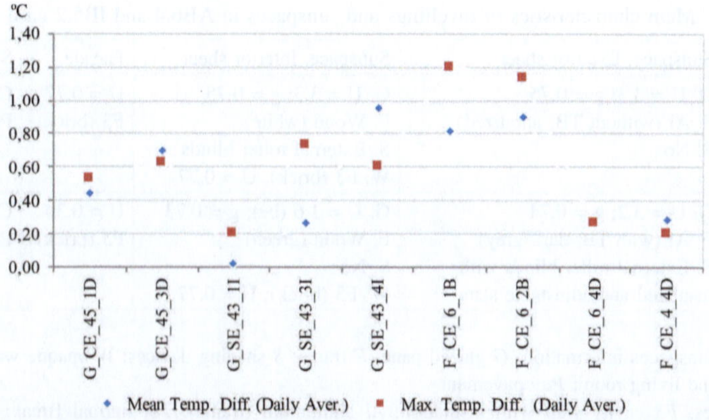

Fig. 7.7 Average difference between several north and south orientated rooms/spaces within a dwelling during daily hours (8–24 h), 17–19 July 2016. GORRI (G) and FERRO (F) case studies

7.1.2.2 Monitoring Sunspaces in Pamplona (Spain)

Sunspaces are mixed elements of solar gain mainly for winter conditions. Nonetheless, a correct functioning must be secured during the year, which does not produce damaging effects in summer conditions. The sunspace works as a buffer or intermediate space between exterior and interior conditions and need adequate ventilation and a shading device in summer, mainly outside the exterior façade sheet.

The general outcome of the research where the monitoring of different sunspaces exposed in this section took place was related to the way they must be designed and used, providing an adequate ventilation (at least a 25% of the external sheet), and a shading device for the external sheet preferably on its exterior, and bearing in mind future climate change scenarios (Monge-Barrio & Sánchez-Ostiz, 2015). Other research of sunspaces prototypes also highlights the importance of shading and an *active* and adequate use of the systems (Sánchez-Ostiz, Monge-Barrio, Domingo-Irigoyen, & González-Martínez, 2014).

Monitoring of two of the sunspaces coincided with two heatwaves that affected Pamplona, 19–21 August 2011 and 17–22 August 2012. Main characteristics of dwellings and their sunspaces are attached in Table 7.2. Both dwellings are situated in the second floor of a collective residential building, AB6.4 dwelling being in a T3 Typology (tower) and IB5.2 dwelling in a T1 Typology (linear block). The first was built in 1970 in Pamplona, and the second in 2004 in *Sarriguren Ecocity*, a town next to Pamplona (Fig. 7.8).

Results of the monitored sunspace AB6.4 without shading on the external sheet during 19–21 August 2011 heatwave in Pamplona are shown in Fig. 7.9. High temperatures are reached in the sunspace (higher than exterior temperatures), also due to a lack of proper ventilation (5% opening of external surface). However, living room temperatures are kept in acceptable adaptive comfort margins except for Category I. It must be highlighted that two older retired people live there, and

Table 7.2 Main characteristics of dwellings and sunspaces in AB6.4 and IB5.2 case studies

	Sunspace. Exterior sheet	Sunspace. Interior sheet	Façade	Slab
AB6.4	G. U = 3.3; g = 0.75 F. Al (without TB, anodized) S. No	G. U = 3.3; g = 0.75 F. Wood (white) S. External roller blinds W. F3 (brick); U = 0.77	U = 0.77 F3 (brick)	Concrete Pav: wood
IB5.2	G. U = 3.2; g = 0.74 F. Al (with TB, dark grey) S. External roller blinds with insulated and adjustable slats	G. U = 1.6 (be); g = 0.73 F. Wood (green) S. No W. F3 (brick); U = 0.77	U = 0.34 F3 (brick)	Concrete Pav: wood

Notes

Codes in sunspaces information: *G* glazed pane; *F* frame; *S* shading devices; *W* opaque wall between sunspace and living room; *Pav* pavement

Other codes: *F3* cavity wall with insulation; *Al* aluminium (frame); *TB* thermal Break; *U* thermal transmittance (W/m^2 K); *g* solar factor

Fig. 7.8 Sunspaces images of AB6.4 and IB5.2 case studies

normally in the summer, they stay in their cottage in the country because of the high temperatures reached in the dwelling.

Monitoring results of sunspaces with solar shading on their external sheet (IB5.2), during the 17–22 August 2012 heatwave in Pamplona, are shown in Fig. 7.10. Contrary to the previous case study, the sunspace reaches lower temperatures than the exterior, becoming a *semi-exterior buffer space*, beneficial for the dwelling also in summer, mostly due to the disposition and use of the external blinds on the exterior sheet of the sunspace. All rooms reach temperatures lower

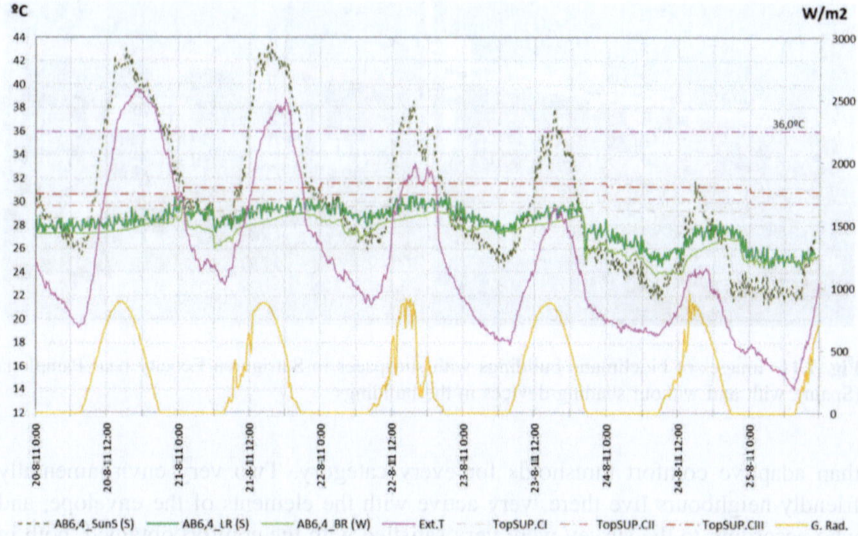

Fig. 7.9 Sunspace facing south and west without shading in the exterior sheet, during the heatwave of 19–21 August 2011 in Pamplona (Spain)

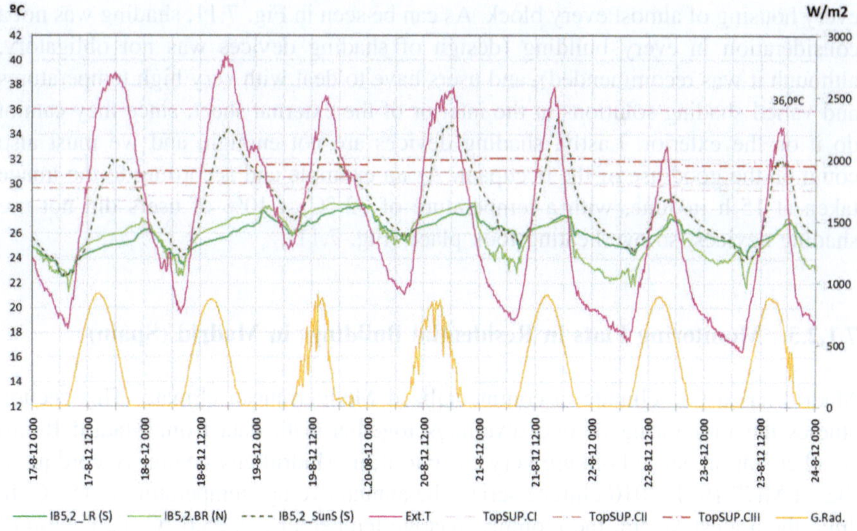

Fig. 7.10 Sunspace facing south with shading devices in the exterior of the exterior sheet, during the heatwave of 17–22 August 2012, near Pamplona (Spain)

Fig. 7.11 Images of bioclimatic buildings with sunspaces in Sarriguren Ecocity near Pamplona (Spain), with and without shading devices in the buildings

than adaptive comfort thresholds for every category. Two very environmentally friendly neighbours live there, very active with the elements of the envelope, and who according to the survey were very satisfied with the comfort obtained, both in winter and in summer. To deal with the high temperatures reached (28 °C), dwellers occasionally use standing fans.

Lastly, we attach images for some sunspaces in Sarriguren Ecocity near Pamplona (Metropoli, 2009), a neighbourhood where bioclimatic strategies were implemented according to normative, and therefore, a sunspace was designed in every housing of almost every block. As can be seen in Fig. 7.11, shading was not a consideration in every building (design of shading devices was not obligatory, although it was recommended), and users have to deal with very high temperatures and varied shading solutions in the interior of the external sheet, since they cannot do it on the exterior. Lastly, shading devices are not enough, and we must also count on the good use of the occupant. As an example and according to the image taken at 15 h in June, with a temperature of 32 °C, a 10% of users did not use shading devices, so overheating took place (Fig. 7.11).

7.1.2.3 Monitoring Flats in Residential Buildings in Madrid (Spain)

Madrid has a Csa climate, a continentalized Mediterranean climate. This section studies the monitoring of two dwellings together with data from Madrid Retiro weather station, since both are very close to it, in Madrid city centre. According to the AEMET 1981–2010 climate series, the annual average temperature is 15 °C. In July, the hottest month, the monthly average temperature is 25.6 °C, maximum of 32.1 °C and minimum of 19 °C (AEMET, n.d.-c). Relative annual average humidity is 57%, being 38% in July.

Monitoring of two case studies in building blocks in Madrid was carried out during the heatwave that affected most of Spain from 14 to 18 June 2017. Table 7.3 shows the given average maximum and minimum temperatures (AEMET, n.d.-b),

Table 7.3 Average temperatures reached 11–17 June 2017, together with temperatures according to UNE-EN 15251

Day	Madrid Retiro (AEMET)			UNE-EN 15251. Adaptive comfort			
	T media	T max	T min	Θ_{rm}	$\Theta_{i\ max}$ Cat. I	$\Theta_{i\ max}$ Cat. II	$\Theta_{i\ max}$ Cat. III
11/06/2017	30.00	37.20	22.80	25.45	29.20	30.20	31.20
12/06/2017	30.30	37.00	23.60	26.89	29.67	30.67	31.67
13/06/2017	29.50	36.20	22.80	28.15	30.09	31.09	32.09
14/06/2017	30.40	37.20	23.60	28.79	30.30	31.30	32.30
15/06/2017	31.50	39.00	24.00	29.42	30.51	31.51	32.51
16/06/2017	30.30	37.60	23.00	30.11	n.a.	n.a.	n.a.
17/06/2017	32.80	40.00	25.50	30.32	n.a.	n.a.	n.a.

Notes
Θ_{rm} running mean outdoor temperature (inferior to 30 °C)
$\Theta_{i\ max}$ upper limit of indoor operative temperature
n.a. not available

together with *running mean temperature* (Θ_{rm}) of adaptive comfort UNE-EN 15251, for those days in Madrid (see Chap. 3). Some days, temperatures went over the average of 30 °C, so top limits on the norm are not applicable, and limits have been used considering 30 °C (that is, 30.8 °C; 31.8 °C and 32.8 °C, for the indoor environment Categories I, II and III).

Both dwellings are in buildings built before thermal regulations in Spain (1979) and both of them have air conditioning in at least one room (Fig. 7.12). Families with very young children (under 2) live in them, and they are both under rental contracts.

Dwelling DOOC11.6 is in a high rise with six floors and attic, built in 1958 and the thermal envelope has not been rehabilitated. The façade is built of brick, wood carpentry with rolling blinds and awnings, and being the sixth floor, its ceiling is under the attics terrace. With a built surface of 112 m^2, it is mostly oriented south (living room and main bedroom), having kitchen and two rooms looking into interior patios. Only the living room has air conditioning, the main bedroom and a secondary bedroom having a fan in the ceiling.

Dwelling DOLA9.2 is in a more modern 17-storey high rise built in 1974, and the envelope has not been reformed either. Façades are brick, sliding aluminium frames with non-thermal break, with rolling blinds. It is a small 64 m^2 flat, with a living room and two bedrooms, with a single west orientation. The three spaces have air conditioning.

Monitored data for two spaces in each dwelling are shown in Fig. 7.13: one with air conditioning in the living room and another one without mechanical conditioning, the bedroom in the case of DOOC11.6 and the kitchen in DOLA9.2. The figure includes the top limits for mechanically (27 °C for Category III) and naturally conditioned buildings (TopSUP.CI to CIII)

Fig. 7.12 Images of DOOC11.6 (Left) and DOLA9.2 (Right). Case studies in Madrid (Spain)

In the case of DOOC11.6, the temperature reached during the heatwave exceeded top limits of Cat. I, and the family, although very environmentally friendly, feel a very high thermal sensation in the bedroom, sometimes even having to sleep in the living room, only space with air conditioning. In the case of DOLA9.2, the space without air conditioning overheats excessively in the afternoon due to its west orientation (blinds were not used during 14–15 June). In both cases, the rest of the house is cooled down through mechanically conditioned spaces, and some actions were improved during the heatwave. For example, June 17, with external conditions more severe than in the previous days, having in DOLA9.2 the blinds down and ventilation more controlled, indoor temperatures are lower than the day before. Finally, we can highlight the difficulties that both dwellings encounter in order to do a night cooling ventilation, due to the high night temperatures registered in Madrid those days (Table 7.3), and specially in DOLA9.2 having a single west orientation.

Lastly, mechanically cooled spaces progressively reduce their maximum and average daily temperatures from the 15th (30.42 and 30.19 °C maximum and 28.78 and 29.43 °C average, in DOOC11.6 and DOLA9.2, respectively) to the 16th and 17th (under 29 °C maximum and 27–28 °C average), therefore, from the first days

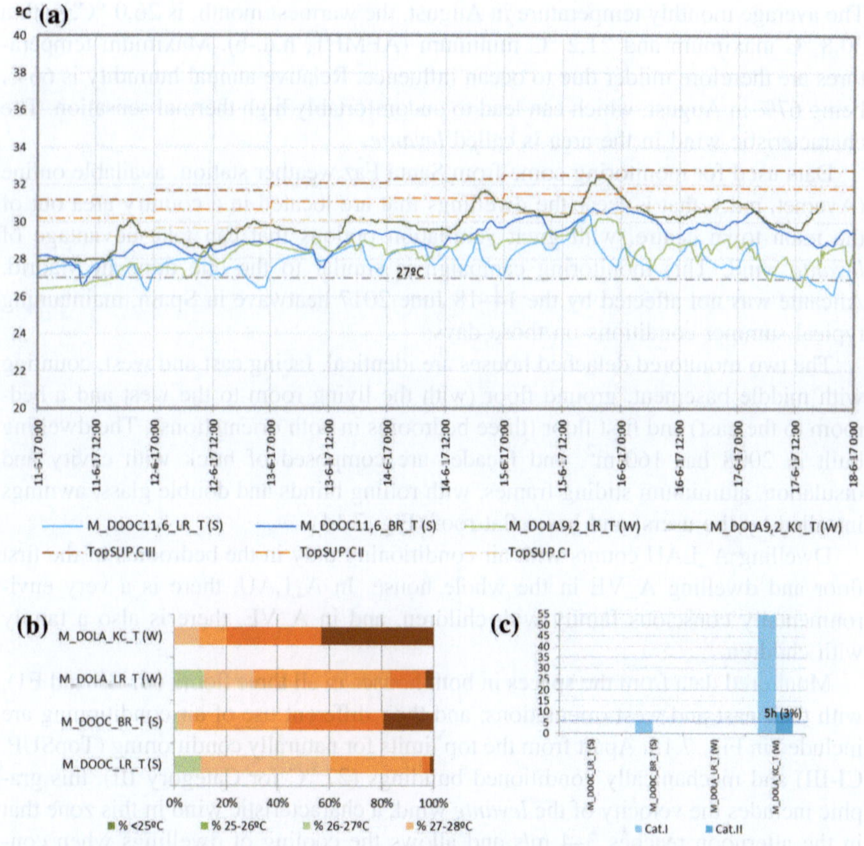

Fig. 7.13 Monitoring of dwellings DOOC11.6 y DOLA9.2 in Madrid, during 14–18 June 2017. Graph of registered temperatures (**a**) temperature frequency (**b**) and hours of discomfort according to UNE-EN 15251 for buildings without mechanic cooling systems (**c**)

of the heatwave to the last ones. In any case, indoor temperatures reached are higher during almost every hour than the upper limit temperature of 27 °C, for Category III (only going below 27 °C during 7 h in DOOC11.6 and 3 h in DOLA9.2, day 17th), threshold temperature for mechanically conditioned spaces, according to UNE-EN 15251 (2008).

7.1.2.4 Monitoring Houses in Alicante (Spain)

Alicante has a BSk climate, dry Mediterranean climate in the south of Europe, and located mainly in the Iberian Peninsula. Average annual temperature is 18.3 °C.

The average monthly temperature in August, the warmest month, is 26.0 °C, with a 30.8 °C maximum and 21.2 °C minimum (AEMET, n.d.-b). Maximum temperatures are therefore milder due to ocean influence. Relative annual humidity is 66%, being 67% in August, which can lead to uncomfortably high thermal sensation. The characteristic wind in the area is called *levante*.

Data used for monitoring come from Santa Faz weather station, available online (Avamet, n.d.), that is, near the dwellings that are located in a country area out of the main town centre, with great ventilation options that can take advantage of *levante* wind. This monitoring campaign is similar to the one done in Madrid. Alicante was not affected by the 14–18 June 2017 heatwave in Spain, maintaining typical summer conditions on those days.

The two monitored detached houses are identical, facing east and west, counting with middle basement, ground floor (with the living room to the west and a bedroom to the east) and first floor (three bedrooms in both orientations). The dwelling built in 2008 has 160 m^2, and façades are composed of brick with cavity and insulation, aluminium sliding frames, with rolling blinds and double glass, awnings installed by the users, and has a flat roof (Fig. 7.14).

Dwelling A_LAU counts with air conditioning only in the bedrooms of the first floor and dwelling A_VE in the whole house. In A_LAU, there is a very environmentally conscious family with children, and in A_VE, there is also a family with children.

Monitored data from the spaces in both houses in all three floors (B, GF and F1), with their east and west orientations, and their different use of air conditioning are included in Fig. 7.15. Apart from the top limits for naturally conditioning (TopSUP. CI-III) and mechanically conditioned buildings (27 °C for Category III), this graphic includes the velocity of the *levante* wind, a characteristic wind in this zone that in the afternoon reaches 3–4 m/s and allows the cooling of dwellings when conditions are adequate.

From A_LAU, we can highlight the strong overheating produced in the top floor (even above outside temperatures), principally in west orientation. At night, in order

Fig. 7.14 Images of A_LAU and A_VE case studies (single-family house) in Alicante (Spain)

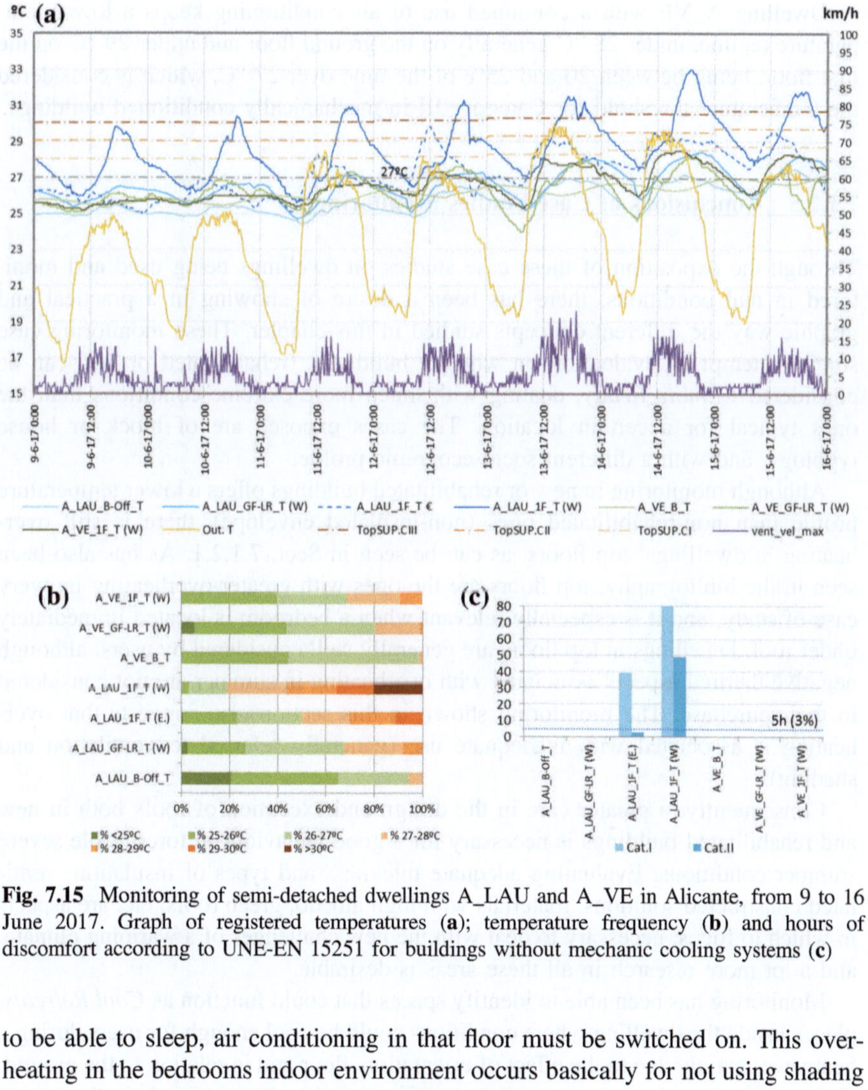

Fig. 7.15 Monitoring of semi-detached dwellings A_LAU and A_VE in Alicante, from 9 to 16 June 2017. Graph of registered temperatures (**a**); temperature frequency (**b**) and hours of discomfort according to UNE-EN 15251 for buildings without mechanic cooling systems (**c**)

to be able to sleep, air conditioning in that floor must be switched on. This overheating in the bedrooms indoor environment occurs basically for not using shading devices in west orientation when opening the windows for ventilation (to take advantage of *levante* wind), and is also a consequence of the bedroom's location under the roof. It is a contrast with ground floor temperatures (GF) where the living room facing west is located. It is well protected from overheating through awnings and a terrace, with opened windows, and without air conditioning, and reaching temperatures under the threshold for adaptive comfort for all three categories. We find lower and more stable temperatures in the semi-basement, generally under 27 °C according to monitoring.

Dwelling A_VE with a continued use of air conditioning keeps a lower temperature regime, under 28 °C generally on the ground floor and under 29 °C on the first floor, being between 20 and 25% of the time over 27 °C, which is considered the temperature threshold for Category III in mechanically conditioned buildings.

7.1.2.5 Conclusions of Case Studies Monitoring

Through the exposition of these case studies on dwellings being used and monitored in real conditions, there has been a desire of showing in a practical and graphic way the different concepts studied in this chapter. These monitoring case studies attempt an evaluation on whether buildings, (rehabilitated or not) can be considered *Climate Ready*, dealing with much more extreme conditions than the ones typical for a certain location. The cases exposed are of block or house typology, and with a different socio-economic profile.

Although monitoring in new or rehabilitated buildings offers a lower temperature profile than non-rehabilitated ones (non-insulated envelope), there is still overheating in dwellings' top floors, as can be seen in Sect. 7.1.2.1. As has also been seen in the bibliography, top floors are the ones with greater overheating in every case of study, and it is especially relevant when a bedroom is located immediately under roof. Dwellings in top floors are generally well considered by users, although negative thermal aspects associated with overheating in summer are not considered in their purchase. The monitoring shown in this section also suggests that overheating is associated with inadequate use (generally referred to ventilation and shading).

Consequently, a greater care in the design and execution of roofs both in new and rehabilitated buildings is necessary for a good behaviour in foreseeable severe summer conditions. Evaluating adequate thickness and types of insulation, ventilated constructed solutions, materials with high albedo, green roofs, etc. are aspects in which to focus, necessary to deal with the new challenges of a warming climate, and a lot more research in all these areas is desirable.

Monitoring has been able to identify spaces that could function as *Cool Retreats*, places inside the dwelling where conditions could be cool enough for users during a heatwave, mainly due to the effect of orientation, floor and in relation to the ground. Houses shows clear temperature stratification, and the ground floor and semi-basements have a cooler temperature regime with lower daily thermal amplitude. They tend to have a greater surface which makes easier the redistribution of spaces in the summer. However, housing in building blocks offers less possibilities of designing a *Cool Retreat*, especially social housing that normally has a smaller surface. Double orientation however, especially in a north-south orientation will help towards this *Cool Retreat*, together with compartmentalized spaces (Saman et al., 2013). Lastly, in zones with very warm summers, to count with air conditioning in at least one room could be essential.

In the study of thermal and energy behaviour in the residential stock, the socio-economic component is of key importance and influences the use of the dwelling. Heatwaves affect both vulnerable and non-vulnerable dwellers, the later normally having more resources to deal with these extreme conditions. If an avoidable overheating takes place during a heatwave (e.g. due to misuse of occupants), and if the dwelling has air conditioning and no economic vulnerability, it will produce only a higher energy consumption. Alternatively, if the dwelling does not count with mechanical systems, whether for being in a milder climatic zone or for being unable to instal or use them, overheating directly affects people's well-being and even their health.

Being environmentally conscious is not enough, you need information and tools. Sometimes the user can act and sometimes this is not possible, for example, trying to ventilate a dwelling with only one orientation, or trying to find a *Cool Retreat* in a dwelling with a single west orientation. Rooted actions like shading or night ventilation have proved to be well known by users, who miss the importance and repercussion of their actions have on overheating and on added energy consumption. At users level it is of great importance to be able to measure in order to make the best use of the dwelling, following the classic idea *if you cannot measure it, you cannot improve it* (Lord Kelvin, 1824–1907). Therefore, a low cost monitoring, or the installation of simple thermometers both indoors and outdoors, would help in gaining consciousness and efficiency. In any case, the economic reduction in the electricity bill from the air conditioning is surely stimulating for the user.

7.2 Winter Measures Are not Enough

Buildings must be built and rehabilitated for the whole year, and in the Mediterranean zone in the south of Europe, they have to deal with both winter and summer conditions that will progressively become more severe in future global warming scenarios. Because in most locations the main energy demand for thermal conditioning is heating, building design and even technical regulations specifically focus on measures related to winter conditions.

Studies consider important improvements having a better energy certification, related with increasing the envelopes insulation (opaque façades and roofs, mostly), improving glass, control of infiltrations and ventilation (e.g. by a heat recovery ventilator system). Buildings with the best energy certifications have lower energy consumption not only in winter but also in summer. Research in Australia, with similar climatic zones, reaches the conclusion that they will also experiment less variation in future energy requirements (Wang, Chen, & Ren, 2010). As determining factors for winter become milder, and summer ones warmer, there will be a need in

Fig. 7.16 Examples of traditional Andalusian patio. Shading and evaporative cooling in Granada (Spain)

some zones for changing the focus of design strategies towards cooling, contrary to present strategies (Karimpour, Belusko, Xing, Boland, & Bruno, 2015).

For this reason, we need to highlight other specific measures for the summer, such as solar radiation control to avoid overheating (contrary to winter's strategy), or giving the building the possibility of ventilation for cooling, which far from optative, are essential. Many other passive strategies that have to do with thermal mass, the albedo (high reflectance), evaporative cooling, etc., offer a wide range of possibilities to reduce overheating in our buildings, or to demand as minimum energy as possible in their thermal conditioning, without compromising occupants' comfort and health. Thanks to these strategies and to occupants' behaviour and use of low-energy equipment, indoor temperatures can be reduced by 2–5 °C or more, according to some authors (Matthies, Bickler, Marin, & Hales, 2008). These measures are well known coming from old traditional architecture in the Mediterranean Region (Menéndez, 2006; Neila González, 2004), from the most simple system to much more sophisticated ones (Figs. 7.16 and 7.17).

All tested and validated strategies and measures applied in severe summer climate zones can be transferred to zones that are starting to be affected with warmer summers, and can affect health because the thresholds are inferior to the Mediterranean ones (Baccini et al., 2008).

Finally, in warm and very warm zones, the elderly represent a population group especially vulnerable to more severe summer conditions, but at the same time, they have lived in times when there was no hegemonic use of air conditioning, and have developed behavioural and environmental strategies to offer the community (Loughnan, Carroll, & Tapper, 2014).

Fig. 7.17 Examples of Mediterranean architecture adapted to the climate. Shading and evaporative cooling. Carré d'Art, Norman Foster (Nimes, France)

7.3 Measures for a Warming Climate Focused on Mediterranean Climates

In order for users of residential buildings to be able to deal with much warmer summers (with higher maximum and minimum temperatures) and longer more extreme and frequent heatwaves, there are three types of measures to be acted upon: related to adaptive and personal behaviour, related to urban spaces and related to buildings.

7.3.1 Adaptive Behavioural Measures

Personal measures that allow users to deal with these very warm situations are frequently quoted from medical and social protection fields as sanitary alerts and are specially directed to children and the elderly (Matthies et al., 2008). Different behaviours will aim to regulate the human body's temperature without health problems.

In residential buildings, there are many possibilities to adapt, from wearing light clothes and light colours to keeping hydrated, not engaging in exhausting physical activities, staying indoors during the central hours of the day, avoiding alcoholic beverages, planning the day according to the weather forecast or spending time in swimming pools or in spaces with air conditioning.

On the other hand, most of the dwellings require manual actions from the occupant. In fact, the efficiency of some of different strategies that follow in Sect. 7.3.3 is directly related to the user. However, it must be underlined that in order to carry out these actions, the dwelling must be designed to allow them. Some of the most usual measures directly linked to user's actions are:

- Shading: use of movable shading devices during the day, when the space is getting direct solar radiation. Shutters, venetian blinds or awnings are very common and regularly used all over Southern Europe.
- Ventilation controlled during the day: close the windows when outside temperatures are higher than inside temperatures. It will be possible to ventilate during the day if there is a cold focus which guarantees adequate temperatures, like interior patios or shaded streets, outdoor spaces with trees, fountains, cool breeze, etc.
- Night ventilation taking advantage of the lowering temperatures or breeze, only if lower than indoor temperatures. Two opposite orientations will be needed in order to be able to have crossed ventilation.
- Moving during the day to a room with air conditioning, or naturally cooler, as a way of *Cool Retreat*. In this respect, the dwelling must be able to offer this opportunity, whether for size, orientation, options to compartmentalize spaces or for having spaces capable of taking advantage of the soil's thermal mass.
- Use of fans, hand or ceiling, to profit from of the refreshing movement of air.
- Use of thermometers and thermostats at home that can orientate on the convenience of applying the previous actions. Also of great interest is the disposition of external temperature sensors, especially with ventilation strategies.
- Disposition of security elements (e.g. window grates in ground floors), protection from insects (mosquito net for windows, etc.) and from outside view, in order to ensure a ventilation in the dwelling by natural means. This is especially relevant to the elderly, very sensitive to security and privacy.

7.3.2 Measures in Urban and Outdoor Spaces

Exterior microclimate is of high importance for naturally conditioned buildings. It is relevant in cities specially if affected by *urban heat island effect*, and in climatic zones with high minimum night temperatures that may not make possible natural night cooling. Exterior pollution and noise (from the outside as well as from the own air conditioning system) can also compromise natural passive strategies in buildings of big cities.

For this reason, at urban level, there are some measures that can help avoid or reduce overheating in dwellings from the outside, such as (Figs. 7.18, 7.19 and 7.20):

- Design of green areas and trees near buildings.
- Light pavements with high albedo (high reflectance).
- Urban design that takes into account buildings and street layout (width, orientation, building height, etc.) in order to naturally shade the area.
- Façades and roofs from the surrounding buildings using materials that do not store the incident solar radiation, such as high albedo materials, green roofs, etc.
- Through the configuration of the urban space, encourage natural ventilation taking advantage of dominant winds, if beneficial.
- Limiting traffic zones

Fig. 7.18 Illustrative image of passive measures in outdoor spaces. Shading, vegetation and evaporative cooling. Nimes (France)

Fig. 7.19 Illustrative image of passive measures in outdoor spaces. Natural and artificial shading. Nimes (France)

Fig. 7.20 Conditioning of semi-exterior spaces. Vents introduce cool air from the basement. Galleria Vittorio Emanuele II. Milan (Italy)

Table 7.4 Review. Part 1. Adapted from Arriazu (Arriazu & Monge-Barrio, 2017)

References	A. Location and climate	B. Climate change scenarios/ heatwaves	C. Methodology	D. Main characteristics of housing and occupants
Karimpour et al. (2015)	Australia Adelaide (BSk)	2070 CC scenario	Energy modelling (AccuRATE) Best design based on minimum heating and cooling energy consumption	Brick veneer house
Wang et al. (2010)	Australia Alice Springs (BSh), Darwin (Aw), Hobart (Cfb), Melbourne (Cfb) and Sydney (Cfa)	General circulation models (GCMs) Based on three scenarios (A1B, A1FI and 550 ppm) from 1990 to 2100	Energy modelling (AccuRATE) Cooling and heating. Base house and modifications to achieve 2.5 and 7 stars	Detached brick veneer residential house. Facades (U = 0.75), Concrete roof tiles, simple glazing, holland blinds; Heating, setpoints 20–18 °C (Living room—bedroom); Cooling, setpoints varies depending on climate zone (23.0–26.5 °C)
Ren et al. (2014)	Australia Brisbane (Cfa) Melbourne (Cfb)	Heatwaves	Energy modelling (AccuRATE) Heatwave scenario: Brisbane (2004) and Melbourne (2009)	Conventional single-family house without air conditioning
Hatvani-Kovacs et al. (2016)	Australia Adelaide (BSk)	Heatwaves	Survey (N = 393)	90% dwellings with cooling systems (all dwelling or a single room); 25% rooms without shading; Mainly without insulation and with light mass Vulnerable population: energy poverty
Brotas and Nicol (2016b)	Athens (Csa), Lisboa (Csa), Rome (Csa) Munich (Cfb), London (Cfb) and Moscu (Dfb)	2020–2050–2080 (CCWorldWeatherGen)	Energy modelling Overheating (CIBSE). –Criterion 1: Hours of exceedance (3%) –Criterion 2: Daily weighted exceedance –Criterion 3: Upper limit temperature	Mid-storey flat facing south and east; $U_{Walls} = 0.18$; $U_{Glass} = 1.4$, g = 0.63 Pattern of use = 24 h; Ventilation rate: 0.3 h^{-1} Night ventilation; Interior blinds

(continued)

Table 7.4 (continued)

References	A. Location and climate	B. Climate change scenarios/heatwaves	C. Methodology	D. Main characteristics of housing and occupants
Barbosa et al. (2015)	Portugal Lisboa (Csa)	M1. 2050–2080 (CCWorld WeatherGen) M2. 2003 heatwave	Energy modelling (energy plus with design builder) Discomfort hours (STAT or ADAPT-15251)	Residential building, four dwellings per floor; each dwelling single orientation (east or west) Façade: brick (high mass) (U = 1.7), double glazed (U = 2.4), PVC frame, concrete slab Vulnerable population: low income
Santamouris et al. (2010)	Greece	2009 summer conditions	Monitoring (N = 214 dwellings)	Mechanically air-conditioned Construction with high thermal mass. U_{Walls} near 0.5, and U_{Roof} near 0.4 Single houses Surface ranges between 55 and 480 m^2
Sakka et al. (2012)	Greece Athens (Csa)	Very hot summer 2007 (hot spells 30–33 °C)	Monitoring (N = 50 dwellings)	Without cooling system Vulnerable population: low income
Pyrgou et al. (2017)	Italy Perugia (Cfb)	Heatwaves 2013	Energy modelling	Residential buildings
Mahdy (2014)	Egypt Alexandria, Cairo, Aswan (BWh)	2020–2050–2080 (CCWorldWeatherGen)	Energy modelling (energy plus with design builder)	2 collective buildings GF + 5F 3 typologies or envelopes (GRC)
Gangolells (2012)	Spain Barcelona (Csa), Valencia (Csa), Madrid (Csa)	2011–2040; 2041–2070; 2071–2100 (A2 and B2, following IPCC)	HDD and CDD	
Saman et al. (2013)	Australia M1: Adelaide (BSk), Amberley (Cfb), Richmond (BSk); M2: Adelaide (BSk), Melbourne (Cfb), Brisbane (Cfa), Hobart (Cfb), Sydney (Cfa), Perth (Csa), Darwin (Aw)	M1: Heatwave M2: 2030, 2050 and 2070	Energy modelling (AccuRATE) M1: heatwaves—overheating (ASHRAE55:2013, adapt): Energy mod. Five case studies M2: Cooling demand: two dwellings	Different typologies: mainly single-family houses and semi-detached houses Vulnerable population: low income and ageing

Table 7.5 Review. Part 2. Adapted from Arriazu (Arriazu & Monge-Barrio, 2017)

References	A Solar radiation control	B Heat transfer control	C Passive measures of heat evacuation and dissipation	D Thermal mass control	E Active measures (systems)	X Control in strategies for all year
Karimpour et al. (2015)	Roof Effective measures: low absorptance (0.1) and foil	Thermal insulation in façade (interior and exterior) Glass: Double glass Low-E Insulation in roof: key role		Thermal mass Pavement (ceramic tiles instead of wood)	Cooling	Heating and cooling. With better combinations, in TMY2070 Cooling: 56.8 instead of 46.5 MJ/m^2
Wang et al. (2010)		Thermal insulation and airtightness: Façades (exterior), Roofs and floor Improvements in frames (from aluminium to wood), double glazed				Most efficient buildings experience less changes in the future Heating and cooling depend on climate: if cooling is now the prevailing demand, in future more total energy demand
Ren et al. (2014)	Shading (exterior) g = 0.77–0.29	Thermal insulation. Roof and floor	Fans			
Hatvani-Kovacs et al. (2016)	Improvement of shading	Improve the insulation mainly in roof			The most popular measure: improve efficiency of the systems	
Brotas and Nicol (2016b)	Exterior shading		Night crossed ventilation			
Barbosa et al. (2015)		Thermal insulation (exterior and interior), better exterior	Night ventilation	Thermal mass Façades		
Santamouris et al. (2010)			Night ventilation			

(continued)

Table 7.5 (continued)

References	A Solar radiation control	B Heat transfer control	C Passive measures of heat evacuation and dissipation	D Thermal mass control	E Active measures (systems)	X Control in strategies for all year
Sakka et al. (2012)	Shading	Airtightness	Night ventilation Limitations: high minimum temperatures during heatwaves	High thermal mass Specially in the first days of heatwaves		
Pyrgou et al. (2017)		Thermal insulation Better insulated buildings, more at risk of overheating and increase in cooling	Ventilation			
Mahdy and Nikolopoulou (2014)		Three envelopes (brick and GRC) Better insulation, better indoor range of temperatures				
Gangolells and Casals (2012)	Shading (fixed or movable), vegetation, awnings, etc.	Thermal insulation	Night ventilation Natural and mixed mode			72% current building stock in Spain is unprotected Decrease in heating: 30% (Barcelona) and 36% (Valencia) Increase in cooling: 107% (Valencia) and 296% (Madrid)
Saman et al. (2013)	Shading Orientation Roof	Roofs Reflectance surfaces and insulation Improvements in windows	Fans Natural ventilation	Soil thermal mass (basements and patios)	Focus not only on reducing total energy demand, but also in peak cooling demand periods. Setpoints	In future scenarios, total energy demand will depend on climate, although an increase in cooling is expected

7.3.3 Measures in Residential Buildings for a Warming Climate. A Review

7.3.3.1 A Review of Mediterranean Climates

Several studies analyse different measures to implement in buildings from the point of view of climate change conditions. This section shows a review of research done in residential buildings in the Mediterranean South of Europe as well as in countries like Australia, with similar climate zones (mostly Csa, Cfa, Cfb and BSk). This review is presented in two tables and is adapted from Arriazu (Arriazu & Monge-Barrio, 2017).

The first table (Table 7.4) shows data from revised articles referring to:

- A. Location and climatic zone, generally referring to Köppen–Geiger classification. Residential buildings in Mediterranean climates have been selected.
- B. Climate change scenarios or heatwave studies. TMY for future scenarios are indicated, or heatwave in relation to which the study is done (mainly from 2003, 2007, 2009 or 2013 heatwaves).
- C. Methodology: energy modelling with heatwave or future scenario, monitoring and/or surveys.
- D. Main characteristics of housing and occupants: Typology, mainly in individual or collective building, and some of them, indicates references to vulnerable people, generally ageing and/or people in fuel poverty.

The second table (Table 7.5) indicates the different measures used in the reviewed studies on the Mediterranean zone. They are:

- A. Solar radiation control: Shading devices, type of glass (g, solar factor), orientation and adequate design.
- B. Heat transference control. Envelopes' transmittance, position of insulation in envelopes and control of thermal bridges, and airtightness.
- C. Passive measures for heat evacuation and dissipation. Natural or mechanic ventilation (the last with very low consumption), night ventilation, stack effect, etc.
- D. Control of thermal mass as energy modulator. Thermal mass versus light construction to avoid overheating.
- E. Active cooling measures. Systems with high energy efficiency, and renovation of inefficient systems, controlling cooling setpoints and using renewable energy resources (mainly photovoltaic, in dwellings).
- X. Annual strategy control. Studies make reference to an annual energy and thermal optimisation with these measures. Reference values are given.

In Figs. 7.21, 7.22 and 7.23, examples of traditional and new passive energy strategies in architecture are shown.

Fig. 7.21 Illustrative images of different types of shading: vegetation, overhangs, shutters, etc. Montpellier (France)

Fig. 7.22 Illustrative images of façade's fixed shading. Seville (Spain)

Fig. 7.23 Illustrative images of passive measures in the Mediterranean area: green roofs, canopies, movable shading, etc. Montpellier (France)

7.3.3.2 Other Studies in Non-Mediterranean Climates

Other articles have been reviewed, and although they do not deal with the Mediterranean zone, they deal with European housing in climate change scenarios and draw interesting conclusions that reinforce the ideas previously exposed in this chapter. Some aspects from three of them are summarized here.

Van Hooff et al. through energy modelling studied three different housing typologies (detached house, terraced house and apartment), and to assess climate change adaptation measures, chose data recorded from de Bilt (the Netherlands) in 2006, known for the occurrence of several heatwaves (Van Hooff, Blocken, Hensen, & Timmermans, 2015). The study found that the number of overheating hours in residential buildings that were built according to the building regulations of 2012 was higher than that for the buildings built in the 1970s, and that increasing the thermal resistance of the envelope resulted in increases in overheating hours. Therefore in the Netherlands, in well-insulated buildings, shading (movable exterior solar shading) or natural ventilation should be provided. It also found differences in the typologies, apartment buildings having the higher overheating rate.

Brotas et al. studied a mid-storey flat in London (United Kingdom) for the 2030, 2050 and 2080 climate change scenarios, and the assessment of discomfort hours was performed through energy modelling, and according to TM54 of CIBSE (Brotas, & Nicol, 2016a). The conclusions were that shading devices were fundamental and should be placed external to glass, night ventilation avoids or minimizes the need for air conditioning and crossed ventilation should be promoted.

Porrit et al., through energy simulation of a Victorian terraced house in London under the 2080 and in heatwave scenario, concluded that the overheating problem could be addressed by purely passive measures, such as wall insulation (better external), window shutters and painting external walls with lighter colours (Porritt, Shao, Cropper, & Goodier, 2011).

As a summary, these studies show how measures widely used in the Mediterranean are starting to be considered important in future warming or heatwave conditions, like shading, night ventilation, consideration of albedo in envelopes or insulation on the exterior side of facades.

7.3.3.3 Conclusions of the Literature Reviewed

A review of passive energy measures related to *popular wisdom*, which can be found in classic passive architecture books, about residential buildings in Mediterranean zones has been carried out, but quantifying the efficiency of the measures according to the location and for a future warming scenario (Fig. 7.24).

Methodology goes from the study through energy modelling of future global warming scenarios to studies of buildings through monitoring during heatwave

Fig. 7.24 Ancient and new architecture in the Mediterranean. On the same way to respond to climate conditions. Nimes (France)

events. The focus is based on cooling energy demand or total energy demand, supposing air conditioning systems are installed in the dwelling, and in overheating that is produced in dwellings, that directly affect occupants comfort and health. The first focus brings into light the special vulnerability of people in energy poverty, and the second, the vulnerability of elderly or ill people and children.

In relation to shading devices, they are ideally placed on the exterior of the glass, movable ones being recommended. Generally, it is well understood their relevance on avoiding overheating and the increase of energy demand.

In relation to heat transfer control, the conclusion is that improvements in façade and roof transmittance (adding insulation) are efficient both in summer and winter. This means that new or rehabilitated buildings with better energy certifications (generally oriented to winter conditions) are also protected in summer conditions. However, this treatment of the envelope is not sufficient, always needing shadings to avoid added overheating, as well as other measures for evacuating and dissipating heat. The most effective insulation goes on the exterior of the envelope in winter, but also in summer, since the interior sheet may have thermal mass, especially in refurbishment, in buildings originally built on brick and stone. To increase efficiency under overheating conditions, this should be combined with night ventilation. Finally, the combination with highly reflective materials are recommended for roofs and façades.

In relation to thermal mass control as energy modulator, it is considered a good solution in everyday use residential buildings, in both summer and winter conditions, as previously mentioned, but solutions that go together with ventilation in order to avoid overheating are preferred.

In relation to passive heat evacuation measures, all studies on this topic agree that night crossed ventilation is the most efficient measure, except under heatwave scenarios or under urban heat island effect (UHI), where this efficiency is questioned due to the high minimum temperatures given at night. It is fundamental to have this in mind when designing the project to give dwellings a double orientation with the windows correctly designed on the façade. These studies also consider other obstacles found mostly in cities like external noise, pollution, etc. The use of fans on the ceiling is also considered an efficient passive measure for its low energy consumption.

7.4 Conclusions

Through the reviewed bibliography and the case studies exposed in this chapter, both focused mainly on locations in the Mediterranean zone, a study is undertaken on the different passive measures to be implemented in residential buildings in the face of the new challenges of a warming climate.

These approximations vary from interviews, monitoring and energy modelling of buildings during heatwaves, to energy modelling in future scenarios of climate change, generally focused on heating and cooling consumption, and on total energy consumption, although some studies were found based on overheating hours. There are different results due to the variety of summer conditions in the Mediterranean locations.

Some research is really concerned about the consequences of these warming conditions in the most vulnerable population, mainly the elderly and people in fuel poverty, so these approaches emphasize the importance of passive architecture and occupants' behaviour, and the relevance of these measures in people's health. Although these measures are well known, they have not been sufficiently quantified and evaluated, and it is worth carrying on research in this area.

We already have warmer summer conditions and important heatwaves, so our designs must respond from now on to these conditions, without unnecessary and avoidable overheating and energy consumption.

Acknowledgements We would like to thank the Spanish Ministry of Economy and Competitiveness for financing the project 'Performance Buildings Envelope Rehabilitation Protocol (prestaRener)' through BIA 2012-38666. Special thanks to Jorge San Miguel and Ana Castillejo, who carried out the monitoring of social housing in Pamplona in 2016 summer, as part of the project prestaRener. Thanks to Ainhoa Arriazu Ramos and to the University of Navarra to grant her research titled *Resilience of the Built Environment to the Climate Change in the Southern European Region. An initial Review*. We also want to recognize the assistance on heatwaves to Cesar Rodriguez Ballesteros from AEMET and Marta Gomez Revuelto from Meteo Navarra, and to Sergi Segura from Avamet, who has provided us with climate data from Alicante. Ultimately, special thanks to all dwellers, who have so kindly facilitated the monitoring of their homes.

References

AEMET. (n.d.-a). Retrieved September 14, 2017 from http://www.aemet.es/es/serviciosclimaticos/vigilancia_clima/resumenes?w=0.

AEMET. (n.d.-b). Retrieved from http://www.aemet.es/es/serviciosclimaticos.

AEMET. (n.d.-c). Retrieved September 1, 2017 from https://opendata.aemet.es/centrodedescargas/productosAEMET?.

AEMET. (2017). *Olas de calor en España desde 1975*.

Arriazu, A., & Monge-Barrio, A. (2017). *Resilience of the Built Environment to the Climate Change in the Southern European Region. An initial Review*. Retrieved from http://hdl.handle.net/10171/44912.

Avamet. (n.d.). Retrieved from http://www.avamet.org/mx-microdades.php?id=c32m014e25.

Baccini, M., Biggeri, A., Accetta, G., Kosatsky, T., Katsouyanni, K., Analitis, A., et al. (2008). Heat effects on mortality in 15 European cities. *Epidemiology, 19*(5), 711–719. (Cambridge, MA) http://doi.org/10.1097/EDE.0b013e318176bfcd.

Barbosa, R., Vicente, R., & Santos, R. (2015). Climate change and thermal comfort in Southern Europe housing: A case study from Lisbon. *Building and Environment, 92*, 440–451. https://doi.org/10.1016/j.buildenv.2015.05.019.

Brotas, L., & Nicol, F. (2016a). *Using Passive Strategies to prevent overheating and promote resilient buildings. PLEA2016 Los Angeles - 32nd International Conference on Passive and Low Energy Architecture. Cities, Buildings, People: Towards Regenerative Environments*.

Brotas, L., & Nicol, J. F. (2016b). *The problem of overheating in European dwellings. Windsor 2016*, (April), 7–10.

Cadot, E., Rodwin, V. G., & Spira, A. (2007). In the heat of the summer: Lessons from the heat waves in Paris. *Journal of Urban Health, 84*(4), 466–468. https://doi.org/10.1007/s11524-007-9161-y.

Díaz, J., Jordán, A., García, R., López, C., Alberdi, J. C., Hernández, E., et al. (2002). Heat waves in Madrid 1986–1997: Effects on the health of the elderly. *International Archives of Occupational and Environmental Health, 75*(3), 163–170. https://doi.org/10.1007/s00420-001-0290-4.

Gangolells, M., & Casals, M. (2012). Resilience to increasing temperatures: Residential building stock adaptation through codes and standards. *Building Research & Information, 40,* 1–20. https://doi.org/10.1080/09613218.2012.698069. (June).

Hatvani-Kovacs, G., Belusko, M., Skinner, N., Pockett, J., & Boland, J. (2016). Heat stress risk and resilience in the urban environment. *Sustainable Cities and Society, 26,* 278–288. https://doi.org/10.1016/j.scs.2016.06.019.

Karimpour, M., Belusko, M., Xing, K., Boland, J., & Bruno, F. (2015). Impact of climate change on the design of energy efficient residential building envelopes. *Energy and Buildings, 87,* 142–154. https://doi.org/10.1016/j.enbuild.2014.10.064.

Laaidi, K., Zeghnoun, A., Dousset, B., Bretin, P., Vandentorren, S., Giraudet, E., et al. (2012). The impact of heat islands on mortality in Paris during the August 2003 heat wave. *Environmental Health Perspectives, 120*(2), 254–259. https://doi.org/10.1289/ehp.1103532.

Loughnan, M. E., Carroll, M., & Tapper, N. (2014). Learning from our older people: Pilot study findings on responding to heat. *Australasian Journal on Ageing, 33*(4), 271–277. https://doi.org/10.1111/ajag.12050.

Mahdy, M. M., & Nikolopoulou, M. (2014). Evaluation of fenestration specifications in Egypt in terms of energy consumption and long term cost-effectiveness. *Energy and Buildings, 69,* 329–343. https://doi.org/10.1016/j.enbuild.2013.11.028.

Matthies, F., Bickler, G., Marin, N., & Hales, S. (2008). Heat–health action plans. WHO Europe, 45. Retrieved from http://www.euro.who.int/__data/assets/pdf_file/0006/95919/E91347.pdf.

Menéndez, H. G. (2006). *Principios y estrategias del diseño bioclimático en la arquitectura y el urbanismo: eficiencia energética.* [Madrid]: Consejo Superior de los Colegios de Arquitectos de España.

METEONAVARRA. (n.d.). Retrieved from http://meteo.navarra.es/estaciones/mapadeestaciones.cfm.

Metropoli, F. (2009). *Sarriguren Ecociudad Ecocity*. Madrid: Gobierno de Navarra, Departamento de vivienda y ordenación del territorio. Retrieved from http://ezproxy.si.unav.es:2048/login?url=http://search.ebscohost.com/login.aspx?direct=true&AuthType=ip,url&db=cat00378a&AN=bnav.b1957324&lang=es&site=eds-live&scope=site.

Monge-Barrio, A., & Sánchez-Ostiz, A. (2015). Energy efficiency and thermal behaviour of attached sunspaces, in the residential architecture in Spain. Summer Conditions. *Energy and Buildings, 108,* 244–256.

Neila González, J. (2004). *Arquitectura bioclimática en un entorno sostenible.* Madrid: Munilla-Lería.

Porritt, S., Shao, L., Cropper, P., & Goodier, C. (2011). Adapting dwellings for heat waves. *Sustainable Cities and Society, 1*(2), 81–90. https://doi.org/10.1016/j.scs.2011.02.004.

Pyrgou, A., Castaldo, V. L., Pisello, A. L., Cotana, F., & Santamouris, M. (2017). On the effect of summer heatwaves and urban overheating on building thermal-energy performance in central Italy. *Sustainable Cities and Society, 28,* 187–200. https://doi.org/10.1016/j.scs.2016.09.012.

Ren, Z., Wang, X., & Chen, D. (2014). Heat stress within energy efficient dwellings in Australia. *Architectural Science Review, 57*(3), 227–236. https://doi.org/10.1080/00038628.2014.903568.

Sakka, A., Santamouris, M., Livada, I., Nicol, F., & Wilson, M. (2012). On the thermal performance of low income housing during heat waves. *Energy and Buildings, 49,* 69–77. https://doi.org/10.1016/j.enbuild.2012.01.023.

Saman, W., Boland, J., Pullen, S., De Dear, R., Soebarto, V., Miller, W., et al. (2013). *A framework for adaptation of Australian households to heat waves*. Retrieved from http://www.nccarf.edu.

au/sites/default/files/attached_files_publications/Saman_2013_Adapting_households_to_heat_waves.pdf.

Sánchez-Ostiz, A., Monge-Barrio, A., Domingo-Irigoyen, S., & González-Martínez, P. (2014). Design and experimental study of an industrialized sunspace with solar heat storage. *Energy and Buildings, 80,* 231–246. https://doi.org/10.1016/j.enbuild.2014.05.031.

Santamouris, M., Sfakianaki, A., & Pavlou, K. (2010). On the efficiency of night ventilation techniques applied to residential buildings. *Energy and Buildings, 42*(8), 1309–1313. https://doi.org/10.1016/j.enbuild.2010.02.024.

UNE-EN 15251. (2008). Parámetros del ambiente interior a considerar para el diseño y la evaluación de la eficiencia energética de edificios incluyendo la calidad del aire interior, condiciones térmicas, iluminación y ruido.

Van Hooff, T., Blocken, B., Hensen, J. L. M., & Timmermans, H. J. P. (2015). Reprint of: On the predicted effectiveness of climate adaptation measures for residential buildings. *Building and Environment, 83,* 142–158. https://doi.org/10.1016/j.buildenv.2014.10.006.

Vandentorren, S., Bretin, P., Zeghnoun, A., Mandereau-Bruno, L., Croisier, A., Cochet, C., et al. (2006). August 2003 Heat wave in France: Risk factors for death of elderly people living at home. *European Journal of Public Health, 16*(6), 583–591. http://doi.org/10.1093/eurpub/ckl063.

Wang, X., Chen, D., & Ren, Z. (2010). Assessment of climate change impact on residential building heating and cooling energy requirement in Australia. *Building and Environment, 45*(7), 1663–1682. https://doi.org/10.1016/j.buildenv.2010.01.022.

Chapter 8
Incidence of Passive Measures in a *Climate-Ready* Architecture. Attending to Energy Demands and Overheating Risks

8.1 Different Approaches for the Study of Residential Buildings in Warming Climates

For studying the performance of residential buildings in actual summer conditions or in future global warming scenarios, two main approaches are considered. In climate zones where summers are traditionally mild and buildings do not normally have air conditioning systems, studies evaluate if the duration of a possible overheating is acceptable. In climate zones with more severe summers and where air conditioning systems are commonly used, studies evaluate cooling demands, as well as peak energy demands (electricity, generally) during severe heat-related events.

Temperature thresholds to evaluate overheating in buildings that are conditioned naturally do not match cooling setpoints in mechanically conditioned spaces. This is mainly because users' expectations are different when having cooling systems than when having natural conditioning spaces. This is reflected in comfort standards, both in European UNE-EN 15251 (2008) and in American ASHRAE55 (2013), as seen in Chap. 3.

In both approaches, residential buildings must incorporate the most appropriate passive measures, so that overheating hours and maximum temperatures are minimal. In case there is an efficient cooling system, since comfort cannot be guaranteed by passive measures alone, it should consume the minimal energy possible. We must be aware that systems or energy supply can fail (especially in peak energy demand situations, for example in prolonged heatwaves), as can the family's capacity to deal with energy costs (energy poverty related to high temperatures). This is why the optimization of passive measures incorporated to architecture is a measure for the short and long term, both at climate and at social level.

© Springer International Publishing AG 2018
A. Monge-Barrio and A. Sánchez-Ostiz Gutiérrez, *Passive Energy Strategies for Mediterranean Residential Buildings*, Green Energy and Technology,
https://doi.org/10.1007/978-3-319-69883-0_8

8.1.1 Limits of Overheating

UNE-EN 15251 proposes different methods to evaluate thermal well-being inside buildings. The simplest one consists in valuing the general conditions of thermal comfort by giving a maximum percentage of hours out of range, specified for each category of indoor environment, and during those hours when buildings are occupied. Residential buildings, since they can have all types of dwellers, must be considered permanently occupied (24 h/7 days per week).

As has been widely discussed in Chap. 3, in residential buildings with different dwellers from all ages, categories of indoor environment can vary from what the norm suggests, having in mind the most vulnerable population (elderly, sick, children or people with reduced mobility) with health problems and their limited capacity for thermoregulation, as well as people in energy poverty who are unable to deal with the economic burden of thermal conditioning systems.

Overheating thresholds are in direct relation with occupants' health because when those thresholds are surpassed in the dwellings, users do not frequently have other options to deal with temperatures inside their homes. So, going over these overheating thresholds directly compromises occupant's health, especially the most vulnerable.

This is why indoor thermal categories for this study are not considered according to the condition of the building, whether new (Cat II) or existing (Cat III), but according to the occupants. In this way if it complies with the highest expectative category, related to the elderly, babies, the sick and handicapped (Cat I), it is having under consideration people of all ages and condition, a characteristic of residential buildings. Considering Category II (normal level of expectative) that is suitable for new buildings and renovations, we could be leaving without protection the most vulnerable ones.

As recommended criteria, acceptable deviations should be under 3% (or 5%) of occupied hours, which is a maximum of 262 h (or 438 h) per year, but no more than 5 h (or 9 h) per week (UNE-EN 15251, 2008).

8.1.2 Reducing Cooling Demand and Peak Loads

The aims in this case are of an energy and economic nature, but not directly related with users health in the Mediterranean area. In fact, air conditioning systems in buildings are often understood as systems for protecting people against severe summer conditions, either due to a certain climatic zone or due to heatwaves. It is often suggested that at least one room in the dwelling should count with a cooling system, as a form of *Cool Retreat*.

In the last years, an important increase has taken place in cooling systems' installation in the Mediterranean area, adding to occupant's perception of them as a basic equipment for the home (in climates with very warm summer). Energy efficiency of air conditioning systems or the substitution of existing ones for more efficient ones, and the use of energy produced in situ or nearby, are essential to reduce energy consumption. Lastly, the design and refurbishment of near-zero energy consumption buildings in summer and winter conditions would contribute to reduce greenhouse effect emissions which contribute to global warming.

8.2 Incidence of Passive Measures in the Built Environment Over Energy Demand: Case Studies in Southern Europe

For studying the incidence of different passive measures in the reduction of energy demand in residential buildings in Southern Europe, different energy modelling in two representative typologies have been carried out. Both heating and cooling demands (H+C-Demand) have been analysed in order to evaluate case studies as they were built (so, without considering any thermal rehabilitation), as well as the impact of various retrofit measures in winter and summer conditions.

The studied residential typologies are T1, collective residential building as a linear block with less than 10 dwellings, ground floor and four floors, and T5, terraced house, with a ground floor and one floor under roof (Fig. 8.1). These typologies are described in Chap. 5.

The proposed Case Studies of each typology consist of an individual or collective residential building built between 1940–80, that has not had any energy rehabilitation (CS1); that has a first combination of measures that supposes the building rehabilitated or new (M1, and variants); and a more demanding combination of measures in the way of a nearly zero energy consumption building, nZEB (M2, and variants). In Tables 8.1 and 8.2, the main characteristics of the envelope are summarized, beside the general patterns of use for heating, cooling and ventilation systems, some of them following the Spanish Building Code (CTE-HE, 2013) and others adapted from it.

The heating and cooling energy demand of these two building typologies has been energy modelled and analysed in ten locations with different climates and severities of winter and summer. Thus, a wide and varied approach is given that covers the complexity of the climate in the Mediterranean Europe, since even having common climatic characteristics, there are important differences between locations.

The selected cities which are shown in Fig. 8.2 are Athens (Greece, GRC), Valencia (Spain, ESP), Madrid (Spain, ESP), Mostar (Bosnia & Herzegovina, BIH), Milan (Italy, ITA), Rome (Italy, ITA), Barcelona (Spain, ESP), Nimes (France, FRA), Portoroz (Slovenia, SVN) and Oporto (Portugal, PRT).

The energy modelling has been developed with DesignBuilder software and EnergyPlus, both well tested and widely recognized. The models have been simulated with IWEC2 climate data of ASHRAE (Huang, 2011). The climate data for

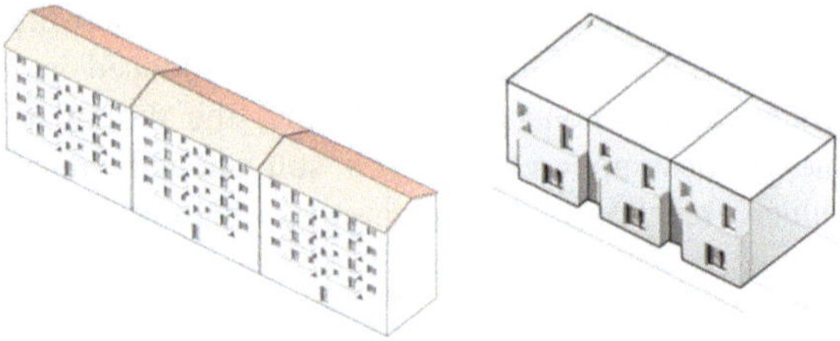

Fig. 8.1 Sketches of studied typologies: T1, linear block and T5, terraced house

Table 8.1 Summary of envelopes. Main characteristics and codes

Façades	
F3	Façade of cavity wall, U_{eq} = 1.95 W/m^2 K
F3+rFei08	F3 with exterior insulation (8 cm), U_{eq} = 0.43 W/m^2 K
F3+rFei08 AL	F3 with exterior insulation (8 cm), with high albedo, U_{eq} = 0.43 W/m^2 K
F3+rFei20	F3 with exterior insulation (20 cm), U_{eq} = 0.15 W/m^2 K
Roofs	
R1	Flat roof, U = 2.65 W/m^2 K
R1+rRei12	R1 with insulation (12 cm) and gravel over slab, U = 0.24 W/m^2 K
R1+rRei20	R1 with insulation (20 cm) and gravel over slab, U = 0.15 W/m^2 K
R3	Pitched roof (with attic without use), U = 2.0 W/m^2 K
R3+rRic12	R3 with insulation in cavity, under the roof (12 cm), U = 0.30 W/m^2 K
R3+rRic20	R3 with insulation in cavity, under the roof (20 cm), U = 0.19 W/m^2 K
Windows	
G1	Wood frame with single glass 6 mm U = 5.7 W/m^2 K
rGLoE	New frame low transmittance + Low Emissivity Double Glazing U_{Glass} = 1.4 W/m^2 K, g-value = 0.61
Ground floor with first floor slab (T1 Typology)	
L2	Reinforced concrete slab U = 2.22 W/m^2 K
L2+rLei08	L2 with insulation under the slab (8 cm) U = 0.4 W/m^2 K
L2+rLei12	L2 with insulation under the slab (12 cm) U = 0.29 W/m^2 K
Ground floor slab (T5 Typology)	
L5	Concrete slab over gravel U = 1.18 W/m^2 K
L5+rLei04	L5 with insulation over the slab (4 cm) U = 0.47 W/m^2 K

Legend
U: Transmittance (W/m^2 K)
U_{eq}: U equivalent considering thermal bridges
g-value: solar factor

Table 8.2 T1 and T5 typologies: summary of envelope characteristics and HVAC conditions

Codes of typologies and measures		Facade	Roof	Windows	Floor 1stF(T1) GF(T5)	Airtightness (h^{-1})	Codes for ventilation rate (h^{-1})	Final codes
T1	**CS**	F3	R3	G1	L2	0.7	0.63/4SN	**CS1.2**
	M1	F3+rFei08	R3+rRic12	rGLoE	L2+rLei08	0.3	0.63/4SN	**M1.2**
							0.63HR/4SN	**M1.2_HR**
							0.4/4SN	**M1.2_0.4**
							0.4HR/4SN	**M1.2_0.4HR**
							0.63HR(NSV)	**M1.2_HR_NSV**
		F3+rFei08AL					0.63/4SN	**M1.2AL**
							0.4HR/4SN	**M1.2_0.4HRAL**
	M2	F3+rFei16	R3+rRic20	rGLoE	L2+rLei12	0.2	0.63/4SN	**M2.2**
							0.4HR/4SN	**M2.2_0.4HR**
T5	**CS**	F3	R1	G1	L5	0.8	0.63/4SN	**CS1.2**
	M1	F3+rFei08	R1+rRic12	rGLoE	L5+rLei04	0.2	0.63/4SN	**M1.2**
							0.63HR/4SN	**M1.2_HR**
							0.4/4SN	**M1.2_0.4**
							0.4HR/4SN	**M1.2_0.4HR**
							0.63HR	**M1.2 HR NSV**
		F3+rFei08 AL					0.63/4SN	**M1.2AL**
							0.4HR/4SN	**M1.2_0.4HRAL**
	M2	F3+rFei16	R1+rRic20	rGLoE	L5+rLei04	0.2	0.63/4SN	**M2.2**
							0.4HR/4SN	**M2.2_0.4HR**

Heating and Cooling Demand (H+C-Demand), Pattern of use:
– Heating: Setpoints (low) 20 °C (8–24 h), 17 °C (0–8 h), from October to May both included (CTE-HE, 2013)
– Cooling: Setpoints (high) 27 °C (0–24 h), from June to September both included
Shading (S): Blinds (8–24 h, June–September) when solar radiation >120 W/m^2, and encoded as "0.2"
Notes and Legend
HR Heat Recovery Ventilation System
AL albedo (high reflectance)
[a]Ventilation (V), Pattern of use:
– 0.63/4SN: 0.63 h^{-1} (CTE-HE, 2013), except in summer (June to September) 1–8 h (4SN: 4 h^{-1} Summer Night). Night ventilation when outdoor temperatures are inferior to the interior ones (2 °C), being the latter higher than 22 °C
– 0.4/4SN: idem, although changing the general ventilation rate to 0.4 h^{-1}
– 0.63HR/4SN: idem 0.63/4SN, but with HR except during summer night
– 0.4HR/4SN: idem 0.4/4SN, but with HR except during summer night
– 0.63HR/NSV: 0.63 h^{-1} all day with HR, but without summer night ventilation

the climate change scenario of 2050 has been generated with CCWorldWeatherGen (Jentsch, James, Bourikas, & Bahaj, 2013), because for the locations of Southern Europe, there is no other common and open climate data source for energy modelling to our knowledge.

In addition to the current state, or building as built, CS1 (Sect. 8.2.1), the incidence and relevance of a proper orientation (Sect. 8.2.2) and the need for the incorporation of different shading devices (Sect. 8.2.3) have been analysed, as well as two main types of retrofit measures, M1 (Sect. 8.2.4) and M2 (Sect. 8.2.5), with different variants specified in each paragraph. Summaries of heating and cooling energy demand per location are finally included in Sect. 8.2.6.

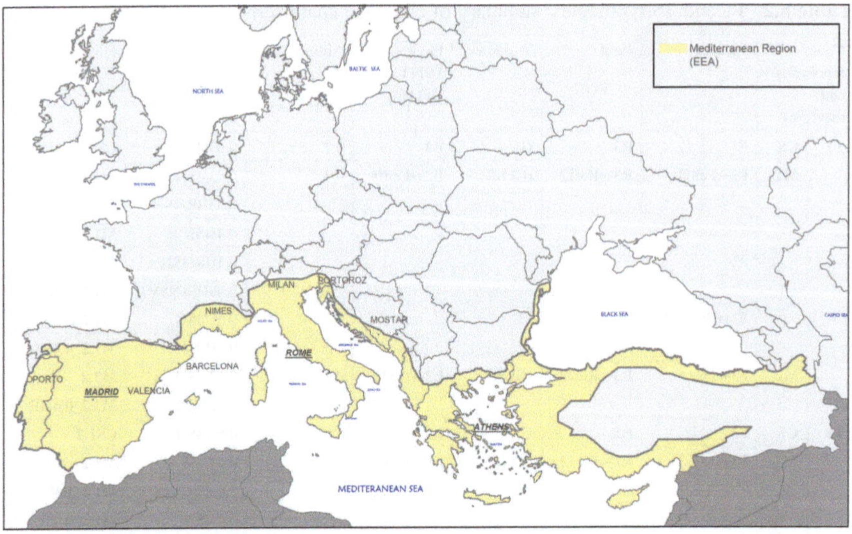

Fig. 8.2 Mediterranean Region of Europe with 10 locations of the research. EEA delimitation (EEA, 2016)

Results show energy demands, first in the floor with less energy heating and cooling demand in the building, second floor in T1 and ground floor (GF) in T5, and second in the floor with more energy demand, fourth floor in T1 and first floor in T5. Heating demand and cooling demand are disaggregated, showing the climate season severity in each location and the variability among locations of the Mediterranean Europe.

The study analyses both the impact of measures on cooling energy demand, and the total energy demand for thermal conditioning (cooling and heating), since one of the most important characteristics of Mediterranean and European locations is that buildings must cope with winter and summer conditions and therefore looks for optimized solutions for the whole year.

8.2.1 *Current State*

Current state (CS) deals with residential typologies built between 1940 and 1980 and without retrofit thermal measures, although incorporating passive summer measures with a common design and use in the Mediterranean Region, as solar shading and night ventilation.

Shading is only included in the Spanish Building Code (CTE) for non-residential buildings. However, in residential buildings, shading devices such as venetian blinds, shutters, overhangs, etc., are commonly designed by architects as part of the architectural project (Fig. 8.3). Occupants widely use and appreciate them, as can be seen during a walk around any town of Southern Europe especially during summer days. T1

Fig. 8.3 Images of shadings in residential buildings, whose design and use is very common and appreciated in the Mediterranean Region. Montpellier (France) and Barcelona (Spain)

Typology has balconies, which also work as overhangs, shading windows immediately below. The fourth floor of this typology has a continuous sloping roof eave.

On the other hand, night ventilation is included in the CTE for residential buildings, although in this study, the pattern of use has been adapted from this Code, as it is detailed in Table 8.2. Current state with the characteristics and patterns of use described in this paragraph is called CS1.2.

All case studies are facing south, the most generally recommended orientation in these latitudes, providing a lower heating demand in winter due to solar gains and being easily shaded in summer. In addition, the construction system of this case study has thermal mass, consisting of brick façades and concrete slabs (see Table 8.1), one of the most typical system in the Mediterranean Region. Thermal mass runs as a temperature modulator and in combination with night ventilation is especially favourable in summer.

In Fig. 8.4, heating and cooling demand of second floor and fourth floor in T1 Typology, and in Fig. 8.5, heating and cooling demand of ground floor (GF) and first floor in T5 Typology, both of Base Case CS1.2, are presented. According to the analysis of results, first in the current scenario and second in the 2050 scenario, it is concluded:

- In all the studied Southern European locations, heating demand is higher than cooling demand, although with different percentages according to the winter and summer severity of each location.
- In all cases, total energy demand is higher in the upper floor than in an intermediate floor in T1 and ground floor in T5. These results are known and coherent with bibliography and monitoring results exposed in Chaps. 5 and 7. In T1 Typology, cooling demand is 48% more on average.
- Locations with higher heating and cooling demand are Milan, Portoroz and Mostar, and locations with lower heating and cooling demand are Valencia, Barcelona and Oporto. As an example, in T1 Typology, Milan's demand, which

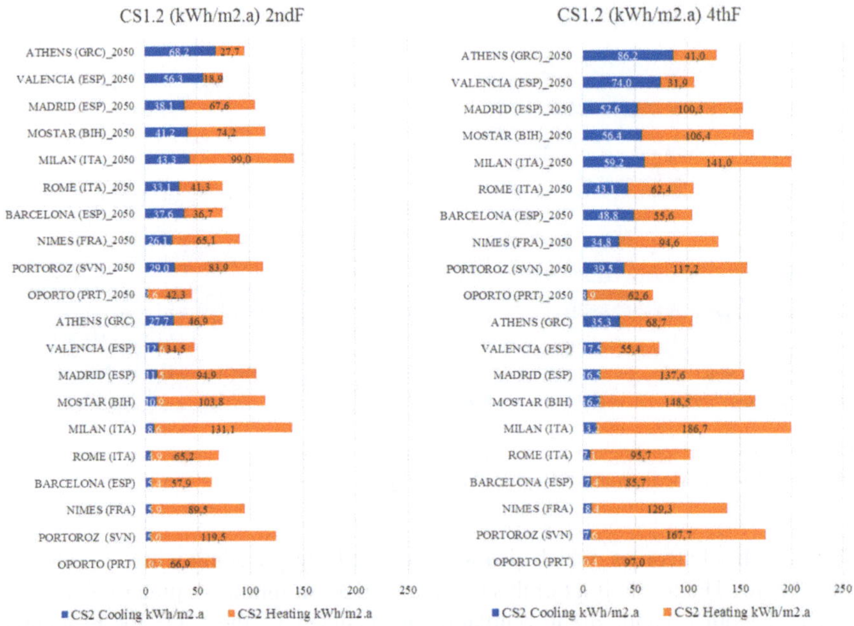

Fig. 8.4 Base Case (**CS1.2**) of T1 Typology (Linear Block). Heating and cooling energy demand in second floor and in fourth floor. Base Case incorporates in summer solar shading in windows and night ventilation

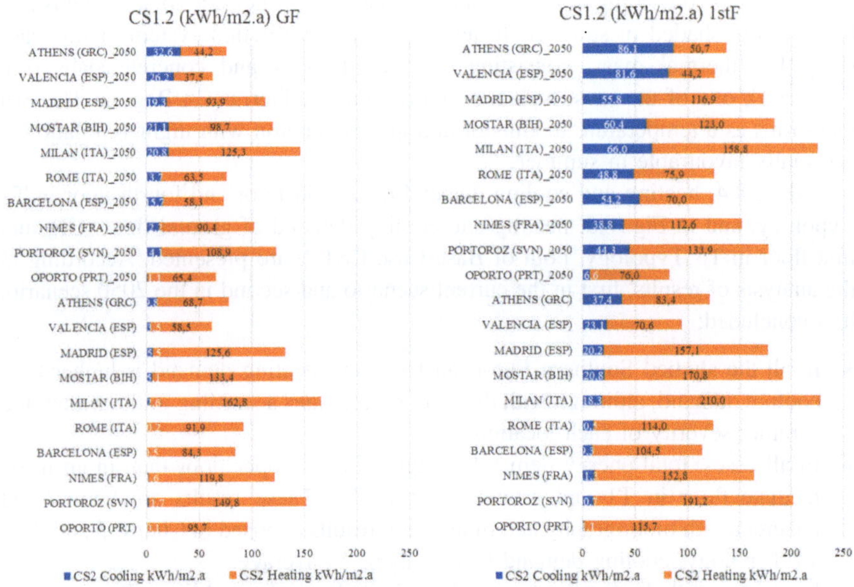

Fig. 8.5 Base Case (**CS1.2**) of T5 Typology (Terraced House). Heating and cooling energy demand in ground floor and in first floor. Base Case incorporates in summer solar shading in windows and night ventilation

is the location with more energy demand, almost triples Valencia's one, the location with the lowest total energy demand.

- Locations with higher heating demand are Milan and Portoroz, and with higher cooling demand are Athens, followed by Valencia, Madrid and Mostar.
- In a future 2050 Scenario, an increase in cooling and a decrease in heating are given in all studied locations as is expected. Cooling demand is higher than heating demand in Athens and Valencia mainly in the T1 Typology (in T5, only in the 1st floor in Athens), producing a shift in the main energy demand.
- In a 2050 Scenario, locations where total energy demand will increase are Valencia (especially in T1, with an increase of 60% in 2nd F and 45% in 4th F) and Athens (especially in T1, with 28% in 2nd F and 22%, in 4th F), because although heating demand is reduced there is a substantial increase in cooling demand.
- In a 2050 Scenario, locations in which total heating and cooling demand will decrease are Oporto, Portoroz and Nimes, although cooling demand has increased specially in the two latter ones. Demand will also decrease in Madrid and Mostar but only when considering Typology T5.
- In a 2050 Scenario, other locations will maintain a similar energy demand depending on typology and floor, although with important variations in the incidence of heating and cooling in the total amount.

8.2.2 Solar Radiation Control: Orientation Incidence

As summer climatic conditions become more severe, control over orientation of windows becomes essential, even having to avoid windows in very unfavourable orientations. Relying on a simple and available solar chart, limits given by orientations to control solar radiation and so preventing overheating can be analysed.

Energy modelling results of the buildings instead of facing south (S), facing west (W-90°) and south-west (SW-45°) are exposed in this paragraph. While Base Case CS1.2 was shown in Figs. 8.4 and 8.5, results of the buildings facing south-west, CS1.2_SW, are shown in Fig. 8.6 and Fig. 8.7, and results facing west, CS1.2_W, are shown in Figs. 8.8 and 8.9. Thereby, incidence of west and south-west orientations effect, in heating and cooling energy demand, is appreciated. It occurs in all locations, and taking into account that all typologies were simulated with movable shading during the day and, in addition T1 Typology has overhangs, as described before.

In all Mediterranean locations[1] and in relation to Base Case facing south (CS1.2), residential buildings facing west are those that have a higher cooling (an increase of 40% in T1 and 21% in T5[2]) and heating energy demand. These values suppose an average increase in cooling demand of 4.5 kWh/m^2 a in T1 and

[1] Oporto is excluded in the mean values of cooling demand in these cases, due to its low demand.

[2] In the current scenario and in T5 Typology, only cooling demand percentages in first floor are considered because in the ground floor, demand increase is very variable. However, both floors are considered in mean energy demand values (kWh/m^2 a).

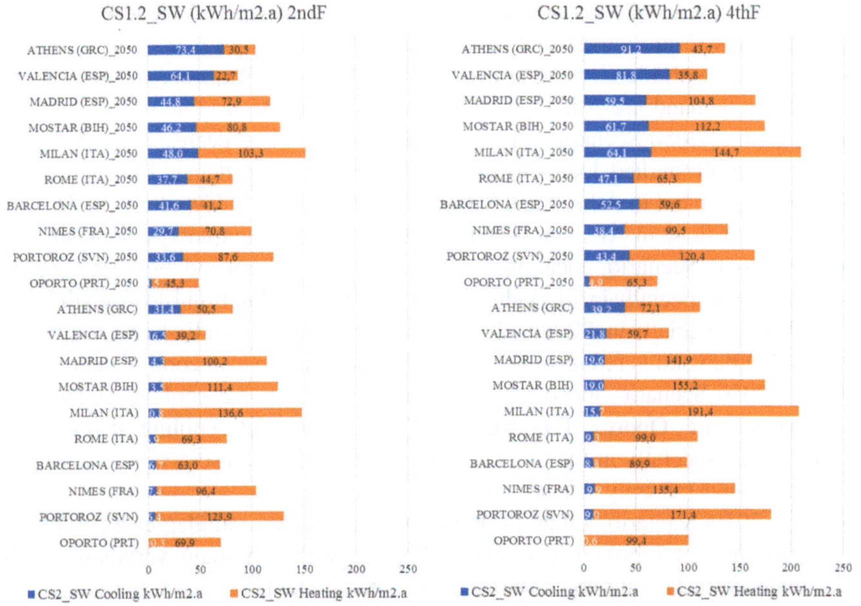

Fig. 8.6 Base Case of T1 Typology (Linear Block), but facing SW-45° (**CS1.2_SW**). Heating and cooling energy demand in second floor and in fourth floor. Base Case incorporates in summer solar shading in windows and night ventilation

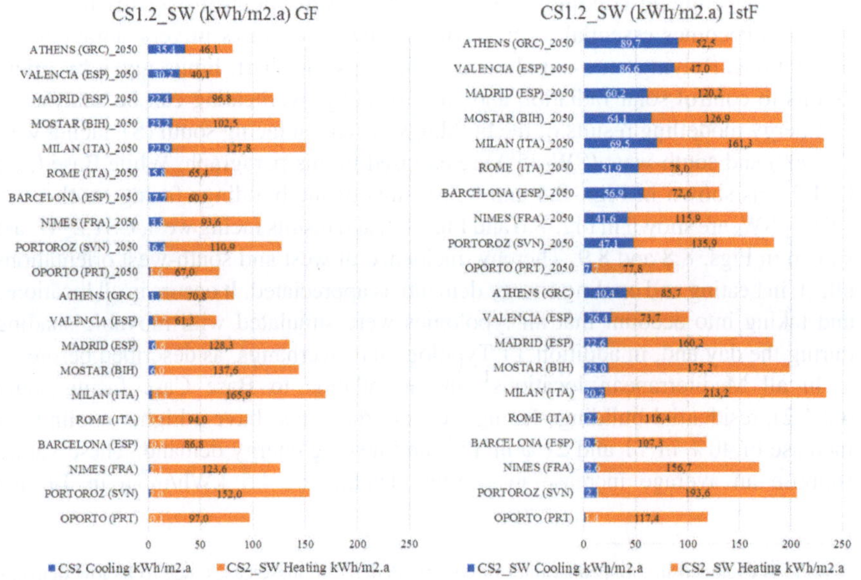

Fig. 8.7 Base Case of T5 Typology (Terraced House), but facing SW-45° (**CS1.2_SW**). Heating and cooling energy demand in ground floor and in first floor. Base Case incorporates in summer solar shading in windows and night ventilation

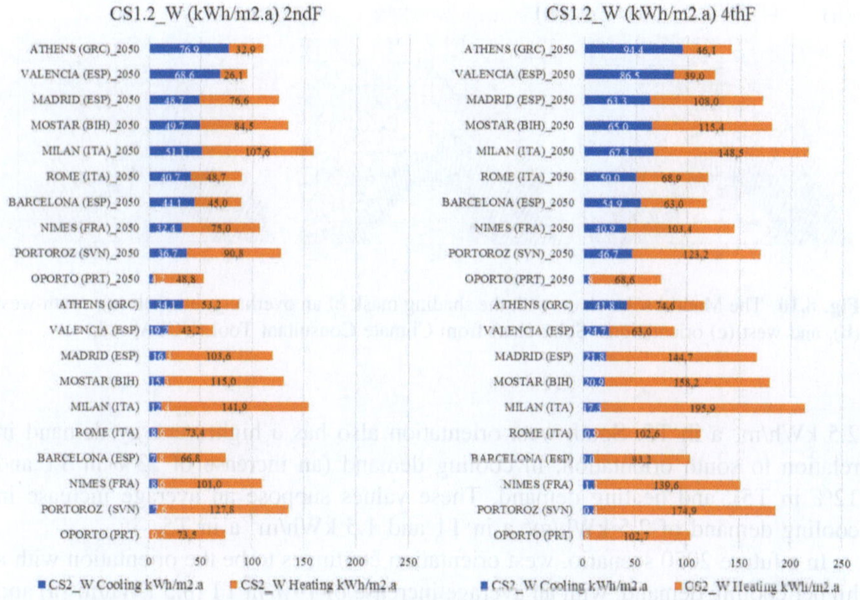

Fig. 8.8 Base Case of T1 Typology (Linear Block), but facing W-90° (**CS1.2_W**). Heating and cooling energy demand in second floor and in fourth floor. Base Case incorporates in summer solar shading in windows and night ventilation

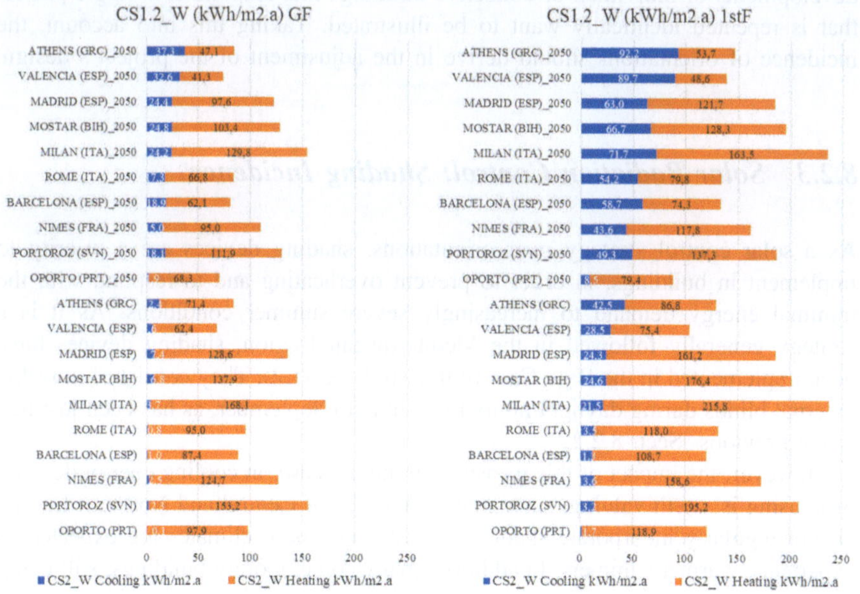

Fig. 8.9 Base Case of T5 Typology (Terraced House), but facing W-90° (**CS1.2_W**). Heating and cooling energy demand in ground floor and in first floor. Base Case incorporates in summer solar shading in windows and night ventilation

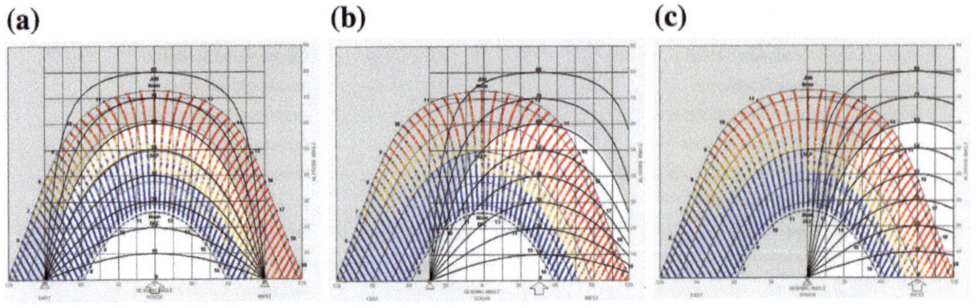

Fig. 8.10 The Madrid solar chart, with the shading mask of an overhang, in south (**a**), south-west (**b**), and west (**c**) orientations. Solar chart from Climate Consultant Tool (UCLA, n.d.)

2.5 kWh/m^2 a in T5. South-west orientation also has a higher energy demand in relation to south orientation, in cooling demand (an increase of 23% in T1 and 12% in T5), and heating demand. These values suppose an average increase in cooling demand of 2.5 kWh/m^2 a in T1 and 1.5 kWh/m^2 a in T5.

In a future 2050 scenario, west orientation continues to be the orientation with a higher cooling demand, with an average increase of 19% in T1 (8.5 kWh/m^2 a) and 16% in T5 (5 kWh/m^2 a), and a higher increase in heating demand. In 2050 Scenario, south-west orientation supposes an increase in cooling demand of 11% in T1 (5 kWh/m^2 a) and 9.5% in T5 (3 kWh/m^2 a).

With these results, differences given in energy demand in big residential urban developments of individual or collective buildings that respond to a single project that is repeated identically want to be illustrated. Taking this into account, the incidence of orientations should derive in the adjustment of the project's design.

8.2.3 Solar Radiation Control: Shading Incidence

As a solar control strategy over orientations, shading devices are a priority to implement in buildings, in order to prevent overheating and to respond with the minimal energy demand to increasingly severe summer conditions. As it is a strategy generally followed in the Mediterranean Region, shading devices have been implemented in the Base Case of this study (CS1.2). They consist of movable external blinds during daylight hours from June to September, as has been justified in the previous Sect. 8.2.2.

However, the impact of this measure on the increase on cooling energy demand of housing is worth valuing, since many countries of Central and Northern Europe do not regularly incorporate shading devices, and these climates are expected to experience warmer summers. In addition, many contemporary buildings, still being

Fig. 8.11 Examples of shading in the Mediterranean from traditional (Palma, Spain) to new designs (Montpellier, France)

located in Mediterranean climates, are being designed without shading relying only on active cooling systems, the achievement of users comfort. On the other hand, and as has been seen in the monitoring paragraph of Chap. 7, solar shading is normally performed manually by occupants in residential buildings, and some of them may not be aware of the incidence of the increase in cooling energy demand, underestimating its relevance (Figs. 8.10 and 8.11).

Graph of Case Study CS1.1 for T1 Typology (linear block) is shown in Fig. 8.12, being the same model as CS1.2, but without movable solar shading. Case Study CS1.3 is shown in Fig. 8.13, and is a CS1.1 model without the balconies that worked as fixed overhangs, shading windows situated just below them. Last, results of Case Study CS1.1 for T5 Typology (terraced house) without retractable solar shading are presented in Fig. 8.14.

In these latitudes, the use of typical overhangs facing south couldn't be sufficient as the only system of solar shading for all summer months. In addition, being a fixed solar protection system, the building receive less solar radiation in winter if it is not adequately designed, and therefore can result in a higher heating demand. On the other hand, solar chart and a shading mask of overhangs in windows facing west and south-west are shown in Fig. 8.10 in order to explain the inefficiency of overhangs in those orientations.

Graphs of Case Studies without retractable solar shading, facing west and in the upper floor of the building typologies studied in this research are included in Fig. 8.15, so it is possible to evaluate the global incidence of some of the main factors that contribute to overheating by solar radiation, as are the orientation, the lack of solar shading and the location in the upper floor of a building.

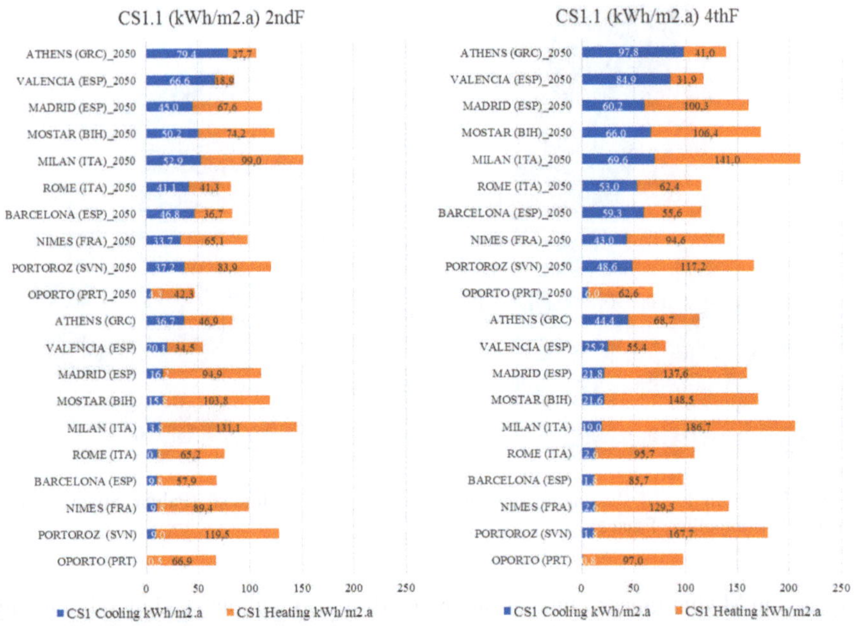

Fig. 8.12 Base Case of T1 Typology (Linear Block), but without blinds as solar system protection (**CS1.1**), in second and intermediate floor (2nd F), and in fourth and last floor (4th F)

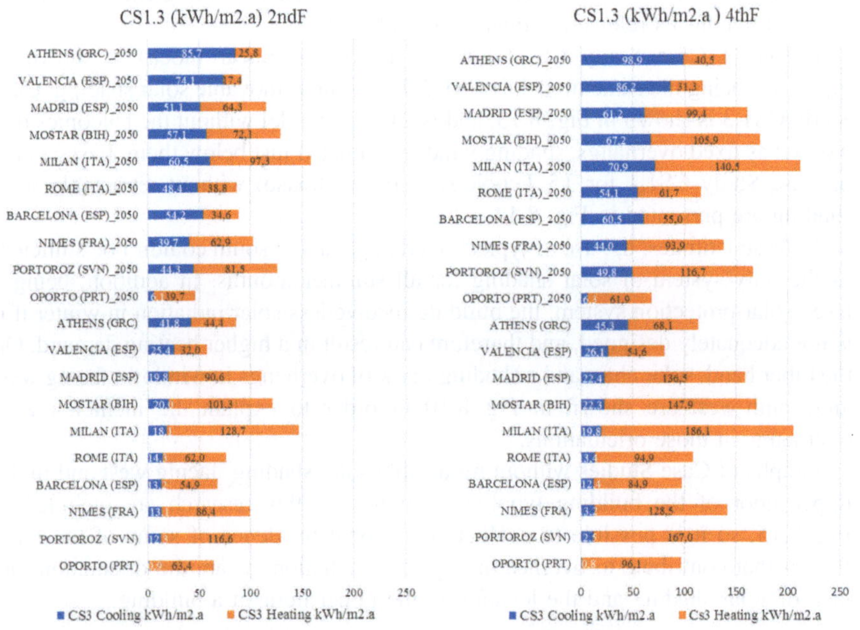

Fig. 8.13 Base Case of T1 Typology (Linear Block), but without blinds and overhangs as solar system protection (**CS1.3**), in second and intermediate floor (2nd F), and in fourth and last floor (4th F)

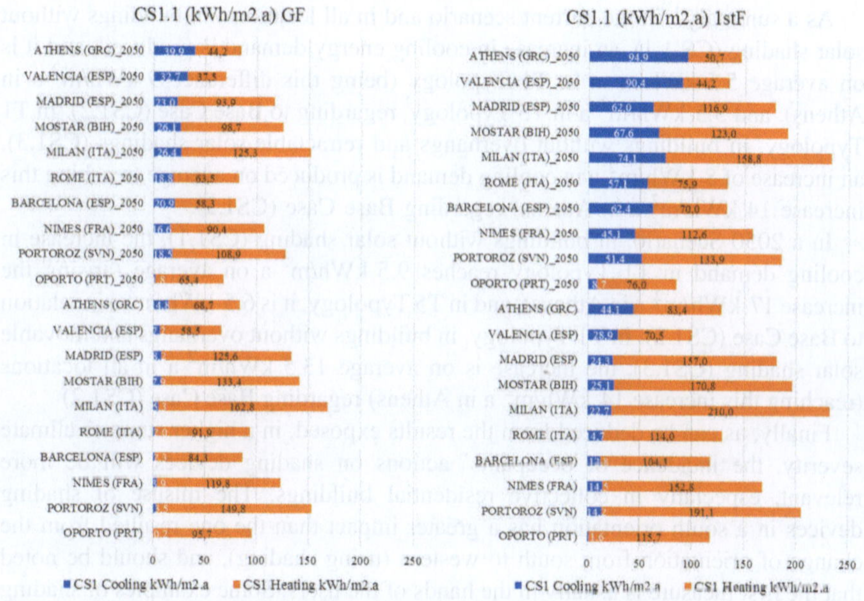

Fig. 8.14 Base Case of T5 Typology (Terraced House), but without blinds as solar system protection (**CS1.1**), in Ground Floor (GF) and in first and last floor (1st F)

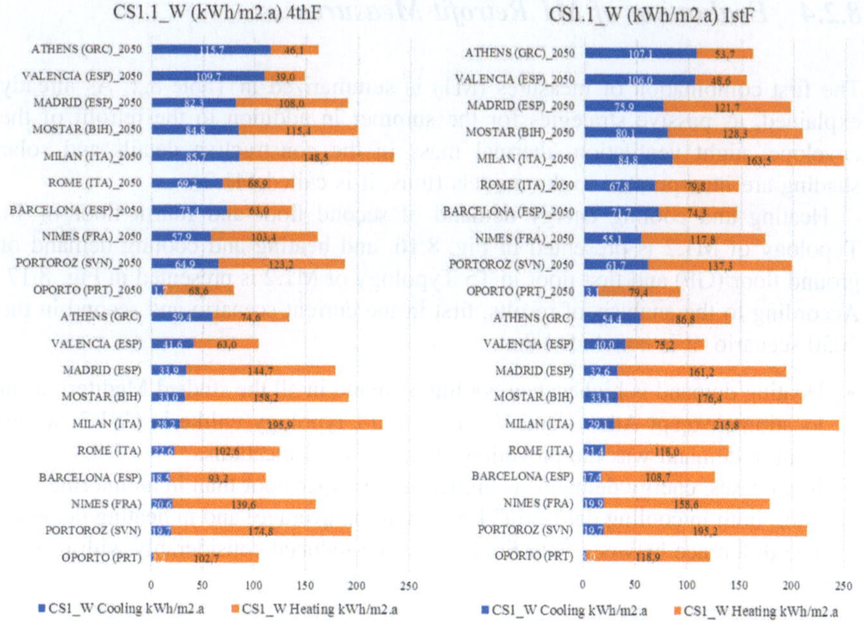

Fig. 8.15 Base Case of T1 Typology (Linear Block), but without blinds and facing west (**CS1.1_W**), in fourth and last floor (4th F), and Base Case of T5 Typology (Terraced House), but without blinds and facing west (**CS1.1_W**), in first and last floor (1st F)

As a summary[3], in the current scenario and in all locations, in buildings without solar shading (CS.1.1), an increase in cooling energy demand is produced, and it is on average 5.5 kWh/m^2 a in T1 Typology (being this difference 9 kWh/m^2 a in Athens), and 3.5 kWh/m^2 a in T5 Typology, regarding to Base Case (CS1.2). In T1 Typology, in buildings without overhangs and retractable solar shadings (CS1.3), an increase of 8 kWh/m^2 a in cooling demand is produced on average (reaching this increase 14 kWh/m^2 a in Athens) regarding Base Case (CS1.2).

In a 2050 Scenario, in buildings without solar shading (CS1.1), the increase in cooling demand in T1 Typology reaches 9.5 kWh/m^2 a on average (arising the increase 17 kWh/m^2 a in Athens), and in T5 Typology, it is 6.5 kWh/m^2 a in relation to Base Case (CS1.2). In T1 Typology, in buildings without overhangs and movable solar shading (CS1.3), the increase is on average 13.5 kWh/m^2 a in all locations (reaching this increase 14 kWh/m^2 a in Athens) regarding Base Case (CS1.2)

Finally, as can be deduced from the results exposed, in a higher summer climate severity, the incidence of occupants' actions on shading devices will be more relevant, especially in collective residential buildings. The misuse of shading devices in a south orientation has a greater impact than the one resulted from the change of orientation from south to western (using shading), and should be noted that the first measure is usually in the hands of the users. Some examples of shading devices are included in Fig. 8.11.

8.2.4 *Evaluation of M1 Retrofit Measures*

The first combination of measures (M1) is summarized in Table 8.2. As already explained, as passive strategies for the summer in addition to the retrofit of the envelope, night ventilation, thermal mass in the constructive details and solar shading are incorporated to the models (thus, it is called M1.2).

Heating and cooling energy demand of second floor and fourth floor in T1 Typology of M1.2 is presented in Fig. 8.16, and heating and cooling demand of ground floor (GF) and first floor in T5 Typology of M1.2 is presented in Fig. 8.17. According to the analysis of results, first in the current scenario and second in the 2050 scenario, it is concluded that:

- Heating demand is higher than cooling demand in all the studied Mediterranean locations, except Athens and Valencia in T1 typology unlike in CS1.2, where heating demand was the prevailing demand in all locations.
- In all cases, energy demand[4] is higher in the upper floor than in an intermediate floor, both in cooling demand (21.5% more on average) and in heating demand. The difference between both floors has been reduced considerably with respect

[3]Oporto is excluded in the average values of cooling demand in these cases, due to its low-energy demand.

[4]Footnote 3 Idem.

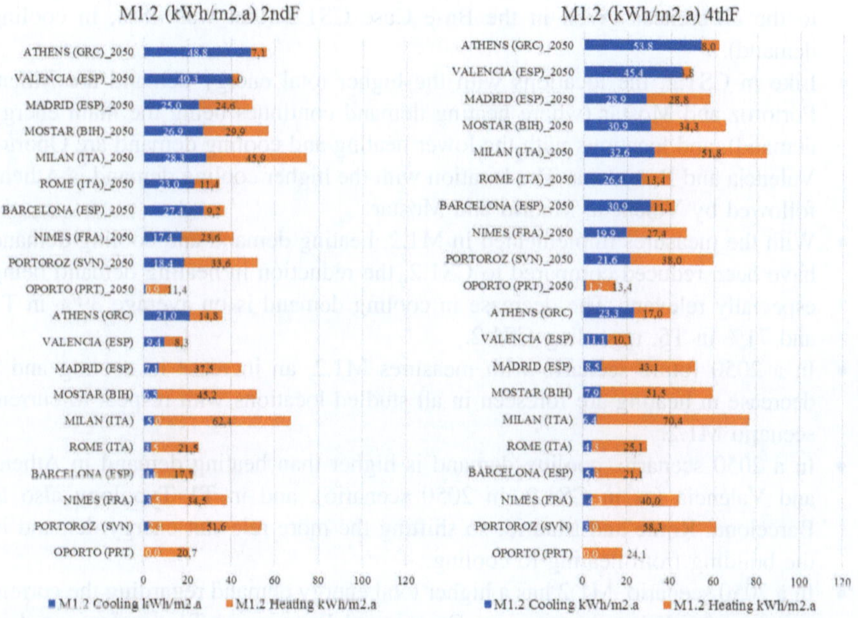

Fig. 8.16 M1.2 Retrofit Measures of T1 Typology (Linear Block). Heating and cooling energy demand in second floor and in fourth floor. Case Studies incorporate in summer solar shading in windows and night ventilation

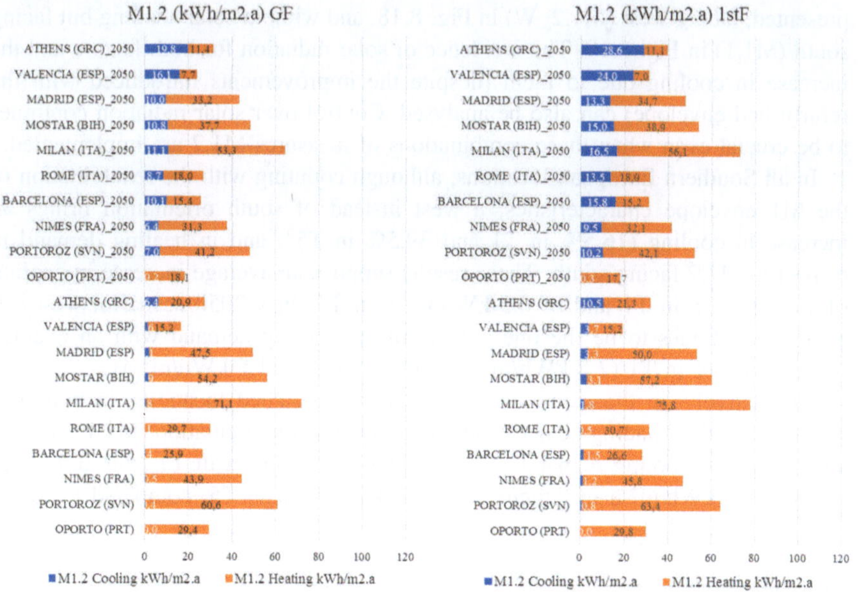

Fig. 8.17 M1.2 Retrofit Measures of T5 Typology (Terraced House). Heating and cooling energy demand in ground floor and in first floor. Case Studies incorporate in summer solar shading in windows and night ventilation

to the differences given in the Base Case CS1.2 (that was 48%, in cooling demand).

- Like in CS1.2, the locations with the higher total energy demand are Milan, Portoroz and Mostar (where heating demand continues being the main energy demand), and locations with the lower heating and cooling demand are Oporto, Valencia and Barcelona. The location with the higher cooling demand is Athens followed by Valencia, Madrid and Mostar.
- With the measures implemented in M1.2, heating demand and cooling demand have been reduced compared to CS1.2, the reduction in heating demand being especially relevant. The decrease in cooling demand is on average 39% in T1 and 71% in T5, regarding CS1.2.
- In a 2050 future scenario with measures M1.2, an increase in cooling and a decrease in heating are foreseen in all studied locations with respect to current scenario M1.2.
- In a 2050 scenario, cooling demand is higher than heating demand in Athens and Valencia (as in CS1.2, in 2050 scenario), and in T1 Typology also in Barcelona, Rome and Madrid, so shifting the more relevant energy demand in the building from heating to cooling.
- In a 2050 scenario, M1.2 has a higher total energy demand regarding the current scenario, in all locations except Oporto and Portoroz in T1 typology, and in Athens and Valencia in T5 Typology. In this future scenario, an increase in cooling demand in Valencia, Barcelona and Athens is especially significant. At the same time, the reduction in heating demand is very relevant

Results of energy demand in the upper floors of the two studied typologies are presented, facing west (M1.2_W) in Fig. 8.18, and without solar shading but facing south (M1.1) in Fig. 8.19. The incidence of solar radiation for both factors and the increase in cooling due to them (despite the improvements introduced with the refurbished envelope) can also be analysed. Control over solar radiation continues to be crucial even when these combinations of measures M1.2 are implemented.

In all Southern European locations, although counting with the rehabilitation of the M1 envelope characteristics, a west instead of south orientation brings an increase in cooling (16.5% in T1 and 39.5% in T5[5]) and in heating demand in respect to M1.2 facing south. These results suppose an average increase in cooling of 1 kWh/m^2 a in T1 and of 0.5 kWh/m^2 a in T5. In a 2050 scenario, west orientation continues to be the one with a higher cooling demand with an average increase of 9% in T1 (2.5 kWh/m^2 a) and 10% in T5 (1.5 kWh/m^2 a).

Alternatively, in the actual scenario and in all studied locations, in buildings without movable shading devices and maintaining south orientation (M1.1), there is an increase in cooling demand of an average 6 kWh/m^2 a in T1 (this difference reaching 10 kWh/m^2 a in Athens), and 3.5 kWh/m^2 a in T5, compared to M1.2.

[5]In the current scenario and in T5 typology, only first floor cooling demands are considered, since on the ground floor the increase in demand is quite variable. In absolute values, both floors are considered.

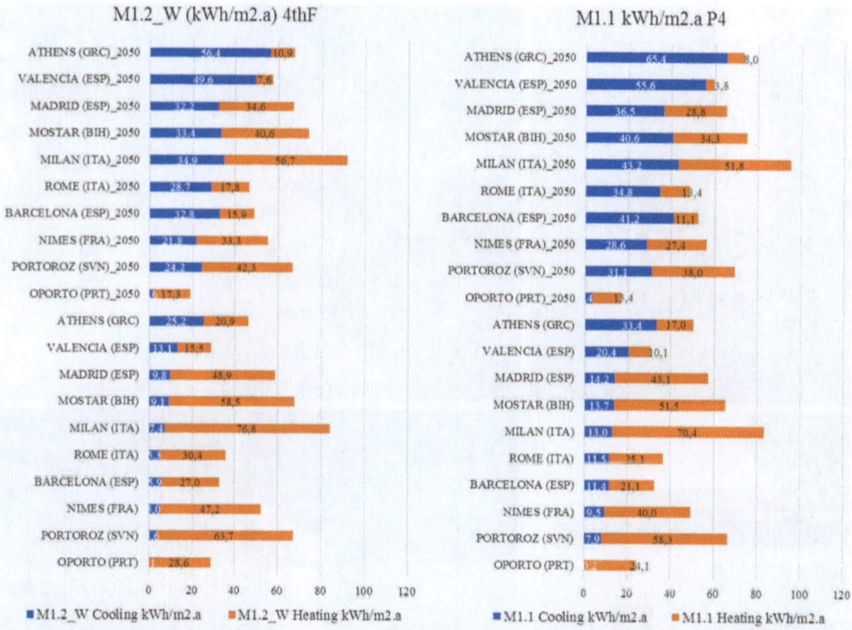

Fig. 8.18 M1 Retrofit Measures of T1 Typology (Linear Block), with solar shading but facing west (M1.2_W), and without solar shading but facing south (M1.1). Heating and cooling energy demand in fourth floor. Case Studies incorporates in summer night ventilation

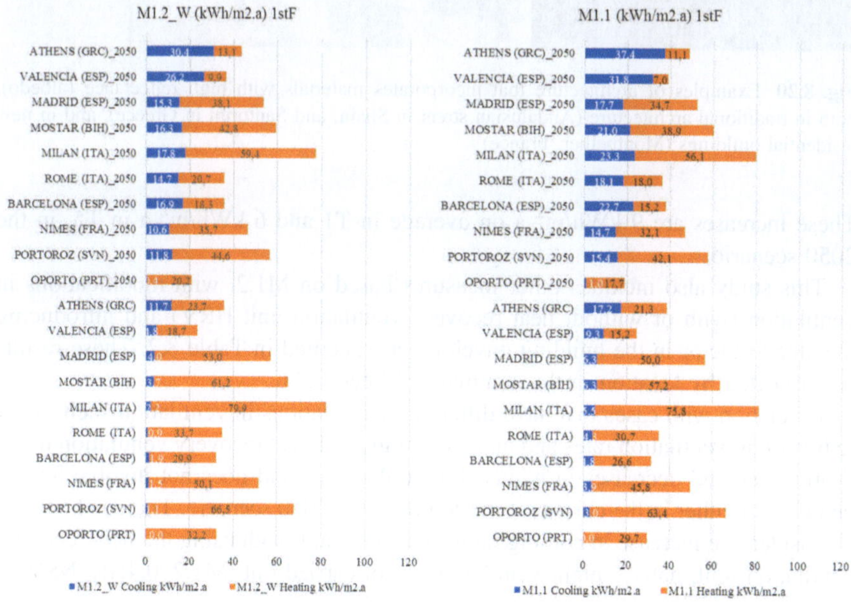

Fig. 8.19 M1 Retrofit Measures of T5 Typology (Terraced House), with solar shading but facing west (M1.2_W), and without solar shading but facing south (M1.1). Heating and cooling energy demand in first floor. Case Studies incorporates in summer night ventilation

Fig. 8.20 Examples of architecture that incorporates materials with high reflectance (albedo), both in traditional architecture (Andalusian street in Spain, and Santorini in Greece), and in new residential buildings (Montpellier, France)

These increases are 9 kWh/m^2 a on average in T1 and 6 kWh/m^2 a in T5, in the 2050 scenario.

This study also includes other measures based on M1.2, with modifications in ventilation (with or without heat recovery ventilation unit HRV) and introducing the albedo factor in the building envelope, as specified in Table 8.2. These results are collected by location in the summary of Sect. 8.2.6.

In relation with cases that have different modifications in ventilation such as the reduction in ventilation rates and an incorporation of heat recovery ventilation units, heating demand decreases in winter, but cooling demand does not involve important drops neither in the current scenario nor in the 2050 scenario. There is however a considerable increase in cooling demands when, although using the heat recovery ventilation unit, natural night ventilation is not carried out (M1.2_0,4HR_NSV).

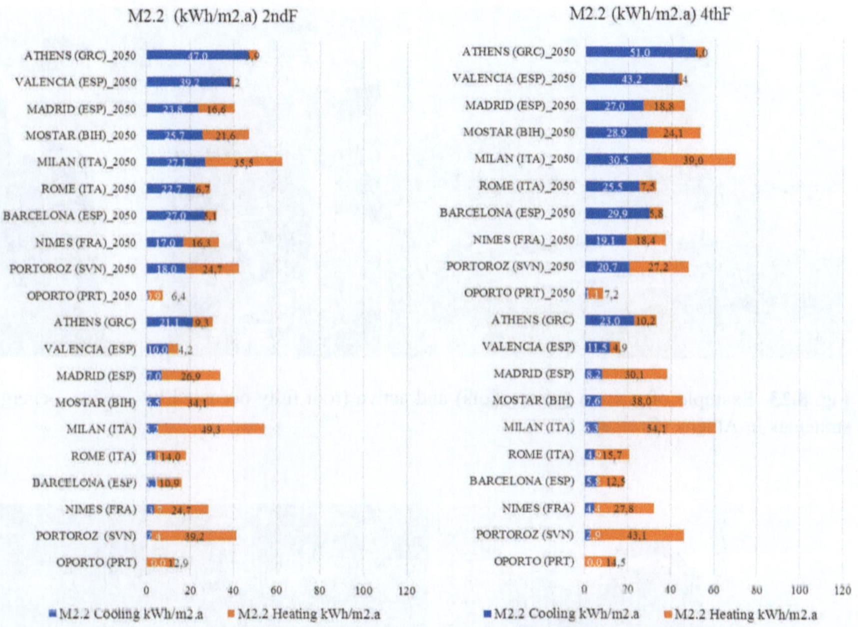

Fig. 8.21 M2.2 Retrofit Measures of T1 Typology (Linear Block). Heating and cooling energy demand in second floor and in fourth floor

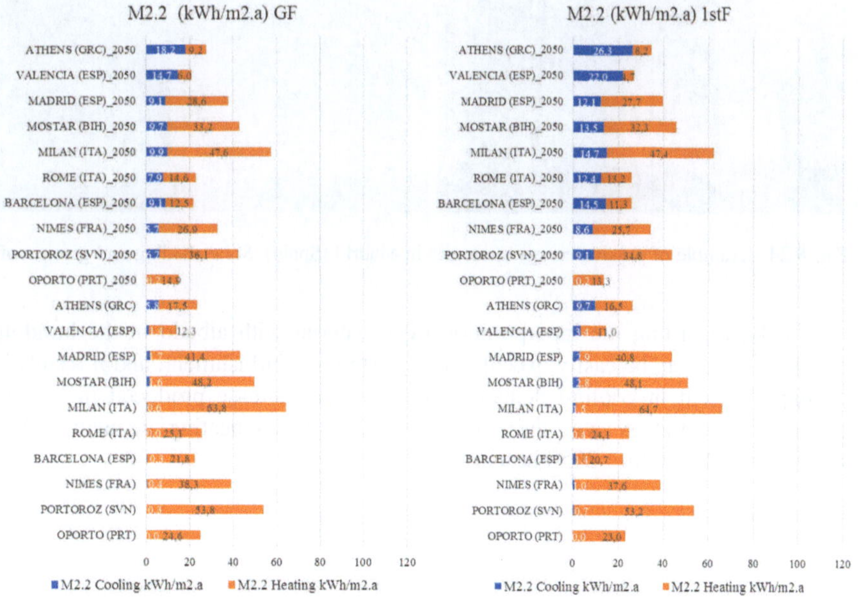

Fig. 8.22 M2.2 Retrofit Measures of T5 Typology (Terraced House). Heating and cooling energy demand in ground floor and in first floor

Fig. 8.23 Example of passive (green roofs) and active (roof fully occupied by services) energy strategies in Athens (Greece)

Fig. 8.24 Example of passive energy strategies in Madrid (Spain). Solar shading and green roofs

Finally, regarding the incorporation of a material with albedo in the building envelope[6], this can be easily executed in an exterior refurbishment, and it results in an improvement in cooling demand similar to the increase produced in heating demand. The strategy will therefore be relevant when heating demand is less necessary and in future 2050 scenarios.

[6]In T1 Typology, with a sloped roof with ceramic tiles, variation in roof's albedo has not been considered since it is not a usual solution.

Fig. 8.25 Example of passive energy strategies Milan (Italy) Movable shading and green facades in high rise buildings

Fig. 8.26 Example of passive energy strategies Rome (Italy) movable shading

8.2.5 Evaluation of M2 Retrofit Measures

M2 measures involve a more demanding rehabilitation of the building envelope, and its characteristics are defined in Table 8.2. Results exposed in this section include the use of movable solar shading devices (so are called, M2.2), and are shown in Fig. 8.21 and Fig. 8.22 in typologies T1 y T5, respectively. Like in M1.2,

Fig. 8.27 Example of passive energy strategies Barcelona (Spain)

Fig. 8.28 Example of passive energy strategies Nimes (France)

Fig. 8.29 Example of passive energy strategies Oporto (Portugal)

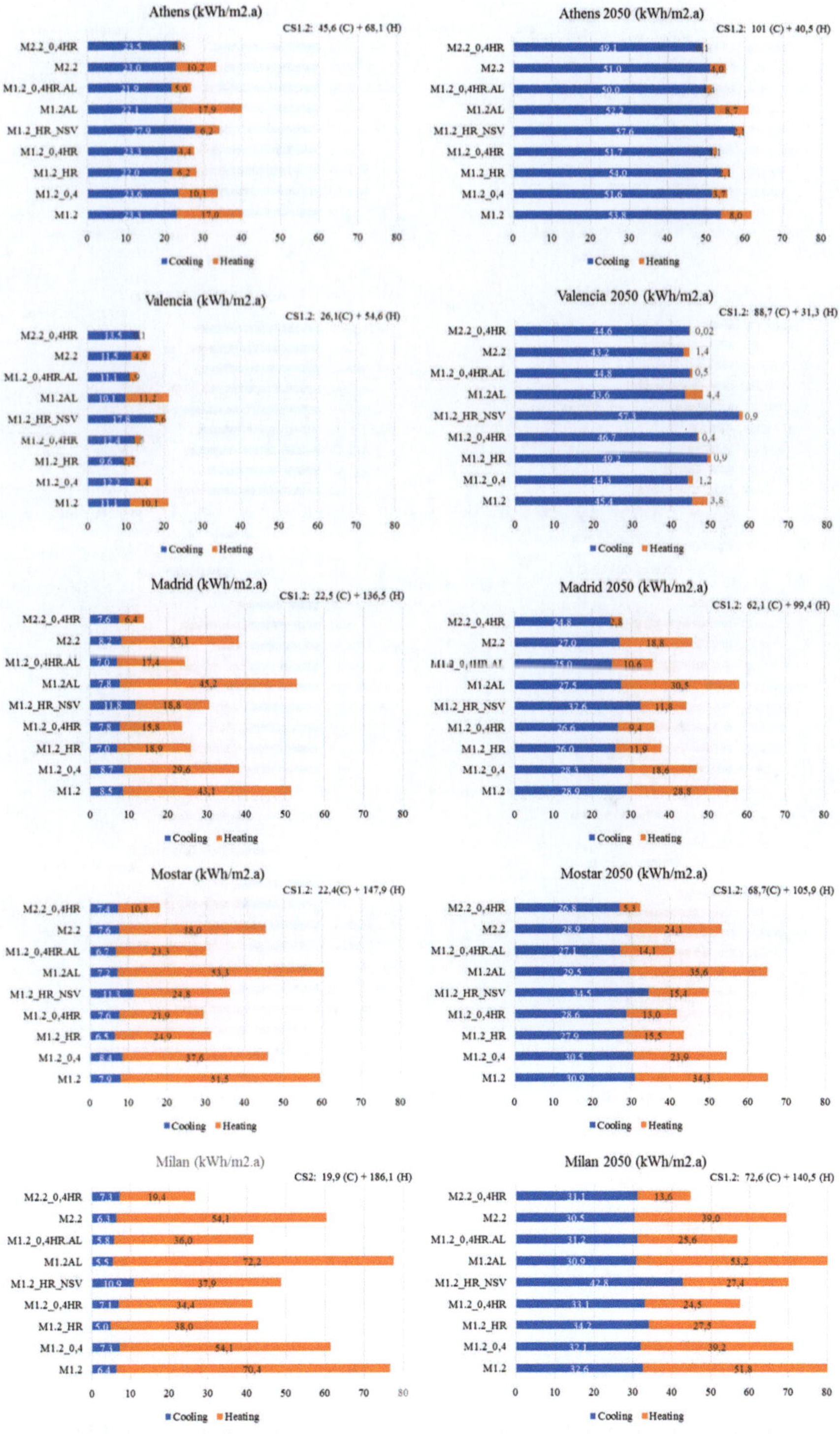

Fig. 8.30 Summary of applied measures M1 and M2, in the current scenario and 2050 scenario, and in T1 Typology (4th floor), in: Athens, Valencia, Madrid, Mostar and Milan

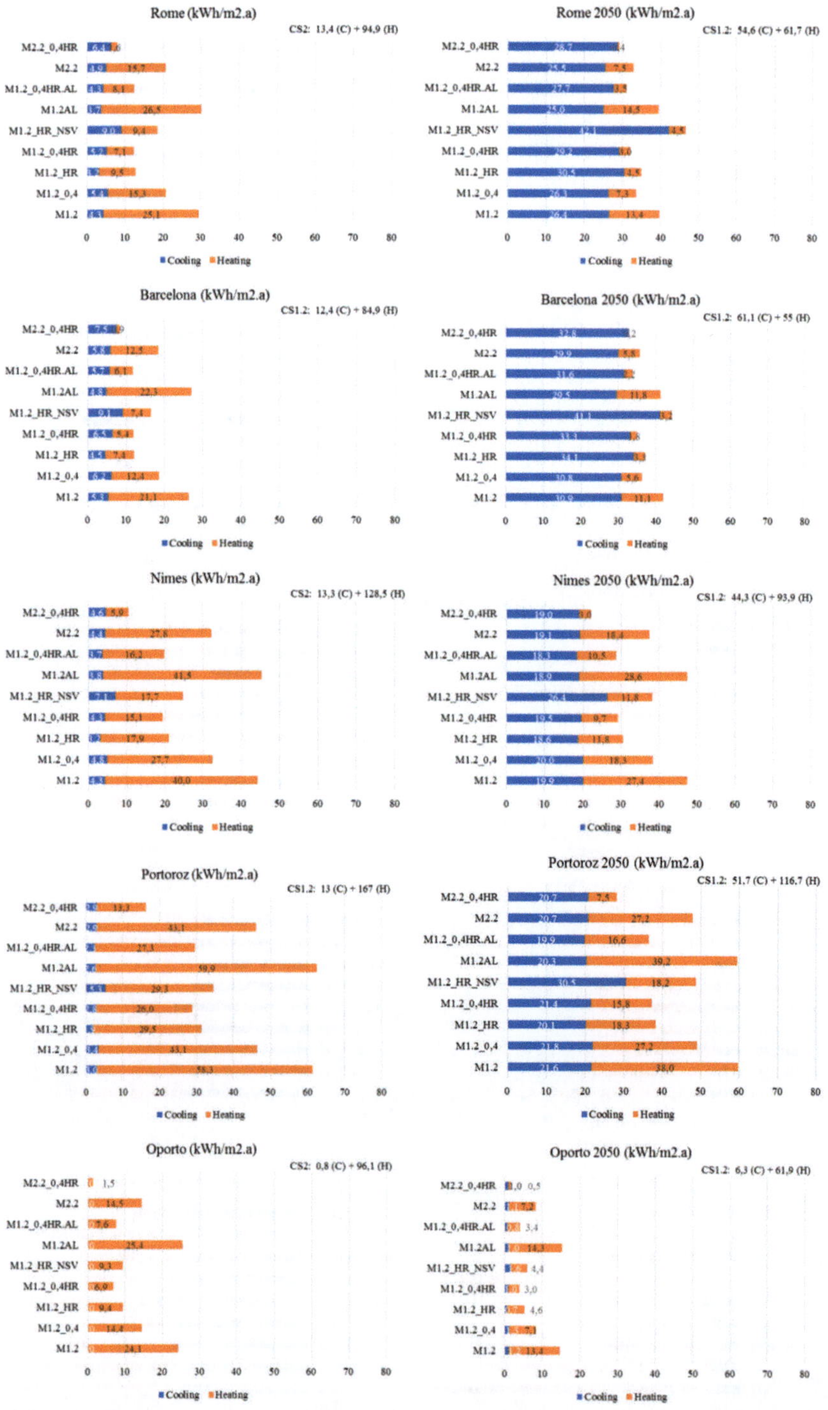

Fig. 8.31 Summary of applied measures M1 and M2, in the current scenario and 2050 scenario, and in T1 Typology (4th floor), in: Rome, Barcelona, Nimes, Portoroz and Oporto

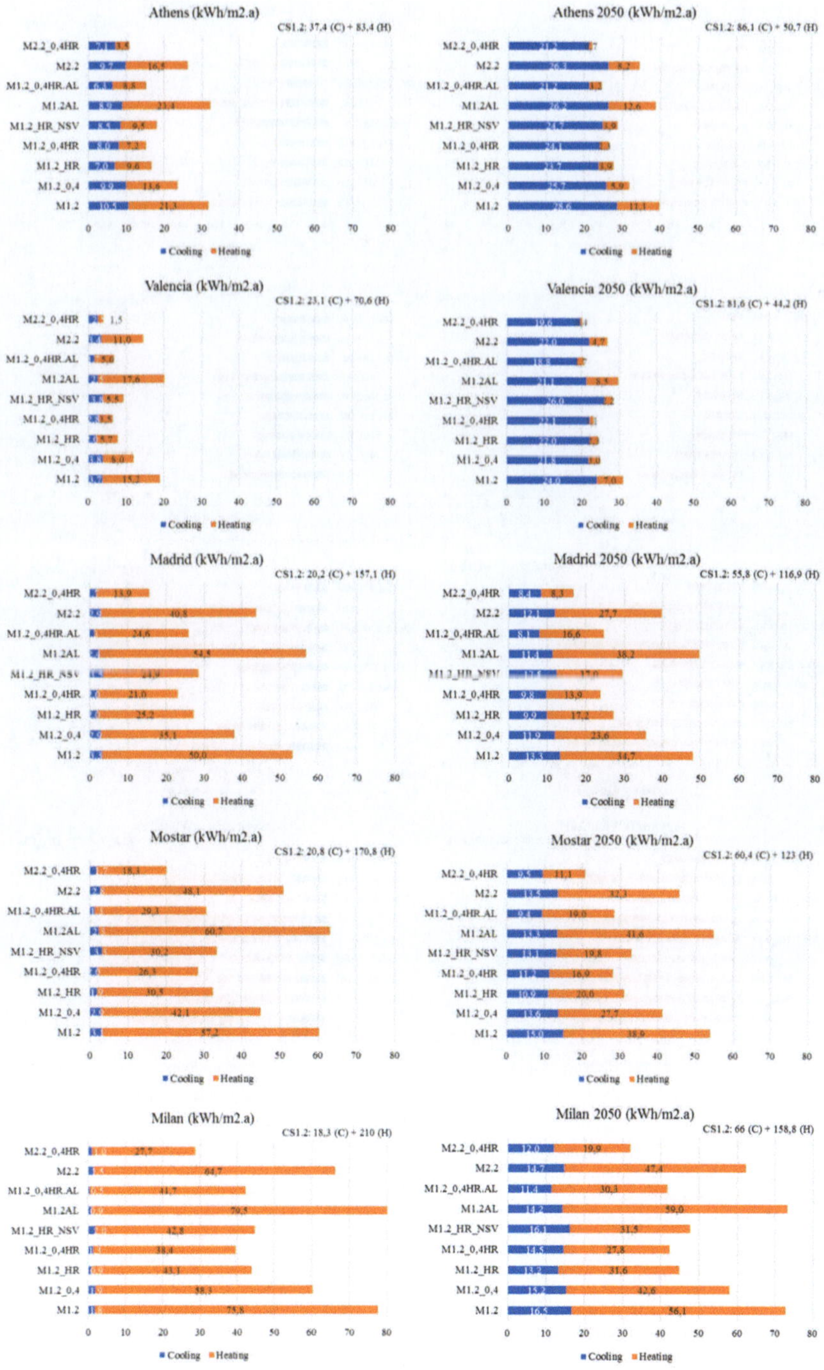

Fig. 8.32 Summary of applied measures M1 and M2, in the current scenario and 2050 scenario, and in T5 Typology (1st floor), in: Athens, Valencia, Madrid, Mostar and Milan

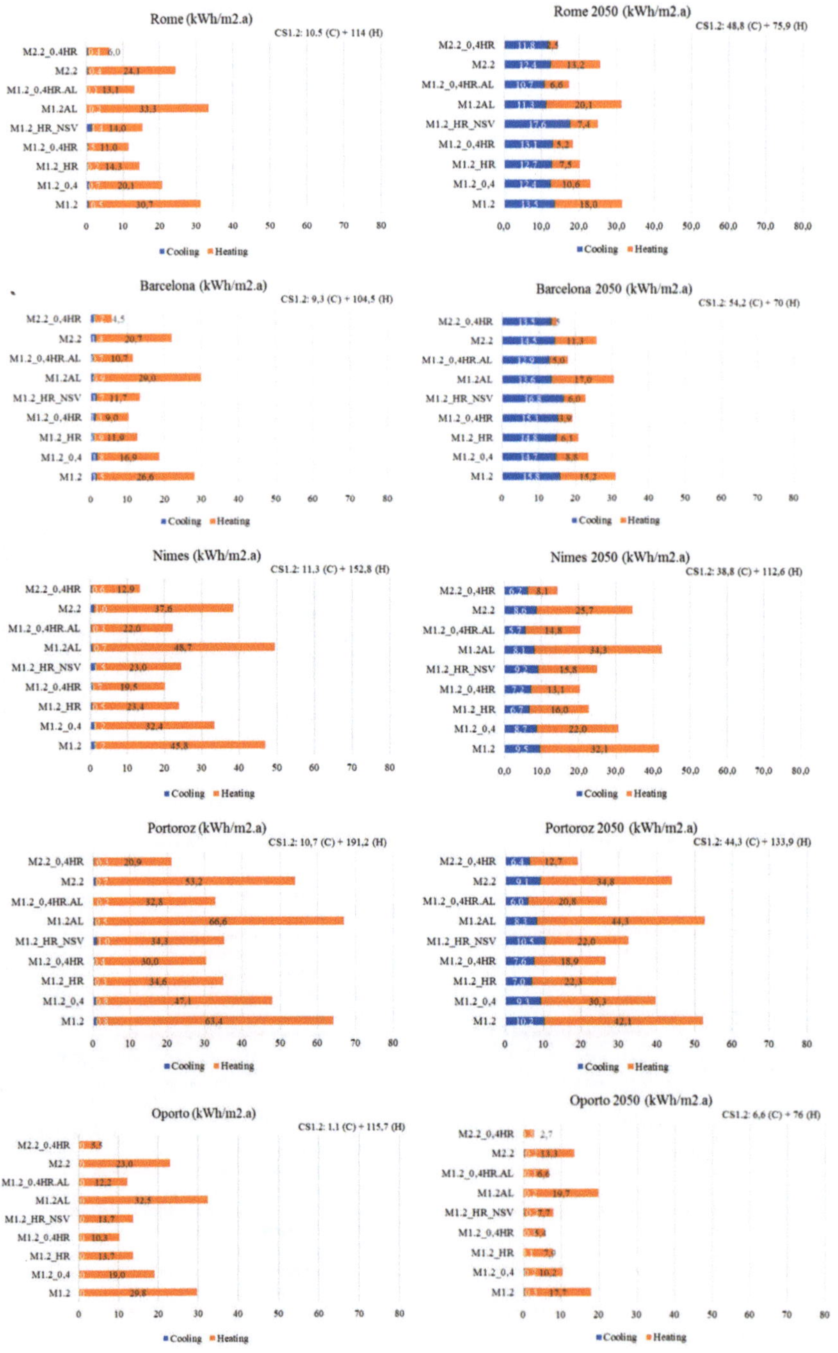

Fig. 8.33 Summary of applied measures M1 and M2, in the current scenario and 2050 scenario, and in T5 Typology (first floor), in: Rome, Barcelona, Nimes, Portoroz and Oporto

the pattern of use in summer also includes night ventilation (with adequate exterior conditions), and thermal mass in the interior, considering the rehabilitation of the envelope by the exterior. According to the analysis of results, first in the current scenario and second in the 2050 scenario, it is concluded that:

- Due to rehabilitation measures being principally aimed at winter conditions, heating demand is strongly reduced in all locations and continues being higher than cooling demand in all locations, except Athens and Valencia. Reduction compared to CS1.2 is 78% in T1 and 73% in T5. Cooling demand is reduced only slightly with respect to M1.2
- Energy demand on the top floor is superior to the intermediate floor, although differences between floors are being reduced, both in cooling (16.5% more on average in T1, and very varied in T5) and in heating demand.
- Locations with a higher total energy demand continue to be Milan, Portoroz and Mostar, due to the incidence of heating, and those with a lower total energy demand continue to be Valencia, Barcelona and Oporto. Cities with a higher cooling demand are Athens, followed by Valencia, Madrid and Mostar.
- In a future 2050 scenario, cooling demand is superior to heating demand in all locations except Porto, Portoroz and Milan in T1, and in Athens and Valencia in T5, shifting the most relevant energy demand in the building from heating demand to cooling demand, in more locations than in M1.2.
- In a future 2050 scenario, with the rehabilitated building M2.2 in T1, all locations have a higher total energy demand with respect to the current scenario rehabilitated, except Oporto with a lower demand (43% in both floors). In T5, only Athens and Valencia have a higher demand, while Oporto (43%) and Portoroz (20%) have a lower demand.

Finally, in the summary of results by location in Sect. 8.2.6, results for M2.2_0,4HR, the same combination of measures that M2.2, but including heat recovery and a reduced rate of 0.4 h^{-1}, (see in Table 8.2) are included. Heating demand continues to be reduced in all cases, but both in the current and in the 2050 scenario, cooling demand has even a slight increase in half of T1 buildings (mainly in Valencia, Barcelona and Rome) and in all T5 buildings.

8.2.6 Summary of Energy Demand by Location

Implemented strategies by cities (Examples in Figs. 8.23, 8.24, 8.25, 8.26, 8.27, 8.28 and 8.29), are shown as a summary for T1 typology (Figs. 8.30 and 8.31) and T5 typology (Figs. 8.32 and 8.33). They have been energy modelled for the current and the 2050 scenario. In every graph, Base Case without rehabilitation is also included (CS1.2), in order to analyse the reduction in energy demand in each case.

Apart from what has been specified in the previous section, these graphs show how far we can go in the reduction of heating, cooling and total energy demand with both levels of rehabilitation measures in each location and with each typology.

A reduction in heating demand is clear in all locations, scenarios and typologies, with the different measures applied (see Chap. 5), but the efficiency of a reduction in cooling is not so clear though, and can even be counterproductive. There has been an aim in showing the incidence of night ventilation on the reduction of cooling energy demand (showing a case without night ventilation M1.2_HR_NSV), and the incidence is clear in all locations

Cooling demand in the actual scenario and at least with a rehabilitation M1.2 in linear block typology (T1) has a cooling demand inferior to 5 kWh/m^2 a in Oporto, Portoroz, Nimes and Rome, and inferior to 10 kWh/m^2 a in Barcelona, Milan, Mostar and Madrid. In Athens, values are over 20 kWh/m^2 a. In terraced house typology (T5), even on the first floor which is the least favourable, cooling is under 5 kWh/m^2 a, in every location except Athens where it is around 10 kWh/m^2 a. However, in 2050 Scenario, in T1 Typology, all locations except Oporto widely exceed 10 kWh/m^2 a of cooling demand, and in T5 Typology, only Portoroz and Nimes, with values next to 10 kWh/m^2 a, together with Oporto, have less cooling demand.

Therefore, although the proposed rehabilitation measures reduce current heating consumption and also future cooling demand, these reductions are not considered enough for the challenges of future warming conditions.

Finally, the terraced house, T5 typology, is much more *Resilient* than the collective building, T1 typology, and offers more possibilities of adaptation on a warming climate, consuming less energy. In addition, the terraced house may only need cooling on the upper floor, and therefore users have the possibility of adaptation inside the dwelling swapping the use of the rooms, in summer without any cooling energy consumption.

8.3 Passive Measures and Overheating: Case Studies in Southern Europe

8.3.1 Methodology

A different and complementary approach to the solely energy one is the study of overheating according to comfort standards. Not all studied locations have severe enough summers for residential buildings to have cooling systems installed. In addition, and as it happens in winter, there is vulnerable population who cannot deal, at a certain moment, with the costs of installing a cooling system and energy required. From here emerges the importance of studying if the passive measures implemented will make buildings' overheating hours lower than those established in comfort standards (Saman et al., 2013).

In the study carried out by energy modelling, the European regulation and its adaptive approach (UNE-EN 15251, 2008) will be followed, so that if the building cannot allow an acceptable thermal environment in this way, cooling by mechanical systems will be necessary to guarantee occupants comfort and health.

As stated in Chap. 3, assigning existing buildings a Category III or new buildings a Category II, in residential buildings where population is intergenerational and where there can be sick or disabled people, there is a percentage of people whose indoor environment conditions should be evaluated according to a more demanding category (Category I). In addition to this, existing buildings with worse features, (either for being cheaper rentals or acquisitions), is where a higher percentage of vulnerable people could live. This is the reason why this study analyses results under Categories I and II of indoor thermal environment conditions.

In the ten locations included in the study, and in the current and future 2050 scenario, the time frequency for interior operative temperatures has been calculated in different ranges for the summer months (June to September included), and hours of overheating according to Annex A.2 of UNE-EN 15251. The limit from which comfort in this space cannot be guaranteed only by passive measures (related to building and use) is a 3% of occupied hours, and if dwellings are occupied 24 h, this supposses 262 h (Annex G).

The study has been carried out with both typologies studied in this chapter, T1 (linear block, multifamily house) and T5 (terraced house), where two floors are distinguished with a differentiated thermal behaviour: the most favourable floor (second floor in T1 and ground floor in T5), and the least favourable that is always the one under the roof (fourth floor in T1 and first floor in T5). In this way, measures' problematic and efficiency can be evaluated by typology. The code of modelling results includes also the considered floor.

Codes and characteristics of the different energy modelling are in Table 8.2, although they are summarized here in order to facilitate the interpretation of the graphs:

- CS1.1: Base Case without rehabilitation, without solar shading devices.
- CS1.2: Base Case without rehabilitation, with solar shading devices.
- M1.2: combination 1 of envelope's rehabilitation with shading, 0.63 h^{-1} ventilation rate, except summer nights with natural ventilation (NV).
- M1.2_0,4: combination 1 of envelope's rehabilitation with shading, 0.4 h^{-1} ventilation rate and NV.
- M1.2_HR: combination 1 of envelope's rehabilitation with shading, heat recovery ventilation unit (0.63 h^{-1}) and NV.
- M1.2_HR0,4: combination 1 of envelope's rehabilitation with shading, heat recovery ventilation unit (0.4 h^{-1}) and NV.
- M1.2_HR_NSV: combination 1 of envelope's rehabilitation with shading, without passive night ventilation in summer (SNV), and a constant ventilation rate with heat recovery ventilation unit (0.63 h^{-1}).

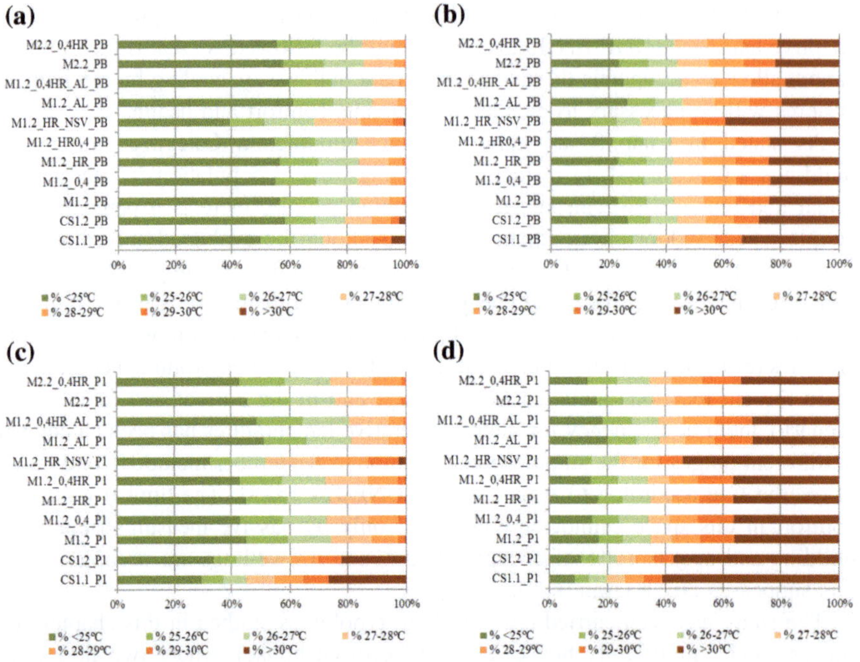

Fig. 8.34 Indoor Temperature Frequency in T5 Typology (Terraced House), in Madrid (**a** Ground floor; **b** First floor) and Madrid 2050 (**c** Ground floor; **d** First floor)

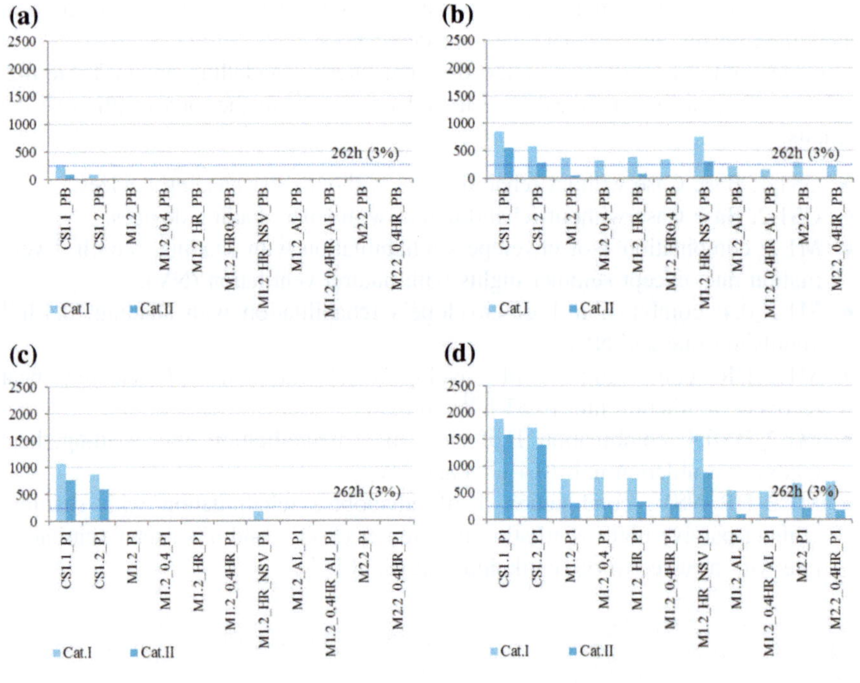

Fig. 8.35 Overheating (hours) over adaptive threshold (UNE-EN 15251) in T5 Typology (Terraced House), in Madrid (**a** Ground floor; **b** First floor) and Madrid 2050 (**c** Ground floor; **d** First floor)

- M1.2_AL: combination 1 of envelope's rehabilitation with shading, 0.63 h^{-1} ventilation rate and NV, façade and roof (the latter only in T5) with albedo (AL).
- M1.2_0,4HR_AL: combination 1 of envelope's rehabilitation with shading, ventilation rate with heat recovery unit (0.4 h^{-1}) and NV, and albedo characteristics (AL).
- M2.2: combination 2 of envelopes rehabilitation measures with shading, 0.63 h^{-1} ventilation rate and NV.
- M2.2_0,4HR: combination 2 of envelope's rehabilitation with shading, ventilation rate with heat recovery unit (0.4 h^{-1}) and NV.

In Figs. 8.34 and 8.35, there is an example of Madrid in T5 typology and in the current and 2050 scenario. The figures with the other ten locations, for T1 and T5 typologies, are attached in Appendix B. In the first figure, there are four graphs (a-d) that show the frequency of indoor temperatures by floor, and for the two temporary summer scenarios. In the second figure, also summarized in four graphs (a–d), we can find overheating hours, according to UNE-EN 15251 for categories I and II, of every simulation, indicating the estimated admissible limit for being able to consider the space conditioned in a natural way and safe for the occupants. All results are given for months with typical summer conditions in the Mediterranean, therefore from June to September.

8.3.2 Conclusions of the Study on Overheating

In the current scenario and in non-rehabilitated buildings (CS1.2), there are various locations where important overheating, over an admissible comfort threshold, takes place (from Athens to Milan). Block typology is especially problematic together with upper floors under roofs, while the ground floor of terraced houses offers possibilities of adaptation inside the dwelling for every category and location. The lack or misuse of shading, shown in CS1.1, illustrates the incidence of solar radiation which can be avoided in summer conditions.

Envelope rehabilitation measures (M1 Retrofit Measures) that is, the first level of energy efficiency improvements showed in this study, suppose a reduction in overheating in summer, compared with the non-rehabilitated buildings studied (CS), in all situations. In T5 typology, all categories of indoor thermal comfort are covered in all locations. However, in T1 typology and in Athens, dwellings cannot be guaranteed an acceptable level of overheating hours. It could not be guaranteed either for Valencia, Madrid or Mostar if Category I is considered, although it could be under Category II.

For a 2050 future scenario, all non-rehabilitated buildings (CS1) suffer important overheating in every category, except those located in Oporto, although ground floor in T5 terraced housing continues offering possibilities of adaptation in the way of *Cool Retreats*, in locations such as Rome, Barcelona, Nimes or Portoroz for both categories, and in places like Milan, Valencia or Athens only for Category II.

M1 Retrofit Measures make terraced houses *Climate Ready* in all locations in a future scenario for a Category II supposing in some locations the specific use of ground floors in the hottest part of the day or during heatwaves. This is not the case however in Category I for the houses located in Athens, Madrid or Mostar. The dwelling could keep an everyday use without serious overheating for every category in Oporto, Portoroz or Nimes.

Although M1 Retrofit Measures give important reductions in heating and cooling energy demand, as just seen in Sect. 8.2, in T1 typology (linear block), there will be important overheating in a future scenario, in every location except in Oporto. This is significant due to the notable part of the stock with social aims in collective residential typologies, although this percentage varies depending on the country (as seen in Chap. 5).

As an example, in Athens (Fig. 8.36) in T1, the measures in the envelope from CS1.2 to M1.2 in current scenario, suppose a decrease in cooling demand from 45.6 to 23.3 kWh/m^2 a, although overheating is almost the same as seen in the graphs in Appendix B. Analysing the frequency in temperatures higher than 30 °C (generally

Fig. 8.36 Athens (Greece) as an example of city affected by Urban Heat Island Effect

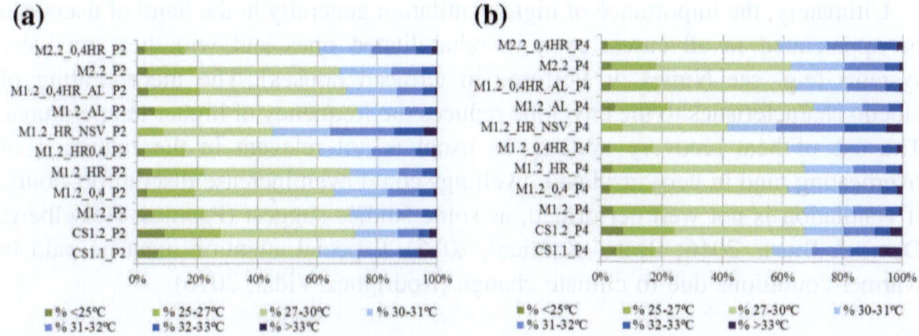

Fig. 8.37 Indoor Temperature Frequency in T1 typology (Linear housing block), in Athens (**a** Second floor; **b** Fourth floor)

out of comfort, as seen in Chap. 3, Figs. 3.7 and 3.8), some incidence in the reduction of these temperatures can be seen (Fig. 8.37). This is the reason why a lot of more research is required in the field of passive cooling and especially in the top floors and directly under roofs, in order to guarantee the health and well-being of occupants notwithstanding their socio-economic condition. As has been discussed, an important part of vulnerable population in residential buildings could be left unprotected in future global warming scenarios.

Fig. 8.38 Extensive use of air conditioning devices without control in the façades. Athens (Greece)

Ultimately, the importance of night ventilation generally in the hand of users can be appreciated in all cases, even in rehabilitated ones and with heat recovery systems (e.g. see Nimes or Portoroz in terraced houses). The incorporating of albedo characteristics to the envelope reduces the frequency of higher temperatures. The use of heat recovery systems as usual is not relevant in the reduction of overheating, and in very insulated dwellings could even increase discomfort hours, if ventilation is not well performed, as some studies suggest (Psomas, Heiselberg, Duer, & Bjorn, 2016; Brotas & Nicol, 2016). Especial attention must be paid in warmer conditions due to climate change (Rodriguez-Vidal, 2016).

8.4 Discussion

In the current climate scenario of Mediterranean Europe, collective and individual residential buildings with energy rehabilitation measures together with specific strategies for summer conditions, can respond adequately considering a typical building category of indoor thermal environment (Cat.II). Therefore, they could be condition in a natural way and without mechanical measures (except in Athens for collective buildings). However, in locations like Athens, together with Valencia, Madrid, Mostar or Milan, dwellings in building blocks cannot guarantee an indoor environment category, for the more fragile and sensitive, like the handicapped, the sick, very young children and the elderly (Cat.I), as is described in the standards. Because of this, since residential buildings may host a percentage of vulnerable population, conditioning must be done through mechanical means to ensure well-being and health of occupants, otherwise users will be unprotected and affected by high temperatures.

Differences in thermal and energy behaviour in residential typologies are clear, as well as the repercussion of spaces under roof in energy consumption and overheating, even with a rehabilitated envelope. In terraced houses, energy demand will be over 5 kWh/m^2 a in Athens, while in the rest of locations it is under 5 kWh/m^2 a. Furthermore, it will be specially reduced if in the central hours of the day, users only occupy the ground floor of dwellings (see Fig. 8.17). However, in dwellings in collective buildings, energy demand will be over 20 kWh/m^2 a in Athens, between 5 and 10 kWh/m^2 a in Mostar, Madrid, Valencia or Milan, and inferior to 5 kWh/m^2 a in Porto, Portoroz, Nimes, Barcelona or Roma, demand going 21% higher in an upper floor under roof (as seen in Fig. 8.16).

In non-rehabilitated dwellings, according to current standards, overheating and the energy demand necessary for maintaining an adequate thermal comfort boost significantly. Terraced houses offer a *Cool Retreat* on the ground floor, and have important overheating in floors under roof in Athens, Valencia, Madrid, Mostar or Milan, which would make it inhabitable without mechanical means. Energy demand in order to be able to use the upper floor of dwellings exceeds 20 kWh/m^2 a from Athens to Mostar (as seen in Fig. 8.5). However, in collective residential buildings without refurbishment, overheating and increases in cooling demand are even higher. Cooling energy demand will be over 20 kWh/m^2 a in Athens, 10–15 kWh/m^2 a in Valencia and Madrid, and between 5 and 10 kWh/m^2 a in the rest of locations except Oporto. In this typology, cooling demand increases on average of 48% in the upper floor of the building. It is worth highlighting that this kind of typology has a big incidence in social dwellings, and precisely, those with the worst conditions will be inhabited by the most socio-economically vulnerable population.

The energy consumption needed for thermal conditioning of residential spaces between the current and the 2050 scenario varies among the locations in the Mediterranean zone due to the variation in climatic severity of summer and winter, building typology and if it is rehabilitated or not, or is built with the current standards of energy efficiency.

In locations such as Oporto, Portoroz or Nimes, total energy demand for heating and cooling will be lower in a future scenario of global warming (both in rehabilitated or new, and non-rehabilitated buildings). In locations like Athens, Valencia or Rome, there will be a higher conditioning energy demand in all non-rehabilitated, rehabilitated and new buildings (considering the same status in the current scenario). However, in buildings rehabilitated or built to energy efficiency standards, total conditioning energy demand in 2050 will be far lower than in non-rehabilitated buildings (Figs. 8.30, 8.31, 8.32 and 8.33).

On the other hand, there is going to be a shift in the building's most important conditioning energy demand in some of the locations of Southern Europe. Cooling demand will become the main demand in Athens or Valencia, and the rest of locations it will depend on typology. Heating demand will continue to be important, and there will be places where both strategies will continue to have a great impact in buildings design, although cooling demand will have a bigger share of the total energy demand. These changes in the percentage of energy demand should have incidence in the way buildings are designed.

In locations where cooling systems are usually installed and used, e.g. Athens or Valencia, this changes in energy demand translate into more cooling consumption, less possibilities of adaptation inside the dwelling, possible peaks in demand that can leave population without supplies (here the importance of in situ renewable energy production), and serious problems on the health of the most socio-economically vulnerable population, who will be affected with energy poverty related to heat. On the other hand, in other cooler locations that do not usually have cooling systems in residential buildings, the increases in temperatures could lead to serious overheating, that could compromise the comfort and health of occupants. These locations must be especially aware in warmer conditions or during heatwave events, since people have fewer opportunities to condition their dwellings and maximum temperature thresholds will be lower (as seen in Chap. 3). Moreover, as conditioning systems are not integrated in the building, their progressive installation under the most severe temperatures will be more inefficient and less integrated in the buildings, both at functional and at aesthetic level (Fig. 8.38).

As in the current, in a future scenario, the differences in the thermal and energy behaviour of building typologies must be highlighted. New or rehabilitated terraced houses with the adequate passive strategy for summer together with a greater possibility of zoning, taking advantage of soil thermal mass (ground floors or basements), open up a possibility of adapting the living spaces during the hours of the day and during the seasons. According to the energy modelling carried out on future warming scenarios, we could consider all terraced houses *Climate Ready* for a Category II (using all the dwelling or only the ground floor, depending on the moment), although for a Category I, Athens, Valencia, Madrid or Mostar would not make it. Energy demand when using cooling is inferior to 5 kWh/m^2 a, in every location except in Athens where it is almost 10 kWh/m^2 a, according to the implemented measure. However, in collective residential typology, important overheating takes place in every location except Oporto. Energy demand to maintain an acceptable indoor environment is under 5 kWh/m^2 a in Oporto, and under 20 kWh/m^2 a in Portoroz and Nimes, this values being widely exceeded in the rest of locations.

On the other hand, non-rehabilitated housing will suffer serious overheating in every typology and for every thermal environment category in 2050. Oporto will be the only location with acceptable overheating, together with Portoroz, Nimes, Barcelona and Roma, but the last ones only in the terraced house typology, for Category II, and supposing only the use of ground floors. The use of air conditioning systems will be mandatory in totally inefficient buildings, with very high consumptions.

In many of the cases studied, passive measures incorporated to design will not be enough by themselves, and natural conditioning of residential buildings will not be possible in order to face future summer conditions. In addition, the most vulnerable population, especially the one that cannot deal with the actual cost of energy, will be totally exposed, because the building will not be capable of providing safe interior conditions without mechanical systems.

This study has been able to test the incidence of either a lack or an incorrect use of shading, as well as the layout of windows in west or south-west orientations, on the increase in cooling and overheating. While a good building design, bearing in mind orientations, as well as design and disposition of shading, is implicit in the project designed by architects, its use is in the hands of occupant, mostly manually and without any economic or energy costs. Shading devices are essential in residential buildings, both now and in future scenarios, and should be included in regulations. They are recommended in climates with severe summer conditions and elsewhere, since they are simple measures that can have a big impact (Fig. 8.39).

Night ventilation or even daytime ventilation when the necessary conditions are given is very efficient as has been tested in the different modelling shown. Buildings must be designed in such a way that the user can efficiently do a crossed ventilation, so, dwellings with a single orientation must incorporate systems that guarantee it. As seen previously in Chap. 7, it is important to highlight that the efficiency of natural ventilation can be reduced in big cities due to the urban heat island effect (UHI), pollution, noise and concerns about safety, mainly in the elderly.

Passive measures are especially relevant in those locations where cooling systems are not implemented in the dwellings, since they will contribute in an

Fig. 8.39 Examples of shading in contemporary buildings: fixed (e.g. in Logroño, Spain) or movable (e.g. in Barcelona, Spain)

important way to the *resilience* of buildings under climate change, whether for warmer temperatures or for more severe heatwaves. In a climate zone where residential buildings normally have cooling systems, greater cooling energy consumption will take place. However in a milder climatic zone, the building itself and the occupant will be in charge of achieving a comfortable indoor environment, without any health risks, for different population groups, and during all the buildings life.

As stated, different measures that contribute to the improvement of buildings' energy efficiency are orientated to heating demand which has proved very effective in winter conditions, but bearing in mind future global warming scenarios, may not be enough. Much more research is needed, in order to condition residential buildings in summer for all the population included the most vulnerable, both with passive measures that minimize energy or system costs, especially in the roofs and in residential collective buildings.

8.5 Conclusions

The energy and thermal behaviour of residential buildings has been studied both in individual and in collective building typology, in summer conditions and for warming future climates. Two approaches have been followed, one considering cooling demands and therefore supposing there are conditioning systems in the dwellings, and another one considering overheating of indoor thermal environment. The incidence of occupants' behaviour and the protection of vulnerable population as one of the main premises of residential buildings, adding to the optimization of passive strategies inherent in their design completes these approaches. Residential buildings have a great potential regarding design and rehabilitation as *Climate Ready,* and great differences have been found between locations and building typologies. Finally, much more research and efforts are needed in every area.

Acknowledgements We would like to thank María Dolores Juri, student in the Master's Degree in Environmental Management and Building Design, MDGAE in University of Navarra (Spain), who carry out the Master's Thesis titled '*Resiliencia al cambio climático en viviendas del sur de Europa de climas mediterráneos*'. Also to Cristina Guell, for her assistance in the elaboration of overheating graphs.

References

ASHRAE55-2013. (2013). Thermal Environmental Conditions for Human Occupancy.

Brotas, L., & Nicol, J. F. (2016). The problem of overheating in European dwellings. *Windsor, 2016*(April), 7–10.

CTE-HE. (2013). Documento Básico CTE-HE Ahorro de energía. *Septiembre*. Retrieved from http://www.codigotecnico.org/images/stories/pdf/ahorroEnergia/DBHE.pdf.

EEA. (2016). Climate change, impacts and vulnerability in Europe 2016 - Key findings. *EEA Report*. https://doi.org/10.2800/66071.

Huang, J. (2011). *ASHRAE Research Project 1477-RP Development of 3,012 typical year weather files for international locations. Final Report.*

Jentsch, M. F., James, P. A. B., Bourikas, L., & Bahaj, A. S. (2013). Transforming existing weather data for worldwide locations to enable energy and building performance simulation under future climates. *Renewable Energy, 55,* 514–524. https://doi.org/10.1016/j.renene.2012.12.049.

Psomas, T., Heiselberg, P., Duer, K., & Bjorn, E. (2016). Overheating risk barriers to energy renovations of single family houses: Multicriteria analysis and assessment. *Energy and Buildings, 117,* 138–148. https://doi.org/10.1016/j.enbuild.2016.02.031.

Rodriguez-Vidal, I. (2016). *Evaluación del estándar de construcción Passivhaus y su aplicación en el ámbito climático de la Comunidad Autónoma Vasca y la Comunidad Foral Navarra. El caso de la vivienda colectiva de protección oficial*. Retrieved from http://hdl.handle.net/10810/17870

Saman, W., Boland, J., Pullen, S., Dear, R. De, Soebarto, V., Miller, W., … Chileshe, N. (2013). *A framework for adaptation of Australian households to heat waves*. Retrieved from http://www.nccarf.edu.au/sites/default/files/attached_files_publications/Saman_2013_Adapting_households_to_heat_waves.pdf.

UCLA. (n.d.). http://www.energy-design-tools.aud.ucla.edu/.

UNE-EN 15251. (2008). Parámetros del ambiente interior a considerar para el diseño y la evaluación de la eficiencia energética de edificios incluyendo la calidad del aire interior, condiciones térmicas, iluminación y ruido.

Appendix A
Other Locations in Southern Europe

In this Appendix A, information about annual and monthly mean temperatures, and heating degree days (HDD) and cooling degree days (CDD), for the current and 2050 scenario are given for other locations sited in Southern Europe and different than those exposed in Figs. 2.5 and 4.8 (Figs. A.1 and A.2).

© Springer International Publishing AG 2018

A. Monge-Barrio and A. Sánchez-Ostiz Gutiérrez, *Passive Energy Strategies for Mediterranean Residential Buildings*, Green Energy and Technology,
https://doi.org/10.1007/978-3-319-69883-0

Location	Country	Koppen-Geiger	Altitude	Latitude	Mean Annual Temp	Mean Annual Temp 2050	Temp of the coldest month	Temp of the coldest month 2050	Temp of the hottest month	Temp of the hottest month 2050
Samos	GRC	Csa	7	37°42'	18,6	20,9	11,4	10,9	28,5	32,2
Messina	ITA	Csa	51	38°12'	19	20,7	12,7	13,4	27,6	30,4
Almería	ESP	BSk	21	36°50'	18,6	21,2	11,8	13,7	26,2	30
Souda (Crete)	GRC	Csa	146	35°31'	18,3	19,9	11,1	12,2	26,5	28,8
Paphos	CYP	Csa	8	34°43'	19,1	21	12,19	13,9	26,4	28,9
Corfu (Kerkyra)	GRC	Csa	4	39°37'	17,3	19,8	9,7	11,6	26,3	30,6
Thesalonika	GRC	Cfa	4	40°31'	15,3	18,1	5,4	7,5	26,3	31
Cáceres	ESP	Csa	405	39°28'	16,3	19,2	7,9	9,7	26,2	30,8
Cerdeña. Cagliari	ITA	Csa	5	39°15'	17,1	18,7	9,5	11	25,9	28,1
Málaga	ESP	Csa	7	36°40'	18,1	21	11,6	13,5	25,7	30,1
Tivat	MNE	Cfb	5	42°24'	15,4	17,9	7,3	9,5	25,7	29,7
Larissa	GRC	Csa	74	39°37'	15,4	17,8	4,6	6,6	25,7	30
Granada	ESP	Csa	570	37°10'	15	17,9	5,9	7,7	25,7	30,3
Andravida	GRC	Csa	14	37°55'	16,8	18,9	9,1	10,9	25,6	29,1
Bari	ITA	Cfa	49	41°7'	16	18,2	8,3	10,1	25,2	28,6
Naples	ITA	Csa	72	40°50'	16,5	18,3	9,2	10,1	25,1	27,8
Alexandroupoli	GRC	Csa	3	40°50'	14,5	17,3	5,2	7,4	25	29,7
Bastia	FRA	Csa	12	42°32'	16,1	18,5	9,6	11	24,6	28,5
Marsella	FRA	Csa	32	43°27'	15,4	17,6	7,9	9,6	24,6	27,6
Cerdeña. Alghero	ITA	Csa	40	40°37'	16,2	17,9	9,6	11	24,5	26,4
Zaragoza	ESP	BSk	258	41°40'	14,9	17,7	6,2	7,7	24,5	29,4
Venice	ITA	Cfb	6	45°30'	13,7	16,6	3,3	5,8	24,4	29,4
Faro	PRT	Csa	8	37°1'	17,8	19,9	11,8	13,4	24,3	26,9
Nice	FRA	Csb	27	43°39'	15,7	17,7	9,1	10,8	24,2	27,4
Girona	ESP	Cfa	129	41°54'	14,7	16,8	7,2	8,9	23,5	27,5
Turin	ITA	Cfb	287	45°13'	12,1	14,6	1,4	3,6	22,8	26,8
Toulouse	FRA	Cfb	154	43°37'	13,6	15,9	5	6,9	22,4	27

Fig. A.1 Locations sited in Southern Europe. Mean annual temperature, mean temperature of the coldest and hottest month for current and 2050 scenario (complementary to Fig. 2.5)

Location	Country	Koppen-Geiger	Altitude	Latitude	HDD (18°C)	HDD (18°C) 2050	CDD (27°C)	CDD (27°C) 2050
Samos	GRC	Csa	7	37°42'	1002	716	184	449
Messina	ITA	Csa	51	38°12'	734	507	67	214
Almeria	ESP	BSk	21	36°50'	755	457	73	260
Souda (Crete)	GRC	Csa	146	35°31'	983	721	95	219
Paphos	CYP	Csa	8	34°43'	735	518	80	227
Corfu (Kerkyra)	GRC	Csa	4	39°37'	1187	842	99	313
Thesalonika	GRC	Cfa	4	40°31'	1829	1418	84	343
Caceres	ESP	Csa	405	39°28'	1563	1140	196	496
Cerdeña. Cagliari	ITA	Csa	5	39°15'	1214	940	83	162
Malaga	ESP	Csa	7	36°40'	901	584	69	314
Tivat	MNE	Cfb	5	42°24'	1738	1299	99	309
Larissa	GRC	Csa	74	39°37'	1891	1499	138	365
Granada	ESP	Csa	570	37°10'	1963	1491	180	430
Andravida	GRC	Csa	14	37°55'	1318	989	82	235
Bari	ITA	Cfa	49	41°7'	1495	1134	62	208
Naples	ITA	Csa	72	40°50'	1363	1058	68	160
Alexandroupoli	GRC	Csa	3	40°50'	2024	1534	70	281
Bastia	FRA	Csa	12	42°32'	1362	976	43	203
Marsella	FRA	Csa	32	43°27'	1639	1263	58	185
Cerdeña. Alghero	ITA	Csa	40	40°37'	1385	1071	74	139
Zaragoza	ESP	BSk	258	41°40'	1880	1456	113	346
Venice	ITA	Cfb	6	45°30'	2203	1693	26	222
Faro	PRT	Csa	8	37°1'	844	560	37	133
Nice	FRA	Csb	27	43°39'	1447	1092	11	86
Girona	ESP	Cfa	129	41°54'	1817	1447	51	176
Turin	ITA	Cfb	287	45°13'	2624	2118	24	139
Toulouse	FRA	Cfb	154	43°37'	2094	1657	40	175

Fig. A.2 Locations sited in Southern Europe. Heating degree days, HDD; and cooling degree days, CDD, for current and 2050 scenario (complementary to Fig. 4.6)

Appendix B
Overheating Study in Southern Europe

In this Appendix B, overheating analysis presented in Sect. 8.3 is exposed for ten locations: Athens (Greece), Valencia (Spain), Madrid (Spain), Mostar (Bosnia and Herzegovina), Milan (Italy), Rome (Italy), Barcelona (Spain), Nimes (France), Portoroz (Slovenia), and Porto (Portugal) (Figs. B.1–B.40).

© Springer International Publishing AG 2018

A. Monge-Barrio and A. Sánchez-Ostiz Gutiérrez, *Passive Energy Strategies for Mediterranean Residential Buildings*, Green Energy and Technology,
https://doi.org/10.1007/978-3-319-69883-0

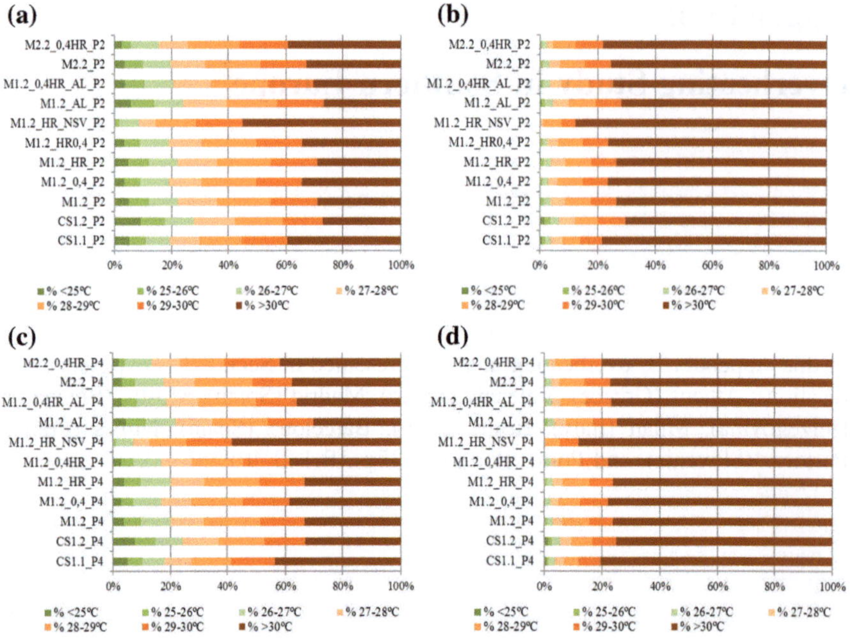

Fig. B.1 Indoor temperature frequency in T1 typology (Linear housing block), in Athens (**a** 2nd floor; **c** 4th floor) and Athens 2050 (**b** 2nd floor; **d** 4th floor)

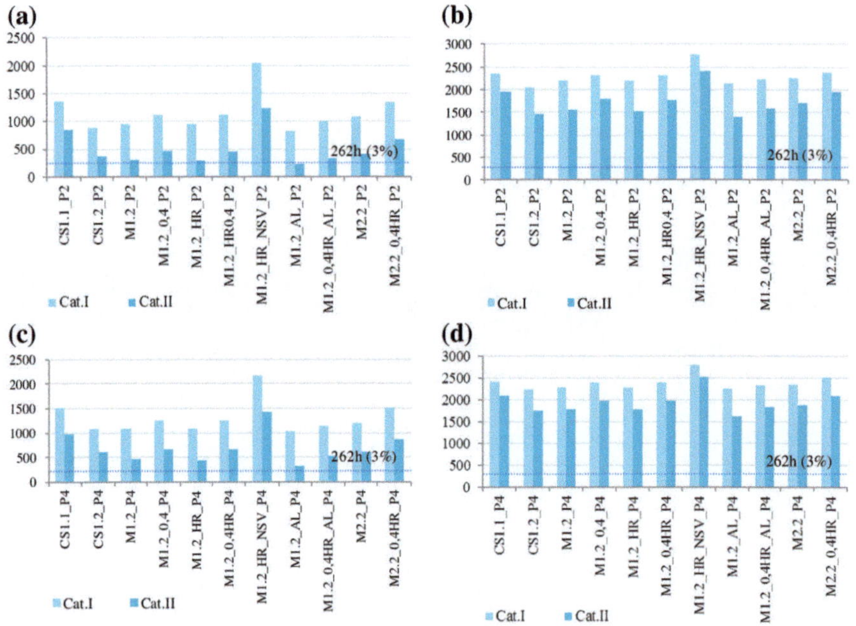

Fig. B.2 Overheating (hours) over adaptive threshold (UNE EN 15251) in T1 typology (Linear housing block), in Athens (**a** 2nd floor; **c** 4th floor) and Athens 2050 (**b** 2nd floor; **d** 4th floor)

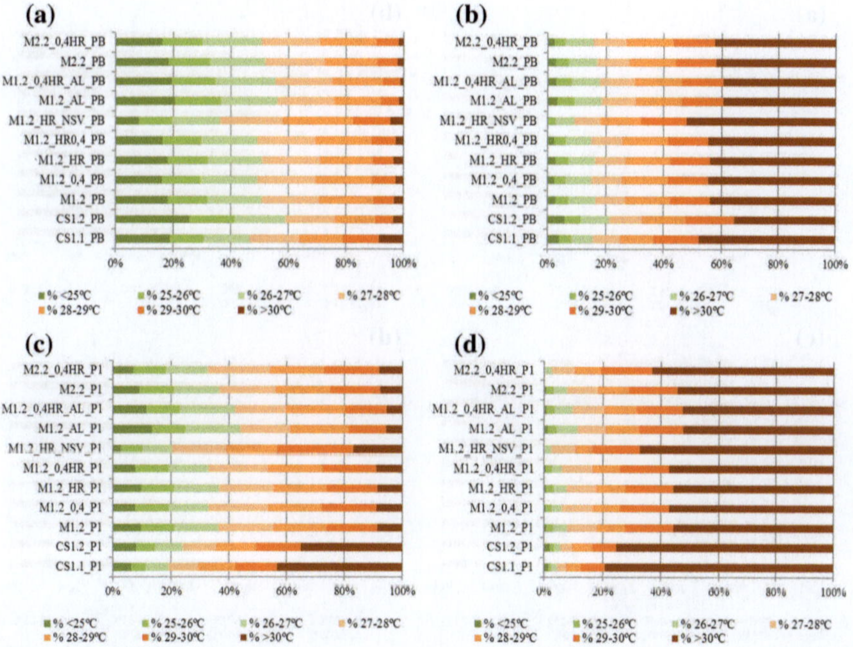

Fig. B.3 Indoor temperature frequency in T5 typology (Terraced House), in Athens (**a** Ground floor; **c** 1st floor) and Athens 2050 (**b** Ground floor; **d** 1st floor)

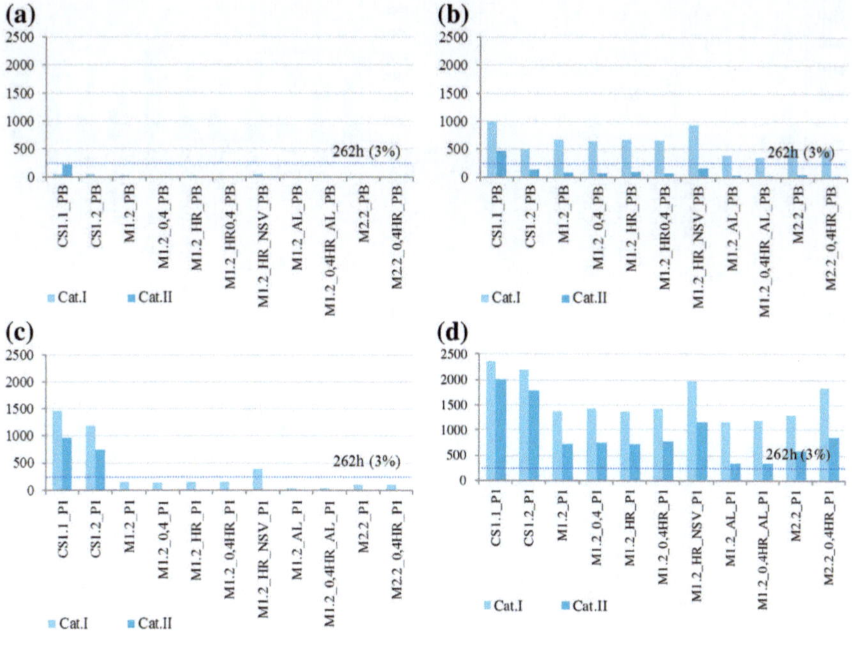

Fig. B.4 Overheating (hours) over adaptive threshold (UNE EN 15251) in T5 typology (Terraced House), in Athens (**a** Ground floor; **c** 1st floor) and Athens 2050 (**b** Ground floor; **d** 1st floor)

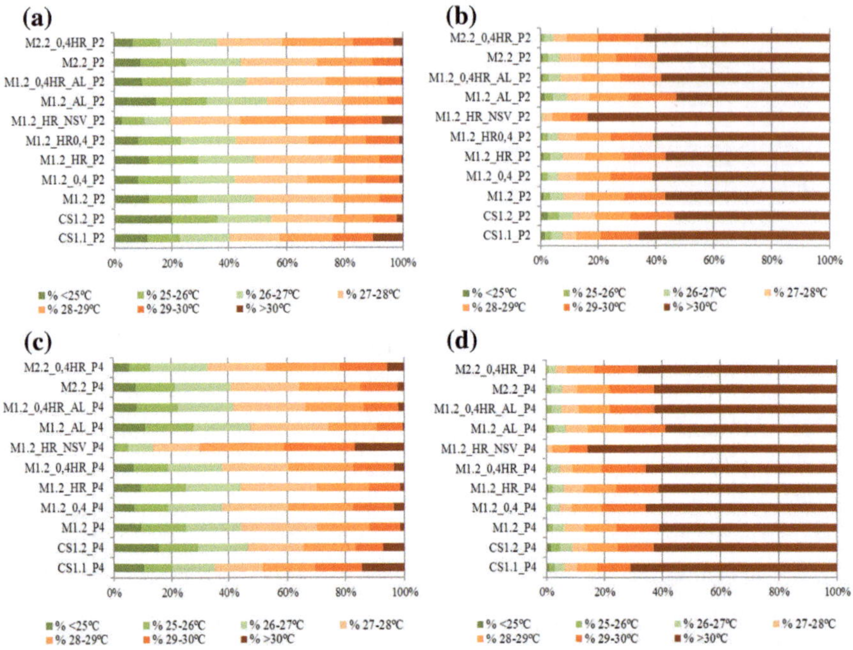

Fig. B.5 Indoor temperature frequency in T1 typology (Linear housing block), in Valencia (**a** 2nd floor; **c** 4th floor) and Valencia 2050 (**b** 2nd floor; **d** 4th floor)

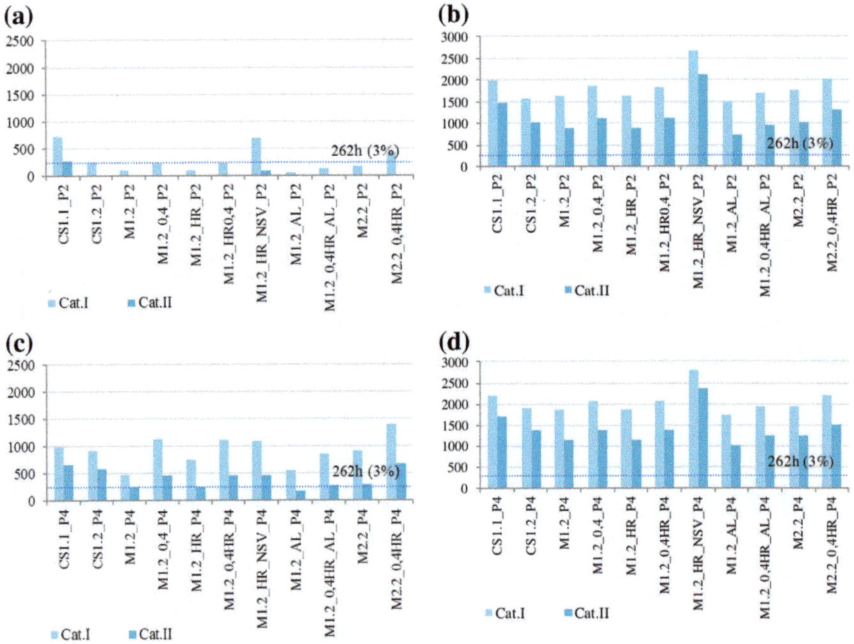

Fig. B.6 Overheating (hours) over adaptive threshold (UNE EN 15251) in T1 typology (Linear housing block), in Valencia (**a** 2nd floor; **c** 4th floor) and Valencia 2050 (**b** 2nd floor; **d** 4th floor)

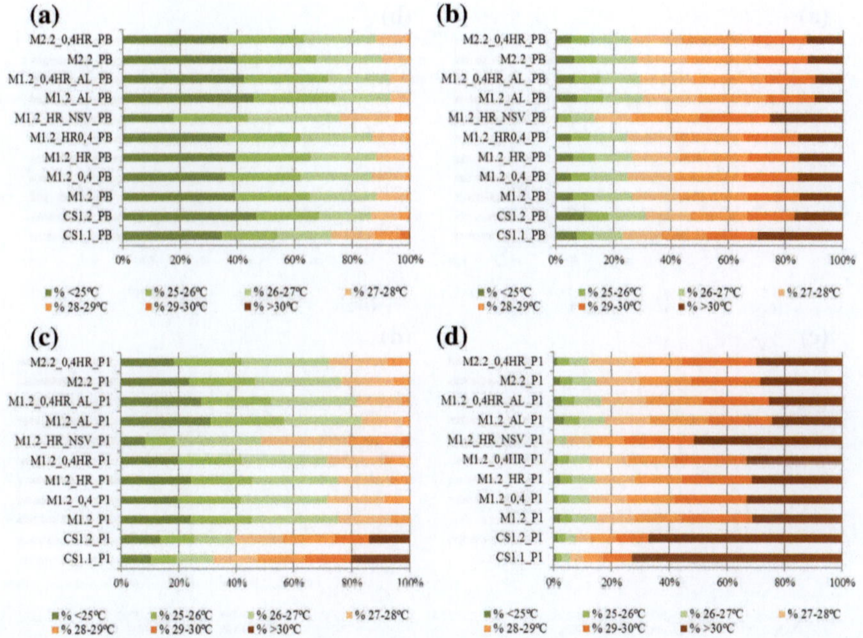

Fig. B.7 Indoor temperature frequency in T5 typology (Terraced House), in Valencia (**a** Ground floor; **c** 1st floor) and Valencia 2050 (**b** Ground floor; **d** 1st floor)

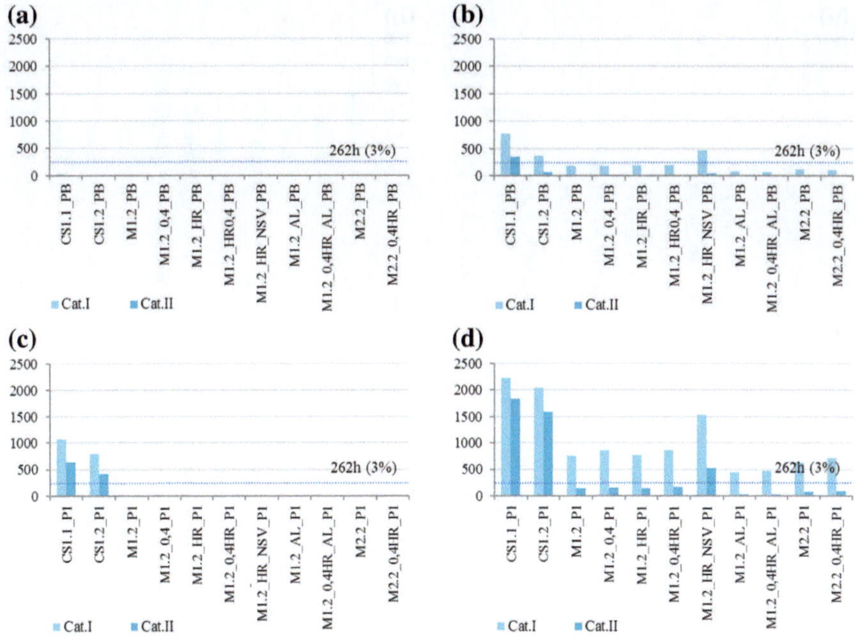

Fig. B.8 Overheating (hours) over adaptive threshold (UNE EN 15251) in T5 typology (Terraced House), in Valencia (**a** Ground floor; **c** 1st floor) and Valencia 2050 (**b** Ground floor; **d** 1st floor)

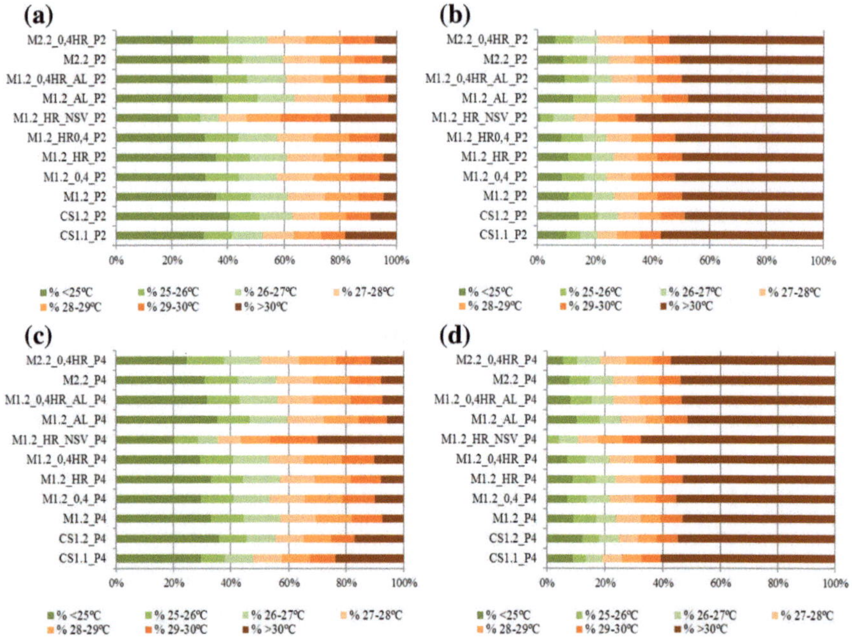

Fig. B.9 Indoor temperature frequency in T1 typology (Linear housing block), in Madrid (**a** 2nd floor; **c** 4th floor) and Madrid 2050 (b 2nd floor; **d** 4th floor)

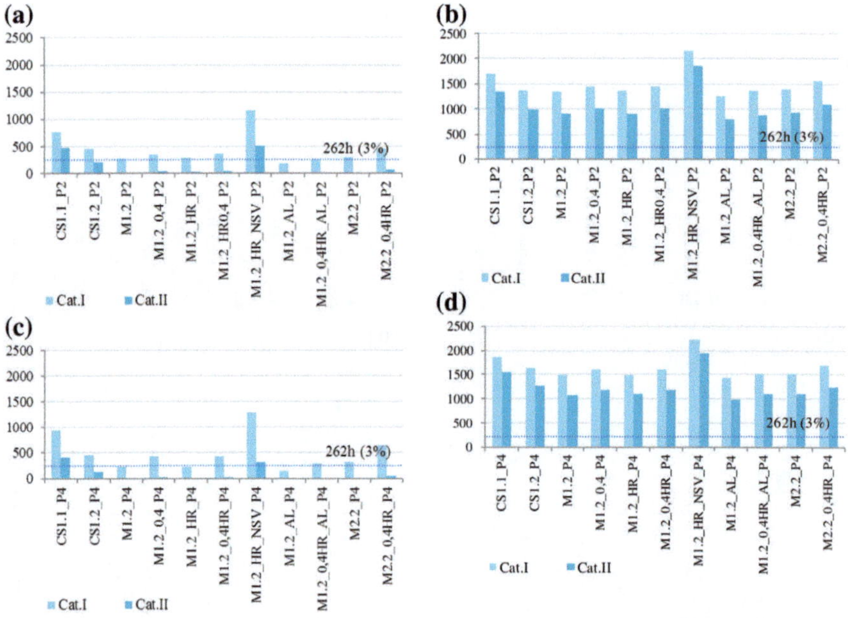

Fig. B.10 Overheating (hours) over adaptive threshold (UNE EN 15251) in T1 typology (Linear housing block), in Madrid (**a** 2nd floor; **c** 4th floor) and Madrid 2050 (**b** 2nd floor; **d** 4th floor)

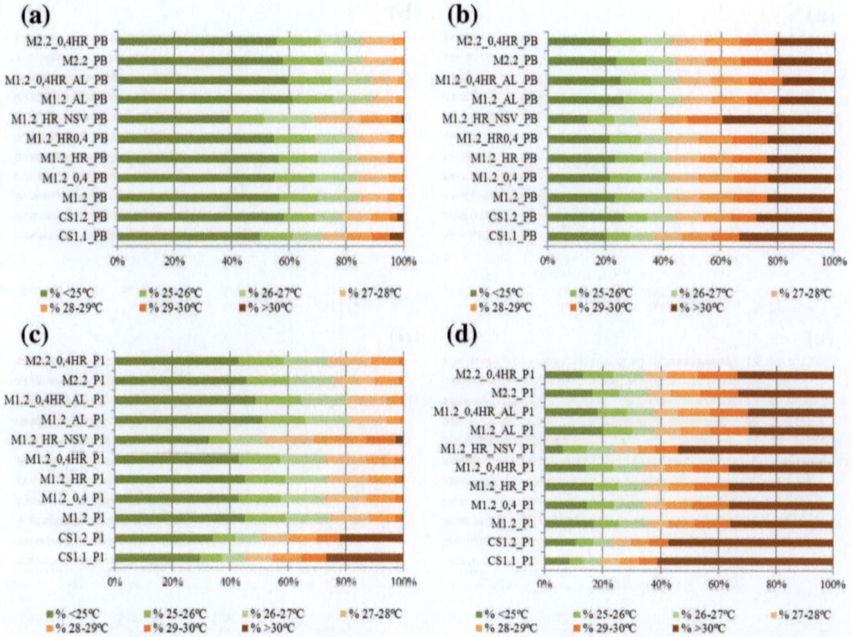

Fig. B.11 Indoor temperature frequency in T5 typology (Terraced House), in Madrid (**a** Ground floor; **c** 1st floor) and Madrid 2050 (**b** Ground floor; **d** 1st floor)

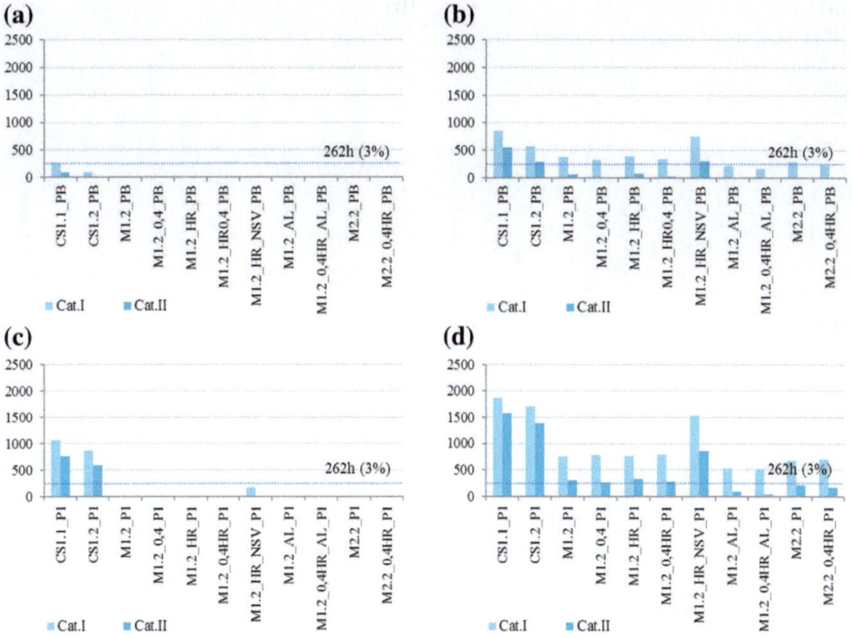

Fig. B.12 Overheating (hours) over adaptive threshold (UNE EN 15251) in T5 typology (Terraced House), in Madrid (**a** Ground floor; **c** 1st floor) and Madrid 2050 (**b** Ground floor; **d** 1st floor)

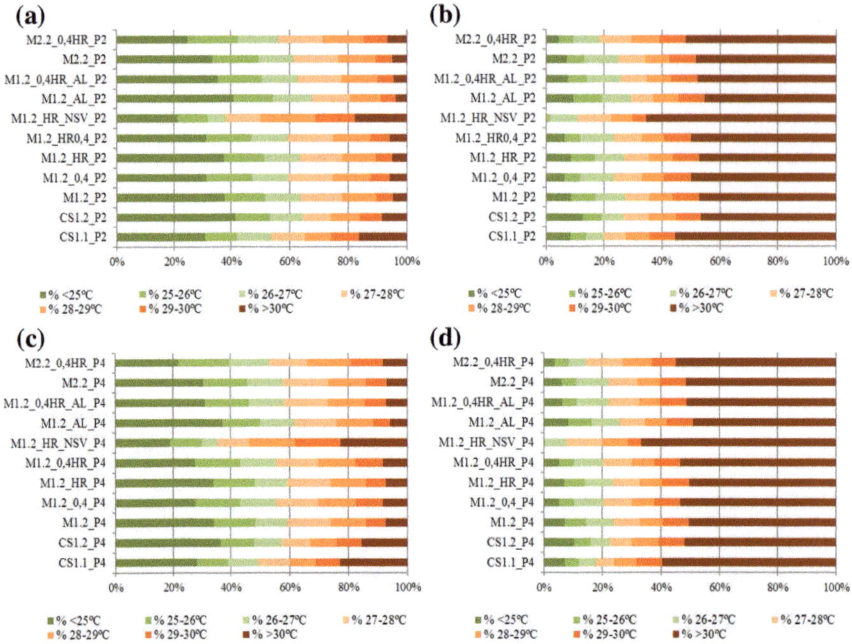

Fig. B.13 Indoor temperature frequency in T1 typology (Linear housing block), in Mostar (**a** 2nd floor; **c** 4th floor) and Mostar 2050 (**b** 2nd floor; **d** 4th floor)

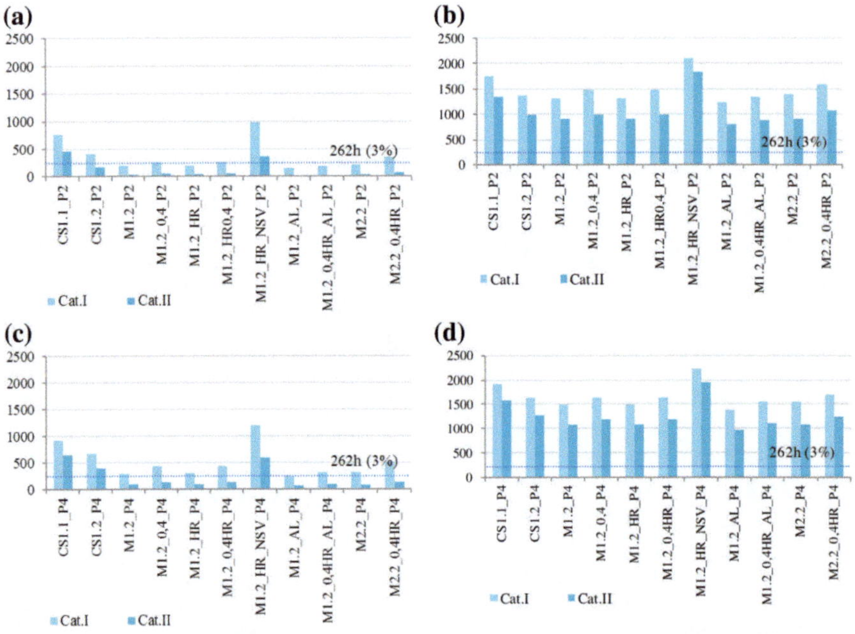

Fig. B.14 Overheating (hours) over adaptive threshold (UNE EN 15251) in T1 typology (Linear housing block), in Mostar (**a** 2nd floor; **c** 4th floor) and Mostar 2050 (**b** 2nd floor; **d** 4th floor)

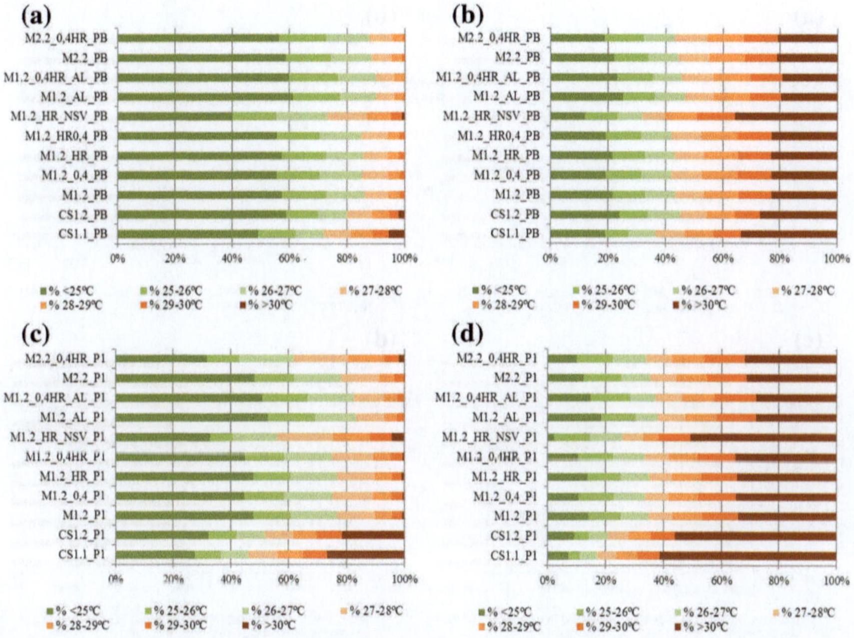

Fig. B.15 Indoor temperature frequency in T5 typology (Terraced House), in Mostar (**a** Ground floor; **c** 1st floor) and Mostar 2050 (**b** Ground floor; **d** 1st floor)

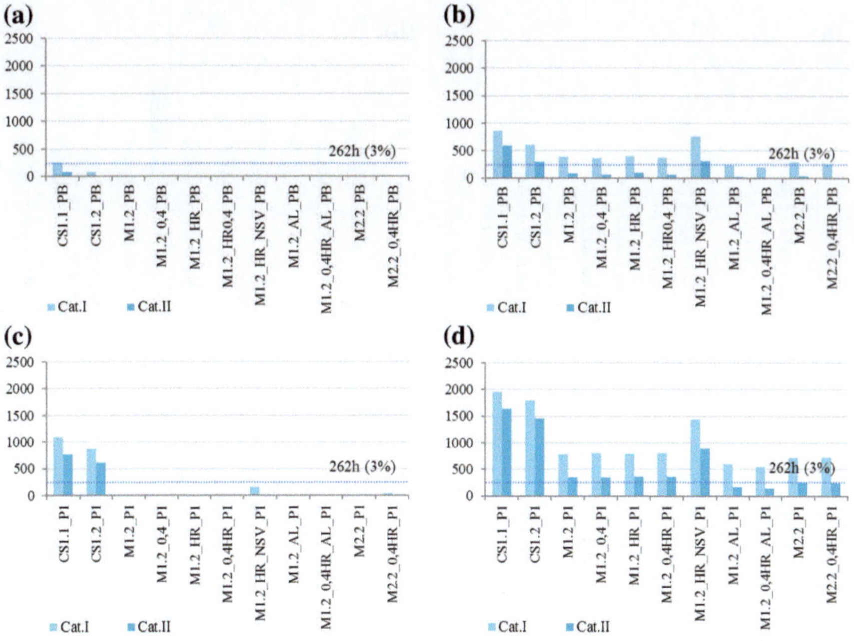

Fig. B.16 Overheating (hours) over adaptive threshold (UNE EN 15251) in T5 typology (Terraced House), in Mostar (**a** Ground floor; **c** 1st floor) and Mostar 2050 (**b** Ground floor; **d** 1st floor)

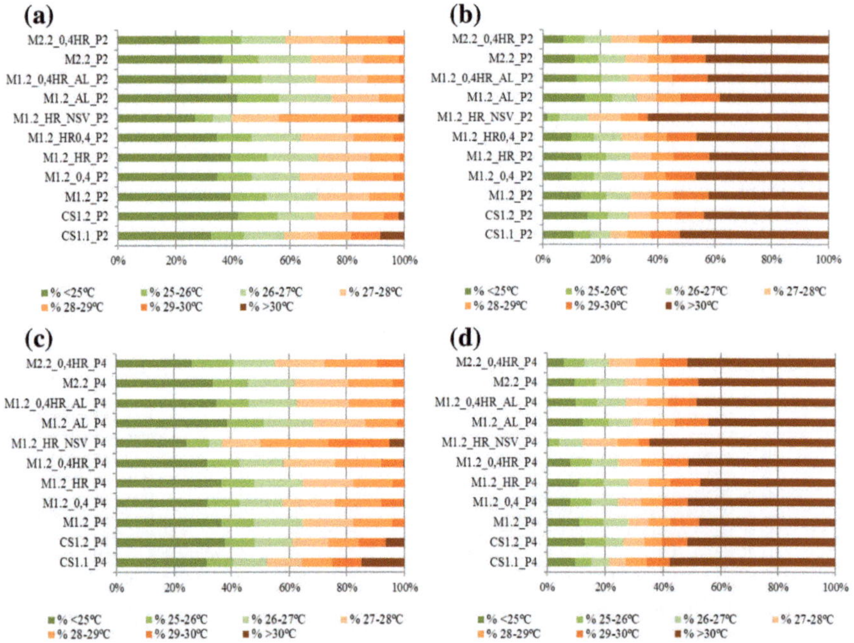

Fig. B.17 Indoor temperature frequency in T1 typology (Linear housing block), in Milan (**a** 2nd floor; **c** 4th floor) and Milan 2050 (**b** 2nd floor; **d** 4th floor)

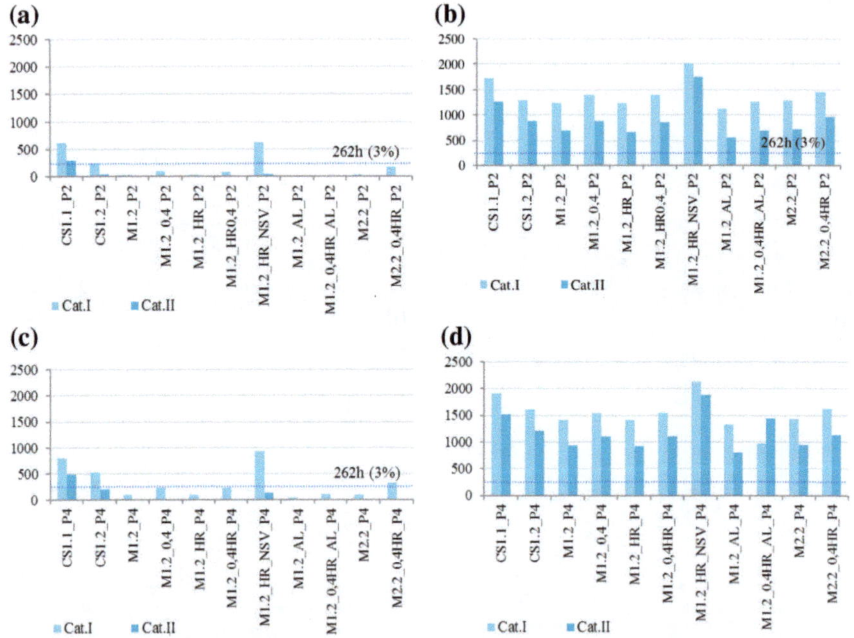

Fig. B.18 Overheating (hours) over adaptive threshold (UNE EN 15251) in T1 Typology (Linear housing block), in Milan (**a** 2nd floor; **c** 4th floor) and Milan 2050 (**b** 2nd floor; **d** 4th floor)

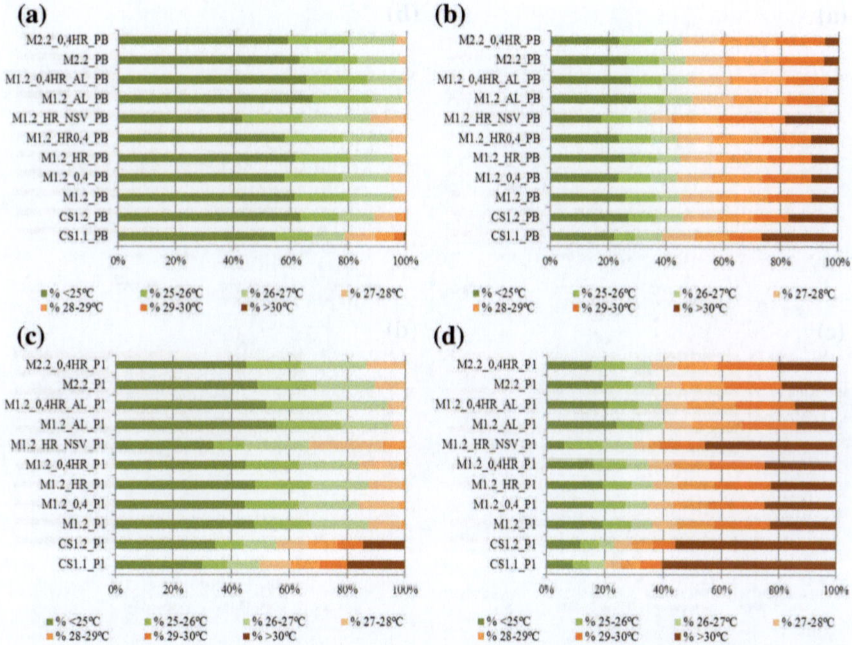

Fig. B.19 Indoor temperature frequency in T5 typology (Terraced House), in Milan (**a** Ground floor; **b** 1st floor) and Milan 2050 (**c** Ground floor; **d** 1st floor)

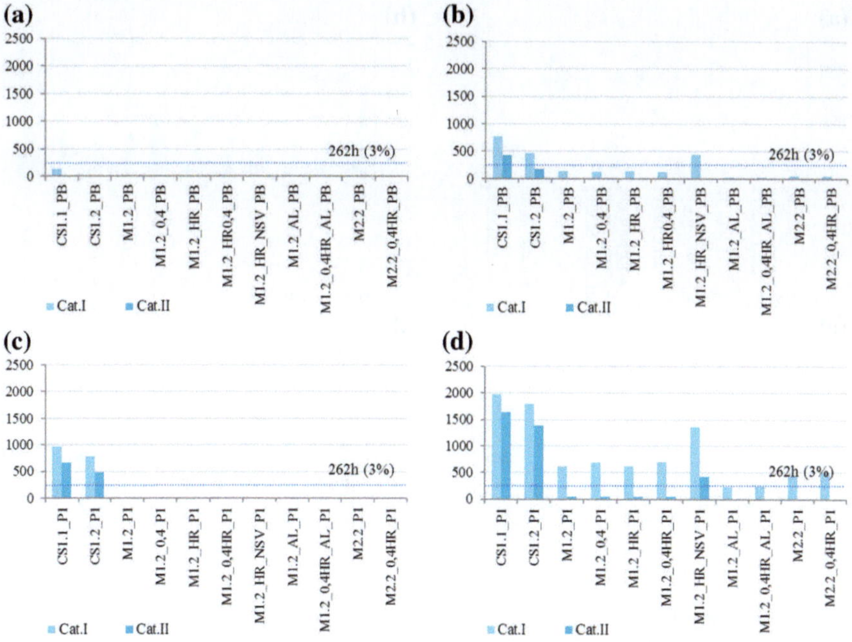

Fig. B.20 Overheating (hours) over adaptive threshold (UNE EN 15251) in T5 typology (Terraced House), in Milan (**a** Ground floor; **c** 1st floor) and Milan 2050 (**b** Ground floor; **d** 1st floor)

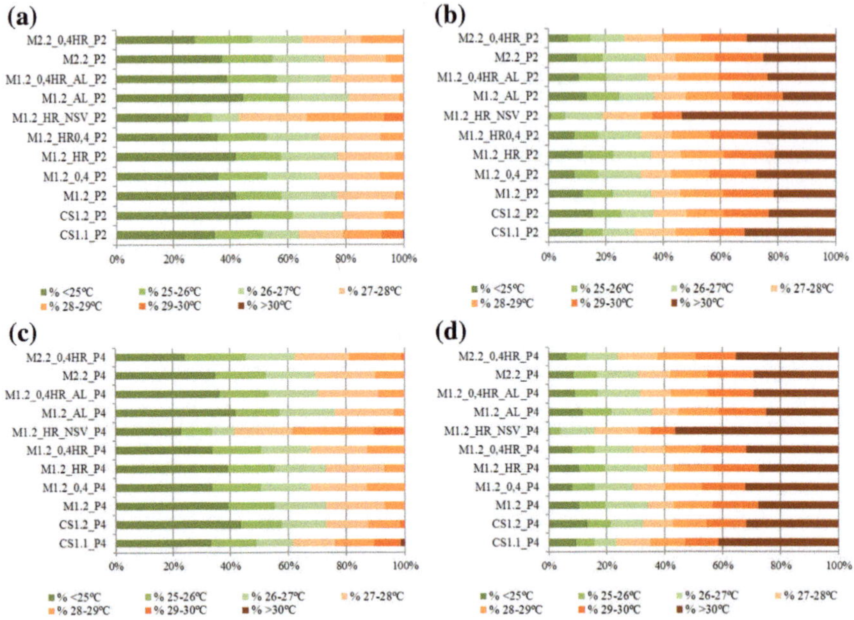

Fig. B.21 Indoor temperature frequency in T1 typology (Linear housing block), in Rome (**a** 2nd floor; **c** 4th floor) and Rome 2050 (**b** 2nd floor; **d** 4th floor)

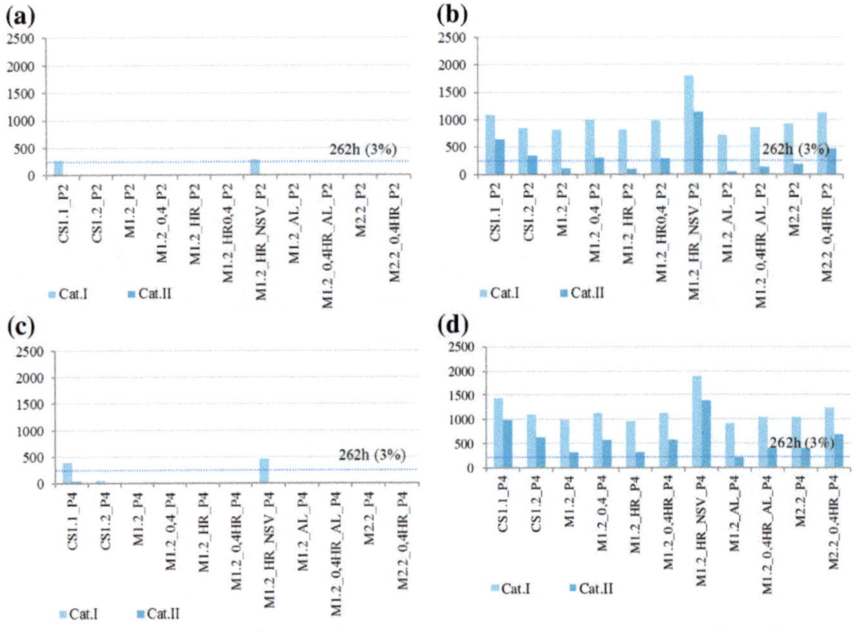

Fig. B.22 Overheating (hours) over adaptive threshold (UNE EN 15251) in T1 typology (Linear housing block), in Rome (**a** 2nd floor; **c** 4th floor) and Rome 2050 (**b** 2nd floor; **d** 4th floor)

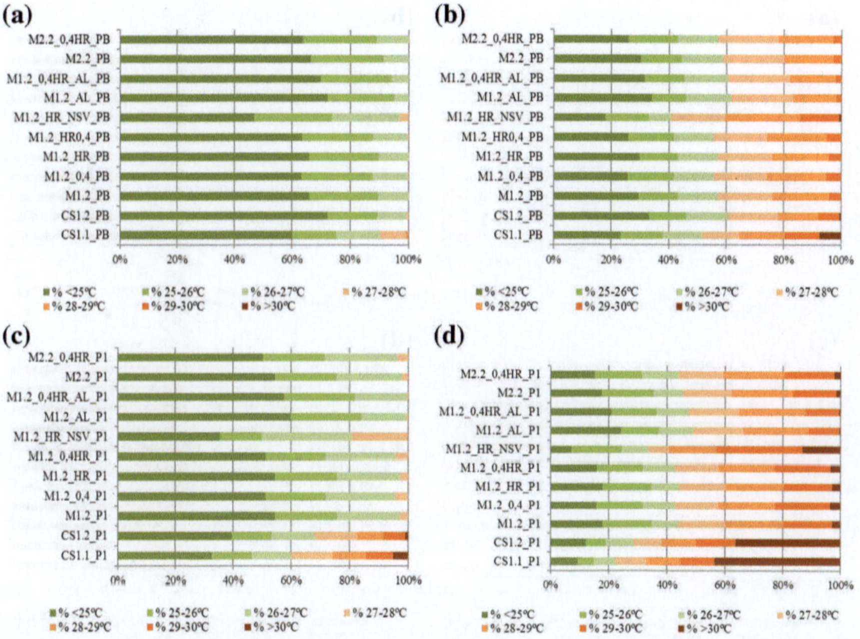

Fig. B.23 Indoor temperature frequency in T5 typology (Terraced House), in Rome (**a** Ground floor; **c** 1st floor) and Rome 2050 (**b** Ground floor; **d** 1st floor)

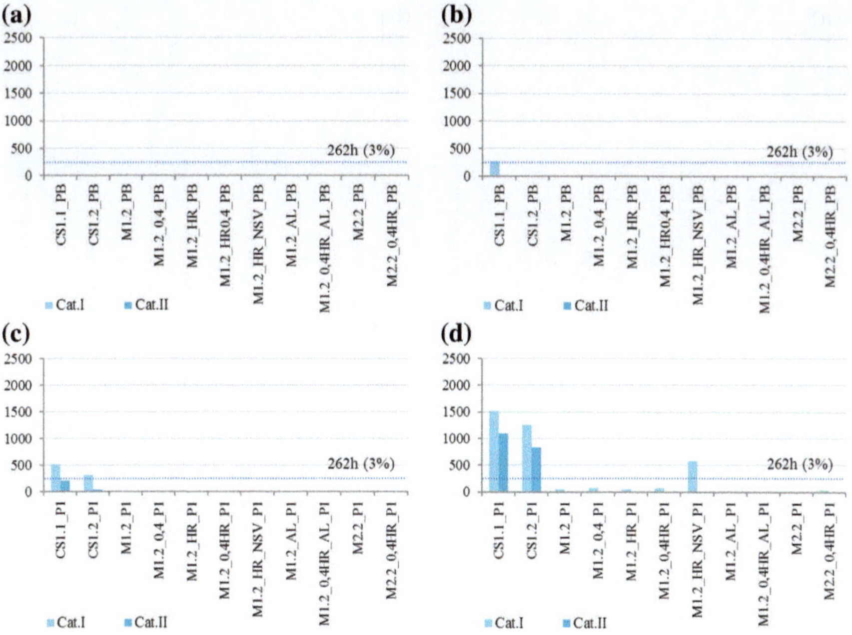

Fig. B.24 Overheating (hours) over adaptive threshold (UNE EN 15251) in T5 typology (Terraced House), in Rome (**a** Ground floor; **c** 1st floor) and Rome 2050 (**b** Ground floor; **d** 1st Floor)

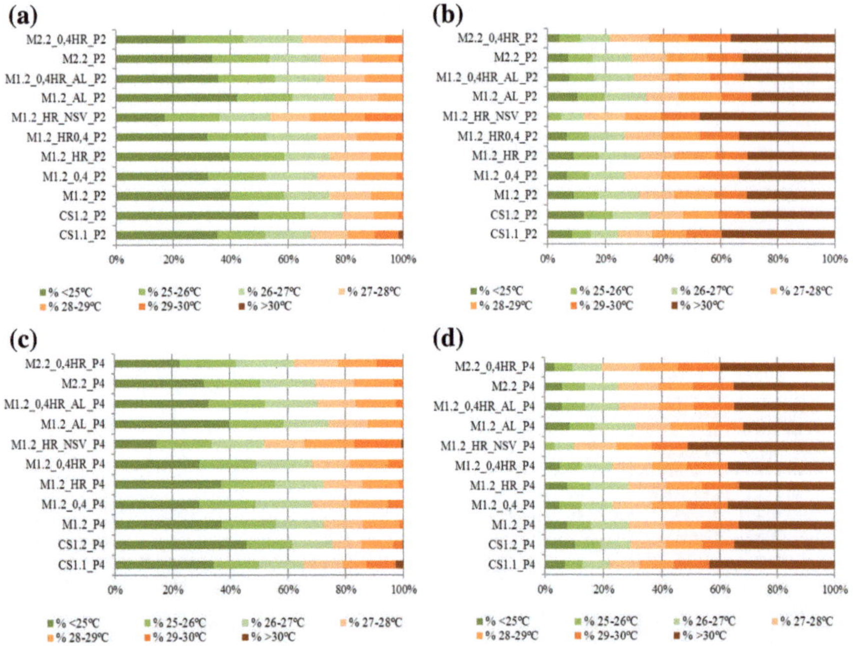

Fig. B.25 Indoor temperature frequency in T1 typology (Linear housing block), in Barcelona (**a** 2nd floor; **c** 4th floor) and Barcelona 2050 (**b** 2nd floor; **d** 4th floor)

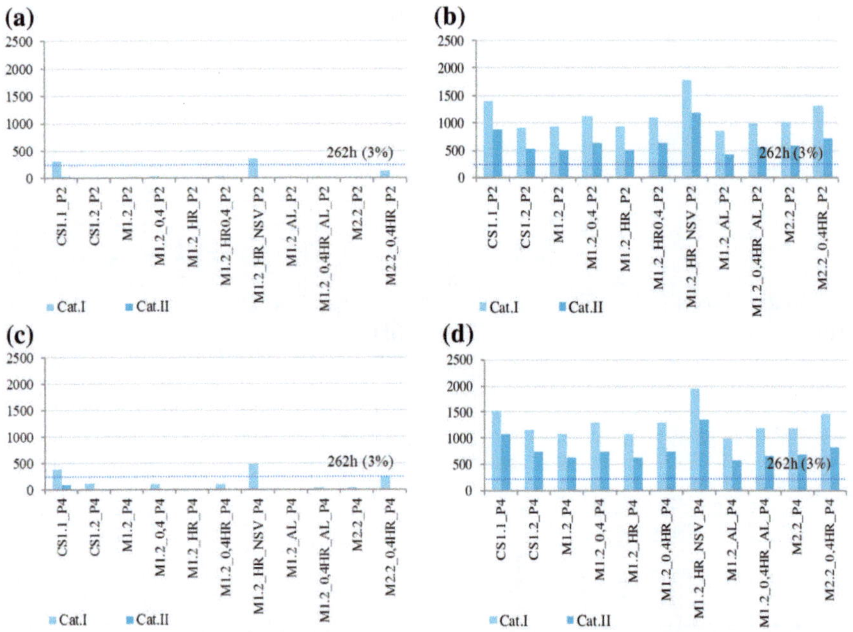

Fig. B.26 Overheating (hours) over adaptive threshold (UNE EN 15251) in T1 typology (Linear housing block), in Barcelona (**a** 2nd floor; **c** 4th floor) and Barcelona 2050 (**b** 2nd floor; **d** 4th floor)

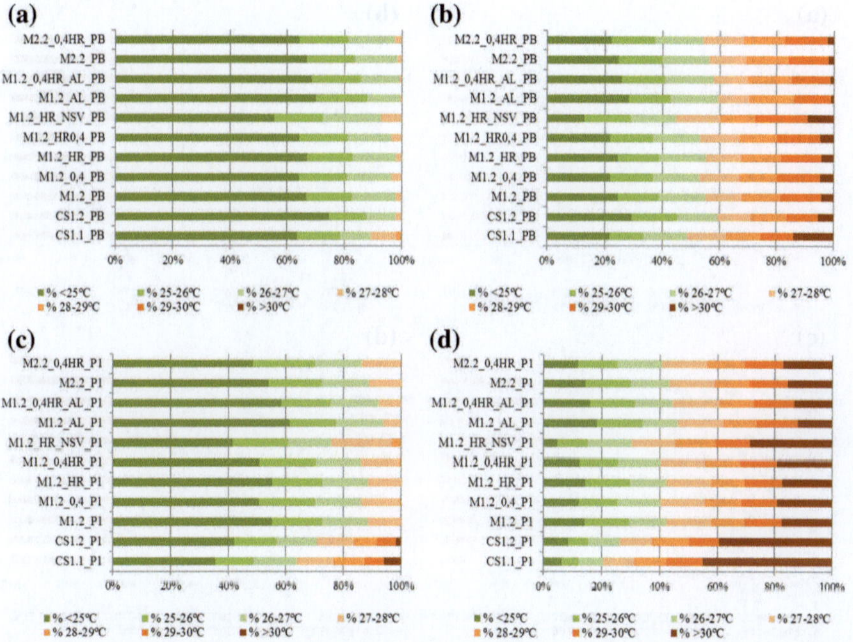

Fig. B.27 Indoor temperature frequency in T5 typology (Terraced House), in Barcelona (**a** Ground floor; **c** 1st floor) and Barcelona 2050 (**b** Ground floor; **d** 1st floor)

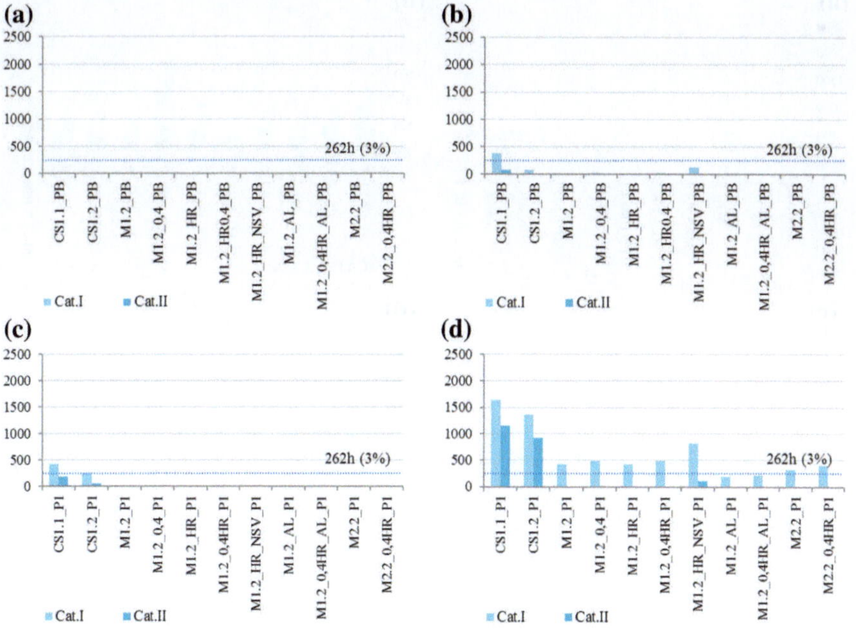

Fig. B.28 Overheating (hours) over adaptive threshold (UNE EN 15251) in T5 typology (Terraced House), in Barcelona (**a** Ground floor; **c** 1st floor) and Barcelona 2050 (**b** Ground floor; **d** 1st floor)

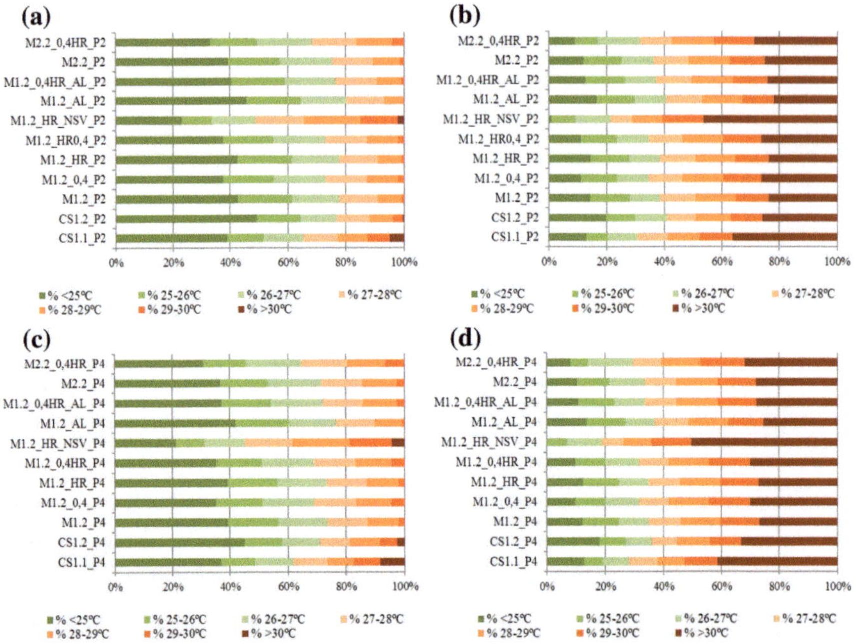

Fig. B.29 Indoor temperature frequency in T1 typology (Linear housing block), in Nimes (**a** 2nd floor; **c** 4th floor) and Nimes 2050 (**b** 2nd floor; **d** 4th floor)

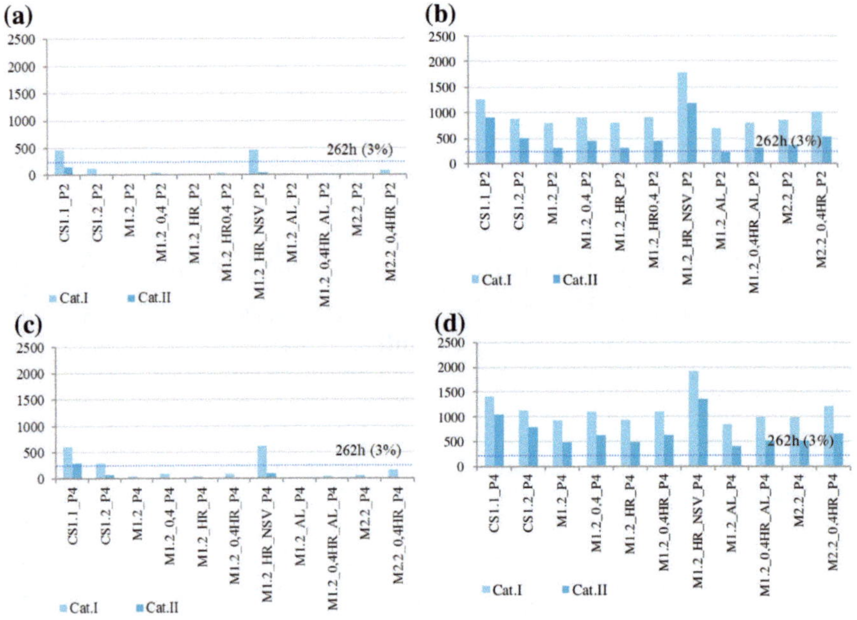

Fig. B.30 Overheating (hours) over adaptive threshold (UNE EN 15251) in T1 typology (Linear housing block), in Nimes (**a** 2nd floor; **c** 4th floor) and Nimes 2050 (**b** 2nd floor; **d** 4th floor)

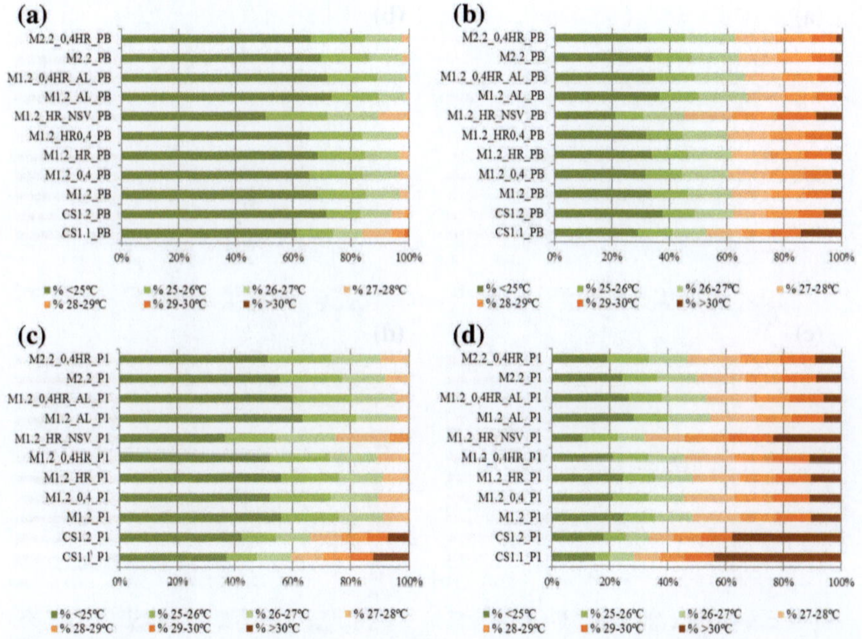

Fig. B.31 Indoor temperature frequency in T5 typology (Terraced House), in Nimes (**a** Ground floor; **c** 1st floor) and Nimes 2050 (**b** Ground floor; **d** 1st floor)

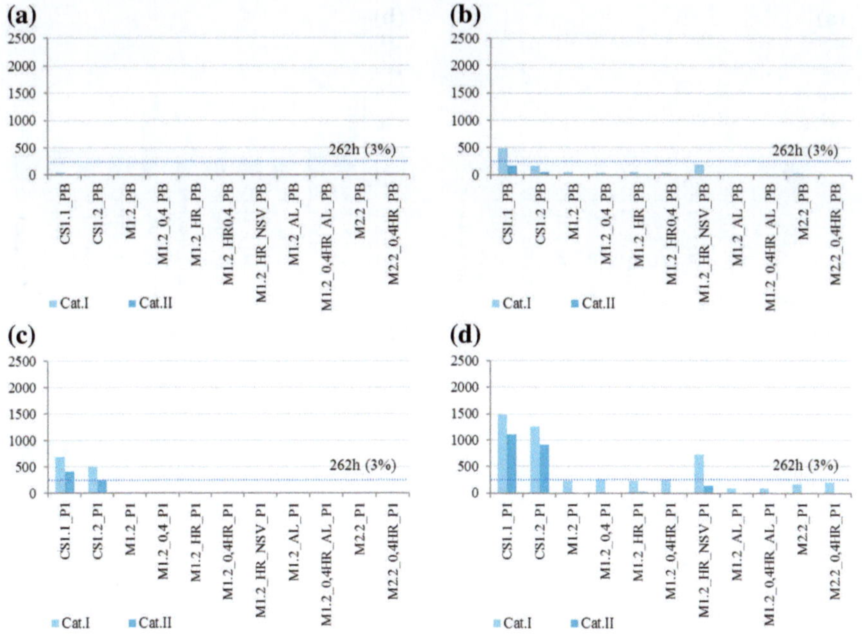

Fig. B.32 Overheating (hours) over adaptive threshold (UNE EN 15251) in T5 typology (Terraced House), in Nimes (**a** Ground floor; **c** 1st floor) and Nimes 2050 (**b** Ground floor; **d** 1st floor)

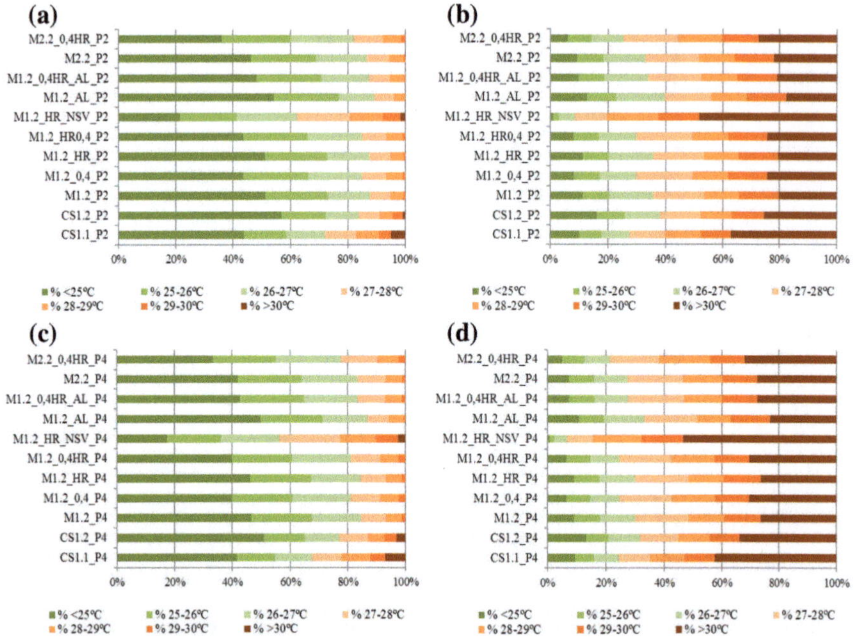

Fig. B.33 Indoor temperature frequency in T1 typology (Linear housing block), in Portoroz (**a** 2nd floor; **c** 4th floor) and Portoroz 2050 (**b** 2nd floor; **d** 4th floor)

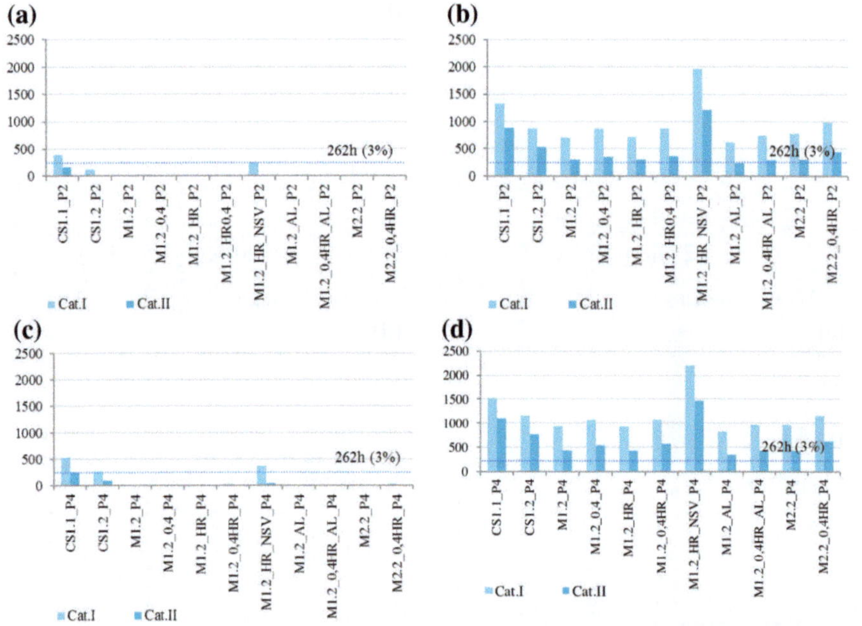

Fig. B.34 Overheating (hours) over adaptive threshold (UNE EN 15251) in T1 typology (Linear housing block), in Portoroz (**a** 2nd floor; **c** 4th floor) and Portoroz 2050 (**b** 2nd floor; **d** 4th floor)

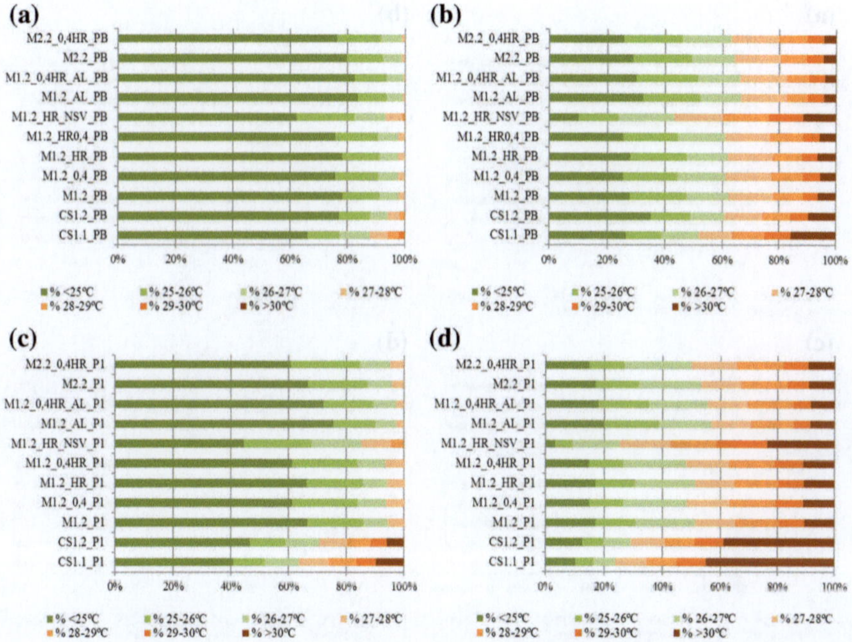

Fig. B.35 Indoor temperature frequency in T5 typology (Terraced House), in Portoroz (**a** Ground floor; **c** 1st floor) and Portoroz 2050 (**b** Ground floor; **d** 1st floor)

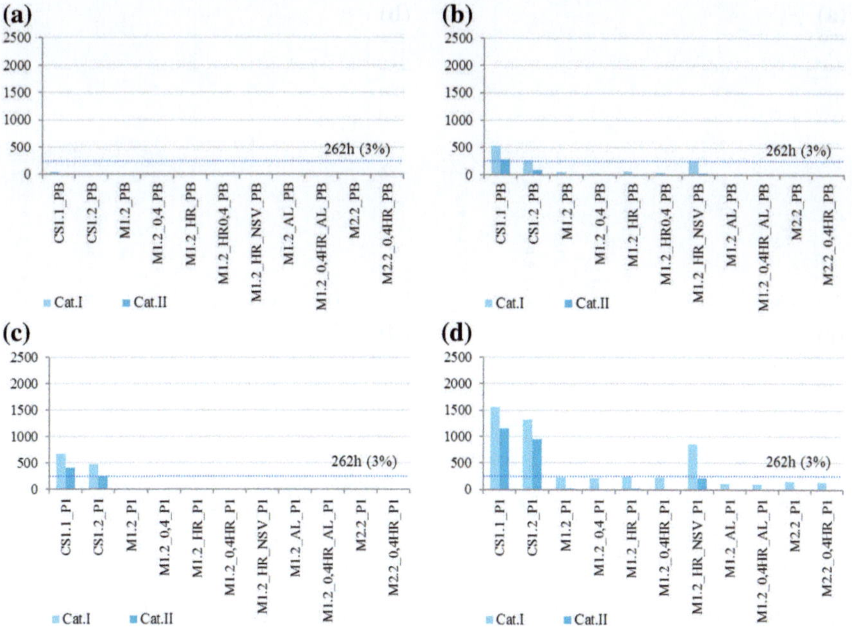

Fig. B.36 Overheating (hours) over adaptive threshold (UNE EN 15251) in T5 typology (Terraced House), in Portoroz (**a** Ground floor; **c** 1st floor) and Portoroz 2050 (**b** Ground floor; **d** 1st floor)

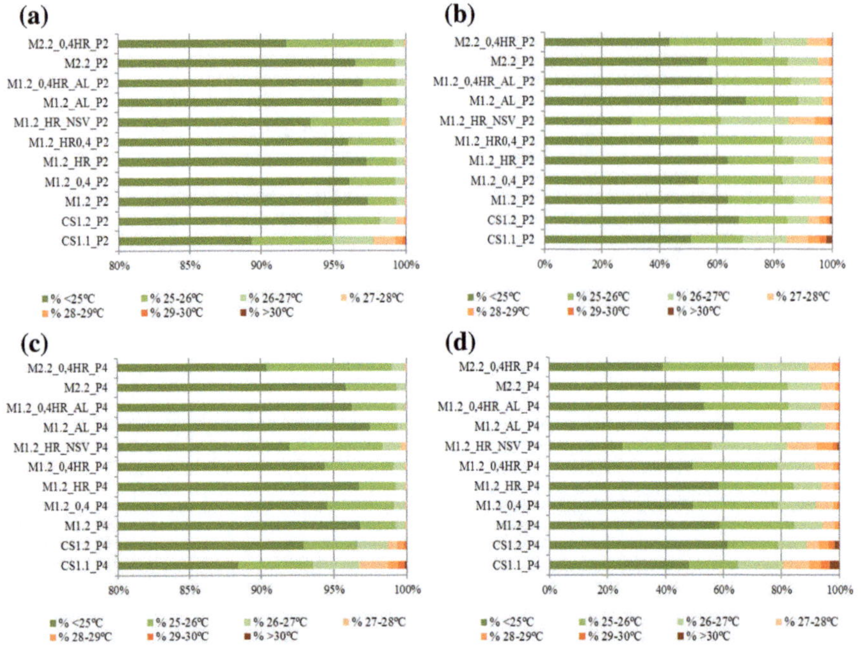

Fig. B.37 Indoor temperature frequency in T1 typology (Linear housing block), in Porto (**a** 2nd floor; **c** 4th floor) and Porto 2050 (**b** 2nd floor; **d** 4th floor)

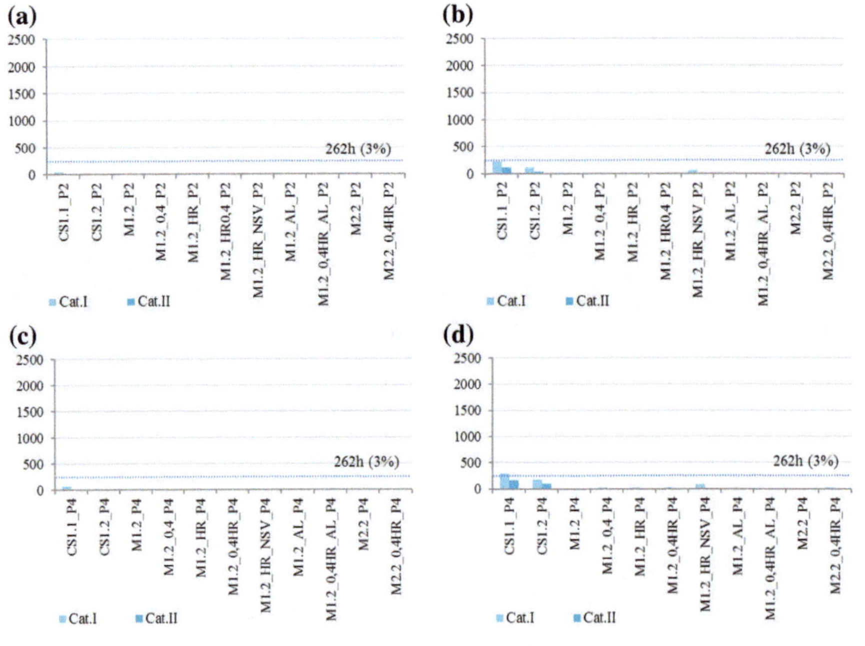

Fig. B.38 Overheating (hours) over adaptive threshold (UNE EN 15251) in T1 typology (Linear housing block), in Porto (**a** 2nd floor; **c** 4th floor) and Porto 2050 (**b** 2nd floor; **d** 4th floor)

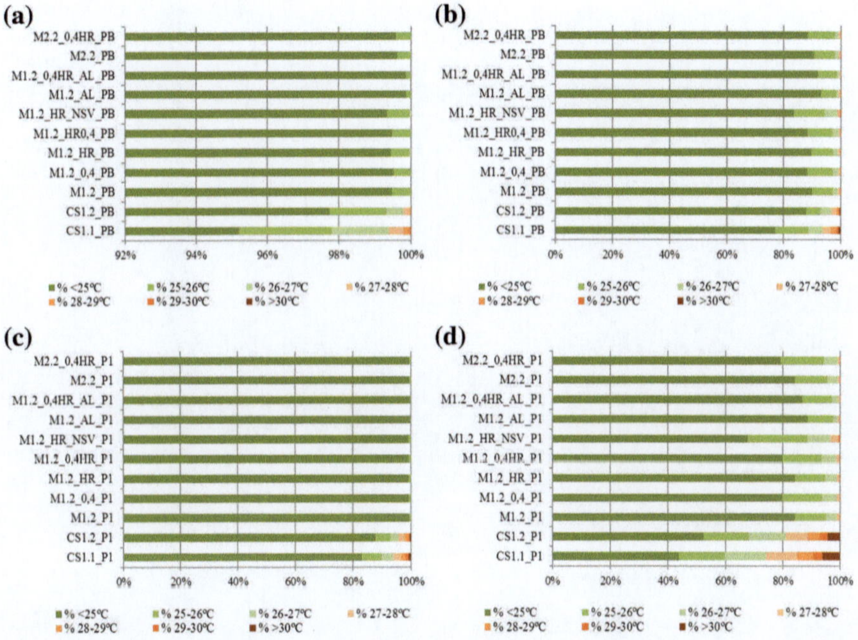

Fig. B.39 Indoor temperature frequency in T5 typology (Terraced House), in Porto (**a** Ground floor; **c** 1st floor) and Porto 2050 (**b** Ground floor; **d** 1st floor)

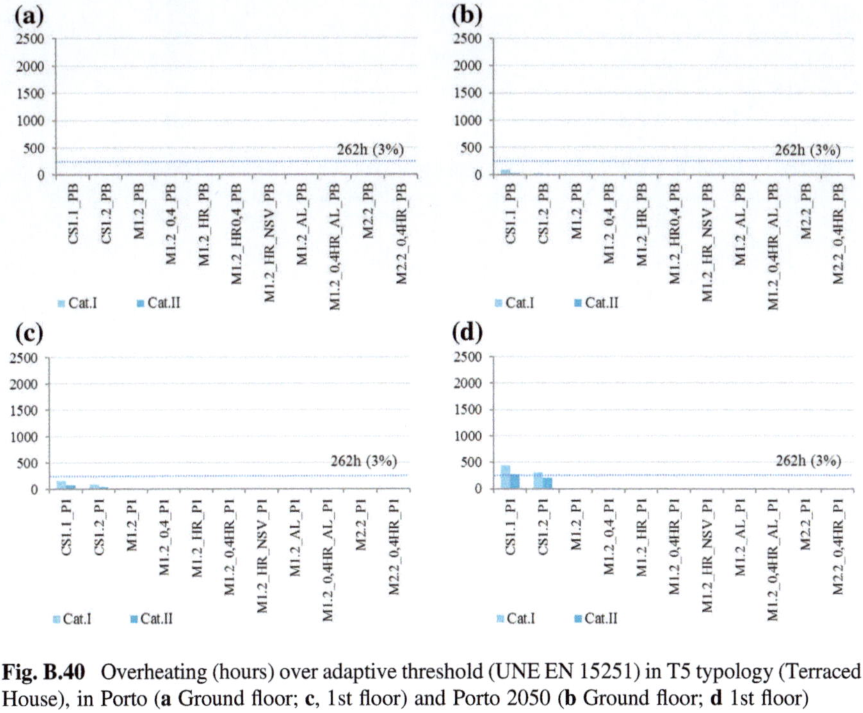

Fig. B.40 Overheating (hours) over adaptive threshold (UNE EN 15251) in T5 typology (Terraced House), in Porto (**a** Ground floor; **c**, 1st floor) and Porto 2050 (**b** Ground floor; **d** 1st floor)

MIX
Papier aus verantwortungsvollen Quellen
Paper from responsible sources
FSC® C105338

If you have any concerns about our products,
you can contact us on
ProductSafety@springernature.com

In case Publisher is established outside the EU,
the EU authorized representative is:
Springer Nature Customer Service Center GmbH
Europaplatz 3, 69115 Heidelberg, Germany

Printed by Libri Plureos GmbH
in Hamburg, Germany